UNIVERSITY OF GLASGOW SOCIAL AND ECONOMIC STUDIES

General Editor: Professor L. C. Hunter

22

LABOUR MARKETS UNDER DIFFERENT EMPLOYMENT CONDITIONS

UNIVERSITY OF GLASGOW SOCIAL AND ECONOMIC STUDIES

New Series

General Editor: *Professor L. C. Hunter*

1. The Economics of Subsidizing Agriculture. GAVIN MCCRONE
2. The Economics of Physiocracy. RONALD L. MEEK
3. Studies in Profit, Business Saving and Investment in the United Kingdom, 1920–62. Vol. 1. P. E. HART
4. Scotland's Economic Progress. GAVIN MCCRONE
5. Fringe Benefits, Labour Costs and Social Security. Edited by G. L. REID and D. J. ROBERTSON
6. The Scottish Banks. MAXWELL GASKIN
7. Competition and the Law. ALEX HUNTER
8. The Nigerian Banking System. C. V. BROWN
9. Export Instability and Economic Development. ALASDAIR I. MACBEAN
10. The Mines of Tharsis, Roman French and British Enterprise in Spain. S. G. CHECKLAND
11. The Individual in Society: Papers on Adam Smith. A. L. MACFIE
12. The Development of British Industry and Foreign Competition 1875–1914. Ed. DEREK H. ALDCROFT
13. Studies in Profit, Business Saving and Investment in the United Kingdom, 1920–62. Vol. 2. Ed. P. E. HART
14. Building in the Economy of Britain between the Wars. Ed. DEREK H. ALDCROFT and HARRY W. RICHARDSON
15. Regional Policy in Britain. GAVIN MCCRONE
16. Regional and Urban Studies. Ed. J. B. CULLINGWORTH AND S. C. ORR
17. Planning Local Authority Services for the Elderly. GRETA SUMNER and RANDALL SMITH
18. Labour Problems of Technological Change. L. C. HUNTER, G. L. REID, D. BODDY
19. Regional Problems and Politics in Italy and France. KEVIN ALLEN and M. C. MACLENNAN
20. The Economics of Containerisation. K. M. JOHNSON and H. C. GARNETT
21. Adam Smith's Science of Morals. TOM CAMPBELL

LABOUR MARKETS

UNDER DIFFERENT EMPLOYMENT CONDITIONS

D. I. MACKAY
D. BODDY
J. BRACK
J. A. DIACK
N. JONES

London

GEORGE ALLEN & UNWIN LTD

RUSKIN HOUSE MUSEUM STREET

First published in 1971

ISBN 0 04 331047 8

Printed in Great Britain
in 10 point Times Roman
by Alden & Mowbray Ltd
at the Alden Press, Oxford

In memory of D. J. R.

PREFACE

Few discussions of economic problems in present-day Britain get very far without coming back to some labour market issue, be it wages, the structure and content of collective bargaining, shortages of skilled labour, labour hoarding, training, institutional restrictions imposed by unions or employees, redundancies, regional disparities in unemployment, or government manpower policy. Yet despite the public attention which has been focused on these issues, labour economics has remained one of the poor relations of economic research. Moreover, much of the work in this field underlines the chief current weakness of empirical analysis, namely a tendency to concentrate on the manipulation of simplistic and highly aggregative models at the expense of detailed investigation of decision-taking at the level of individual employers and employees. A thorough understanding of these processes is necessary if the dichotomy between macroeconomic and microeconomic analysis is to be resolved and if the former is to be rooted in reality. This is particularly relevant in labour market analysis for the key decisions are usually made by individuals operating in a local context.

This volume sets out to describe and explain labour market behaviour by analysing the personnel records of the 75,000 manual workers employed by sixty-six engineering plants over the period 1959–66, and relates this information to wage data, employment statistics and other relevant material. The plants were located in five separate areas chosen to represent different labour market conditions. The majority were based in the cities of Birmingham and Glasgow, where unemployment was respectively much below and much above the national average, but the study also encompassed a dozen recently established engineering plants in an old industrial area, a new town and a small isolated community. We are thus able to see whether employers and employees responded to differences in the economic environment in the manner predicted by labour market theory, and to describe their behaviour in a number of different dimensions which are relevant to public policy.

Four main subject areas are distinguished—wages, labour turnover, labour recruitment and mobility, and personnel and manpower policies—and in each of these areas our discussion proceeds with two main aims in mind. The first is simply to describe and present information on a range of labour market issues presently obscured by a blanket of ignorance. If it achieves nothing else this volume should help provide an improved factual basis on which to erect future discussions. The second and more important objective is to outline the main elements of the competitive theory of labour markets, and to attempt to derive from this predictions which can be tested against our observations. In this manner we can establish more clearly the strengths and weaknesses of existing theory in a manner which may suggest pointers to future investigations.

The project was made possible by the financial support of the Department of Employment and Productivity, and the good offices, advice and encouragement of that Department were of invaluable assistance at all stages of the research. Convention prevents us from naming individuals but we can at least record our appreciation of the assistance we received from the Department's officials at five main centres—Birmingham, Edinburgh, Glasgow, London and Watford. Reports of our findings were prepared for the Department and form the basis of much of the current volume. We would however wish to make clear that the Department is in no way associated with the opinions expressed, for which the authors accept full responsibility.

We are also grateful to the managers of those engineering plants on whose co-operation the success of the project depended. The nature of the research made considerable demands on their time and patience. Both were willingly given. We hope that we will be forgiven if at times in this book it seems as if we have bitten the hand that fed us. Apart from the case-study plants we received encouragement and support from representatives of both sides of industry, most particularly from the Engineering Employers Federation, its two affiliated associations in the West Midlands of England and in Scotland, and the Amalgamated Union of Engineering and Foundry Workers.

Our remaining debts are mainly to our colleagues at the University of Glasgow who have borne the unwanted externalities of our research. The initial idea for the project owed much to Donald Robertson who was a pioneer of the study of plant wage structures, and whose support and advice before his death was a constant source of encouragement to the research team. The formidable task of collecting the data for the enquiry, together with much of the subsequent processing, was carried through by a team of technical assistants whose main members were Anne Carey, Susan Grier, Mary Kelly, Maureen Robb, Mary Smith and Doris Williamson. Ann Cairns, Gordon Stevenson, Jo Stott and Christine Duff wrote the complex computer programmes required and a reluctant computer was forced to disgorge results by Professor Gilles and Joan

Coverdale. Myra Nelson and Pat Rennie with patience and efficiency typed the various drafts of the manuscript occasioned by the comments and suggestions of Bob McKersie (University of Chicago), Derek Robinson (Oxford), and Tony Thirwall (Kent), and by those of Laurie Hunter, Alan Evans, John Harris, Frank Herron, Peter Norman and Alan Sleeman of the University of Glasgow. Our severest critics were, however, our colleagues Harry Garnett and Graham Reid who laboured mightily to save us from ourselves. Where they have not succeeded the fault is ours.

D. I. MACKAY, *University of Glasgow*
D. BODDY, *University of Glasgow*
J. BRACK, *University of Warwick*
J. A. DIACK, *University of Glasgow*
N. JONES, *University of Glasgow*

CONTENTS

PART 1: INTRODUCTION

CHAPTER 1

THEORY AND EMPIRICAL EVIDENCE

'The whole of the advantages and disadvantages of the different employments of labour . . . must, in the same neighbourhood, be either perfectly equal or continually tending to equality. If in the same neighbourhood, there was any employment evidently any more or less advantageous than the rest, so many people would crowd into it in the one case, and so many would desert it in the other, that its advantages would soon return to the level of the other employments.'

Adam Smith

'One can hardly pick up a new book on labor nowadays without finding the author jumping gleefully on what he thinks is the corpse of demand and supply, or proclaiming with trumpets, "The labor market is dead, long live human relations".'

Kenneth Boulding

The intellectual foundations of labour economics remain those laid down by Adam Smith in 1776. Other branches of economic theory have changed almost beyond recognition since the publication of *The Wealth of Nations*, but the competitive theory of the labour market has no serious rival. This is not to say that it has commanded widespread respect or that it has not been challenged. On the contrary, most empirical studies of labour market behaviour have stressed that the operation of market forces might be seriously impeded by inadequate knowledge, institutional rules, monopolistic practices and so forth. While the proponents of the competitive or 'traditional' theory see order through the operation of demand and supply, the institutionalists and behaviourists see only chaos. One group can identify an integrated wood, the other, using the same evidence, perceives only a set of unrelated trees.

The fundamental assumption underlying labour market theory is that individuals and institutions, such as firms and trade unions, pursue their own interests. Further, it is assumed that these interests are capable of being defined in economic terms or, at least, that economic factors are sufficiently important to determine or strongly influence decisions. Given this, hypotheses can be constructed predicting possible reactions to a given change in economic circumstances. This may be illustrated by reference to labour mobility, the area in which theory is most developed and where the predictions of theory have been most seriously challenged by empirical research. Theory suggests that job choice is determined by the bundle of wage and non-wage conditions attached to different jobs. Current earnings, employment prospects, probable future earnings, the nature and conditions of work and similar factors are all evaluated by the person seeking work, whose ultimate decision between alternative job openings depends on the *balance of net advantages*. It is important to observe that both pecuniary and non-pecuniary factors are held to influence job choice. The theory does *not* predict that individuals will move from lower-paid to higher-paid jobs but only that such movement will occur provided other things are equal.[1]

In the opposite camp stand the sceptics who seem agreed on the following conclusions which are thought to be inconsistent with the assumptions of the competitive model. Manual workers have limited job horizons; they know little in detail about the wage or the non-wage conditions attached to different employments, or even the full extent of the job openings which are available; those in work are often not interested in considering potential alternatives to their current employment, and the job search when it does take place is limited in scope, haphazard and heavily influenced by a restricted circle of relatives and friends. In consequence, the labour mobility observed in the market is often wasteful, leading to disappointed expectations and imposing heavy costs on employees and employers without eroding the large wage differentials which exist even within a local market area.

It is extremely difficult to choose between these alternative views. For example, if we set out to test labour market theory we find that it is so malleable that it can 'explain' virtually any set of observations. It is difficult to refute (or confirm) by reference to empirical data.[2] If labour should move from a low to a high wage-paying plant, this can be held to substantiate the model. Conversely, should labour not move towards the high wage-paying plant it can be argued that this is due to adverse, non-pecuniary factors which offset the inducement of the higher wage. In any situation there are alternative actions each of which may be consistent with the postulate of economic rationality.

[1] S. Rottenberg, 'On Choice in Labor Markets', *Industrial and Labor Relations Review*, Vol. 9, 1955–6.

[2] See, for example, R. Perlman, *Labour Theory*, 1969, esp. pp. 105–23.

Even greater difficulties arise when examining the labour market policies adopted by firms, difficulties which are simply swept under the carpet by the competitive model's assumption of a homogeneous labour supply. When hiring labour to undertake any task, the employer can either hire workers of better quality and pay high wage rates, or he can hire a larger number of workers of lower quality at low wage rates. Either action may be rational in the economic sense. Similar problems arise in all other aspects of a firm's labour market policies. The employer, when faced with a labour shortage, can raise wages, lower hiring standards or pursue a more active recruitment policy. Again, as output falls the employer may attempt to retain his work force by eliminating overtime and adopting short-time working, or he may choose to reduce his wage bill through redundancies. There are, then, considerable difficulties in interpreting observed behaviour so as to evaluate the predictive power of the simplified, theoretical model largely due to the fact that theory is of little help in predicting specific actions in any labour market situation.

The main result of empirical research has been to make economists and other social scientists more sceptical of the strength of competitive forces. No one, not even Adam Smith, believed that the competitive model was a faithful image of the real world. It was simply an abstraction which attempted to isolate the fundamental forces at work. The model did not predict that net advantages would be equalized outside an idealized world but merely that there would be a tendency towards equalization, the strength of which would depend on the extent of imperfections in the market.[1] However, even allowing for this qualification, it is true to say that empirical research has substantially modified views about labour market behaviour. As a result, labour economics remains in a confused state. While the competitive model is difficult to square with important aspects of labour market behaviour, no viable alternative has emerged to take its place.

The present volume does not pretend to provide a satisfactory answer to this problem. Its aim is not to put forward an alternative theory, but to test the elements of traditional theory against observed behaviour. We should also make clear from the outset that our principal objective is to attempt to establish the usefulness of the assumptions which economists normally employ to explain the

[1] Thus the quotation from *The Wealth of Nations* is immediately succeeded by the sentence, 'This at least would be the case in a society where things were left to follow the rational course, where there was perfect liberty, and where every man was perfectly free both to choose what occupation he thought proper; to change it as often as he thought proper.'

actions of employers and employees. As one of our opening quotations suggests, this is a rather unfashionable approach, for many modern students of labour markets seem to believe that economic rationality is of little relevance in explaining such actions. This we believe to be mistaken. Economics is not, of course, the only relevant discipline. Indeed, one of our main tasks is to indicate its shortcomings more clearly and to demonstrate how institutional and other pressures can modify, and override, market forces. Nonetheless the economist does have something to contribute in explaining labour market behaviour. Our interest lies in the extent and the limitations of this contribution.

This is done in what, it is hoped, is a novel and useful manner, by examining how employers and employees react to the labour market environment in which they find themselves. By this means it may be possible to identify those areas where existing theory provides a satisfactory explanation of events and those where further developments may be necessary. We begin by looking at labour market theory and the results of empirical research in greater detail, so that we can derive certain hypotheses concerning labour market behaviour. Without such a groundwork, a description of the case-study plants, while interesting in itself, would provide no basis for generalization and prediction.

While the theoretical underpinnings are of first importance to an understanding of the forces at work, the results of the present volume may also be useful in a more direct and immediate manner. In Britain, very little research on the operation of labour markets has been undertaken despite the fact that many aspects of labour market behaviour are the subject of debate and controversy. It is desirable, even if nothing else is accomplished, to describe and analyse certain features of such behaviour which are of pressing practical importance. There appears to be a growing belief in Britain that labour markets do not operate efficiently; that the traditional methods of collective bargaining between employers and unions ignore the public interest and are not sufficiently responsive to economic pressures; that unions have imposed restrictive practices which impede labour mobility and result in overmanning and a wasteful use of labour; that managements 'hoard' labour in recessions and pay insufficient attention to training needs, so frustrating a more efficient distribution of labour; and that employees act irrationally so that much labour mobility is haphazard and unrelated to economic needs.

One consequence of this growing dissatisfaction with labour market behaviour has been increased state intervention in the operation of the market. The need for such intervention at certain points has, of course, long been recognized, e.g. through the establishment of the employment exchanges, but in recent years the emphasis of legislation has shifted towards more wide-ranging policies designed to improve the working of labour markets. This intervention has attempted to influence the behaviour of employees and labour market institutions by the application of financial inducements or penalties, such as those embodied in the Industrial Training Act, the Redundancy Payments Act, the Selective Employment Tax and the Regional Employment Premium, or it has supplemented private action by public institutions such as Government Training Centres. If state intervention is necessary, and if it is to be effective, it must be based on an accurate view as to how employers, unions and employees behave, since the functioning of the labour market continues to depend very largely on decisions taken at the microeconomic level. At the moment, it is doubtful whether sufficient information is available to allow informed judgments. One commentator has remarked that 'We probably know more about conditions on the surface of the moon than we do about the operations of local labour markets.'[1]

At this microeconomic level there is a large number of processes, each of which is important to the efficient operation of the economy and each of which is, as yet, imperfectly understood. These concern, *inter alia*, wage structure and wage determination; labour turnover; labour mobility within and between plants; personnel and manpower policies relating to manpower forecasting, selection, training, redundancy and labour 'hoarding'; and the role played by agencies such as the public employment exchanges. Issues such as these form the core of the following chapters, which are based on a study of engineering plants[2] operating in different labour market environments over the period 1959–66 inclusive. The fundamental objective is to describe and analyse how the labour market policies of plants and the behaviour of manual employees are influenced by the employment conditions and other characteristics of the labour market areas in which they are established. A sample of engineering plants was taken from five different labour market areas—Birmingham, Glasgow and three other local labour markets in Scotland which will subsequently be referred to as North Lanarkshire, New

[1] D. Robinson, 'Myths of the Local Labour Market', *Journal of the Institute of Personnel Management*, Vol. 1, 1967–8, p. 36.

[2] A plant comprises those manufacturing buildings in the same location and under the same management. It may form part, or the whole, of the activities of a firm.

Town and Small Town[1]—these areas having been selected to represent different structures of industry, employment and unemployment.

The method of investigation and the labour markets selected for examination are described in greater detail in Chapters 2 and 3. In the remainder of this chapter, an attempt is made to describe the elements of labour market theory, to illustrate its assumptions and from these to formulate hypotheses about labour market behaviour. The discussion also considers the relationship between existing theory and the results of previous empirical research to indicate those theoretical and practical issues which are as yet unsatisfactorily resolved. Particular attention is paid to the manner in which differences in the external labour market environment might be expected to influence labour market behaviour. This can then be tested against the experience of the case-study plants in the five labour market areas examined.

Though all aspects of labour market behaviour are likely to be interrelated in greater or lesser degree, the following discussion is selective. It focuses attention on certain subject areas which are considered to be especially important, and where the data at our disposal allow us to examine how the behaviour of management and employees is conditioned by their economic environment. Section 2 deals with inter-plant wage differentials and Section 3 with internal plant wage structures. Sections 4 and 5 are concerned with labour turnover and with labour mobility involving a change of employer. Job-changing within the plant is discussed in Section 6, while Section 7 deals with manpower forecasting, selection procedures, hiring standards, recruitment channels, redundancy and a number of related issues.

2. THE INTER-PLANT WAGE STRUCTURE

The competitive theory of labour markets predicts that within a local labour market there will be a tendency for labour of the *same* quality to obtain the same earnings irrespective of the employer for whom the labour works. Should no such tendency be observable, the only explanation consistent with the competitive theory is that the wage differentials observed are counterbalanced by offsetting non-pecuniary factors which are greatest in the low wage-paying plants.[2] It

[1] Hereafter these three areas are referred to collectively as the 'Other Scottish Areas'. The names New Town and Small Town have been used to conceal the identity of those plants in these two areas which co-operated in this enquiry.

[2] This, however, has to be demonstrated: otherwise one assumes as true what one sets out to prove.

should be recognized that the model does not predict complete equalization of wages or even of net advantages at any moment of time, since barriers to labour mobility exist so that complete equalization will never be obtained. However, it is held that there will be a *tendency* towards equalization; that substantial wage differentials will not persist over time unless offset by non-pecuniary advantages; and that, given time, labour can recognize and will react to differences in net advantages so that no plant can set wages and other conditions of employment independent of the behaviour of its competitors. This latter factor provides the link between the wage policies of different plants. 'Potential mobility', in response to wage differentials, 'is the ultimate sanction for the interrelation of wage-rates'.[1]

Empirical evidence from a variety of sources has brought into question the extent to which labour mobility is responsive to wage differentials. The existence of substantial differentials within the same labour market appears the common rule rather than the exception.[2] Although much of the information is based on a snap shot at a particular moment of time, there is some evidence that substantial differentials persist over time. One recent study has concluded that widespread ignorance of existing differentials exists among both employers and employees, and that employers do not have the necessary information on competitors' wage levels, so that wage changes are unrelated. Moreover, no apparent relationship appears to exist between changes in earnings and employment at the plant level. Hence 'the labour market does not work in the way generally supposed. It is far more chaotic than even the sceptics have believed.'[3]

Inter-plant wage differentials may arise if, due to imperfect knowledge, an individual is unable to assess a job except by working at it. The process of labour mobility can then be regarded as 'job shopping', the individual moving from job to job until he finds one which is suitable. Much intervening mobility may be haphazard and wasteful, involving costs for both employees and employers. If this is so, there is likely to be a time-lag before labour responds to the emergence of a wage differential. Imperfect knowledge may, then, provide some part of an explanation for the pattern of plant differentials observed at a given moment of time. If, however, this pattern remains largely undisturbed over long periods of time, then one has to

[1] J. R. Hicks, *The Theory of Wages* (1932), p. 79.

[2] R. A. Lester, 'Wage Diversity and its Theoretical Implications', *Review of Economics and Statistics*, Vol. 28, 1946; and 'A Range Theory of Wage Differentials, *Industrial and Labor Relations Review*, Vol. 5, 1951–2.

[3] Robinson, *op. cit.*, pp. 38–9. See also *Wage Drift, Fringe Benefits and Manpower Distribution* (O.E.C.D., 1968).

find other explanations, or else to suppose that labour mobility is even more haphazard and irrational than suggested by previous studies.

A recent empirical investigation of the relationship between employment changes and earnings differentials found that 'there is no evidence of a strong systematic relationship between changes in earnings among individual industries and variations in relative employment'.[1] This is not to suggest that the level of earnings has no influence, for there appears to be a limiting condition that industries which have earnings substantially below the average, or whose relative earnings have shown a substantial decline, will have recruitment difficulties.[2] Where such circumstances do not apply it appears that the wage structure is inflexible and that the redistribution of labour is a function of variations in job opportunities rather than of differences in relative earnings.[3] However studies based on industry aggregates, where the dependent variable is the *net* change in employment, cannot demonstrate that employees are insensitive to wage signals. As we shall see in Section 4, it is possible that employees could respond to wage differences in a manner not revealed by data of this nature.

As soon as we drop the simplifying assumption that labour is homogeneous, we have to recognize that inter-plant differentials may also arise because of the labour market policies adopted by plants. For example, a 'high wage' policy may enable a plant to set stricter hiring standards and reduce labour turnover so that real costs might be cut despite higher money wages. Similarly, collusion between employers through 'no poaching' and anti-pirating agreements and the adoption of seniority rules in determining promotion and demotion may reduce labour mobility and give rise to wage differentials.

We can establish two contrasting hypotheses of the behaviour of inter-plant differentials. If wages are not a key factor in job choice there will be little relationship between plant wage levels. In the extreme case the supply of labour would be price-inelastic and there would be no tendency for a common level of wages to be established. Labour turnover would not be responsive to differences or changes in relative earnings. Wage-setting would depend on the administrative decision of management influenced by union pressure but unaffected by mobility or potential mobility in the labour market. Alternatively,

[1] O.E.C.D., *Wages and Labour Mobility* (1965), p. 16.

[2] *Ibid*.

[3] W. B. Reddaway, 'Wage Flexibility and the Distribution of Labour', *Lloyds Bank Review*, Oct. 1959.

if labour is responsive to wage differentials the plant's wage policy will be influenced by the actions of its competitors and by its experience in recruiting and retaining labour.

Our task, therefore, is to examine and explain the structure, and changes in the structure, of inter-plant differentials. We wish to know, first, whether there is such a thing as a market rate for a particular type of labour, or whether at any point of time substantial inter-plant differentials exist. Second, are such differentials as do arise stable over time, or do plants frequently change their position in the wage hierarchy? Third, does management have the required information on which to base a rational wage policy? In other words, is management generally aware of the level of earnings in rival establishments or, failing this, can management identify and make the appropriate response to recruitment difficulties and changed turnover levels? Fourth, is the extent of inter-plant differentials standard between markets and within the same market over time, or do differences in plant wage levels contract and expand in response to the employment conditions prevailing? These and similar questions are examined in Chapter 4.

3. THE INTERNAL WAGE STRUCTURE

Management, in setting a wage for any occupational group, must consider not only the relationship of occupational earnings in the plant to those prevailing in other establishments but also the relationship between the earnings of different groups *within* the plant. The theoretical model suggests that the earnings of any group will be determined by demand and supply for that particular type of labour and will reflect management's evaluation of the group's contribution to production and the ease or difficulty with which labour of that type can be recruited. A set of earnings differentials will then arise reflecting, *inter alia*, the nature of the work, innate skills, length of training, and the degree of responsibility involved. The function of a wage structure as seen by the economist is to promote an efficient utilization of the available labour. To this end it is necessary that the wage structure should reflect the demand and supply for labour performing different tasks within the unit. In addition, *changes* in demand and supply conditions should be accompanied by an adjustment in relative earnings, so that wage 'signals' assist rather than retard changes in the distribution of employment.

Managements are then supposed to react to recruitment difficulties in the external market through adjusting wage relativities.[1] The

[1] There are other possible actions some of which will be considered later.

question is whether they can recognize recruitment difficulties for certain groups and have sufficient control over the wage structure to bring about an increase in the relative earnings for these groups. If the internal wage structure is inflexible and not susceptible to management control, then any change in customary differentials will give rise to pressure for increased earnings from other groups. Where such 'coercive comparisons' are important two consequences seem to follow. First, a plant which is a high wage-payer for one group of labour will tend to be a high wage plant for all the types of labour it employs. Second, the wage structure will move bodily upwards over time so that *all* groups of workers in a particular establishment will tend to have *either* a small rise *or* a large rise in earnings relative to the average rise in earnings for all plants in the local labour market. If this is so, the relative earnings of the different occupational groups may not reflect the demand for and supply of labour at a particular point in time, nor may they be responsive to changes in demand and supply over time.

It may be supposed that the wage structure in engineering would be sensitive to the external market environment, for, as we shall see later,[1] the institutional framework confers considerable freedom on individual plants, enough it would seem to adapt wage structures to local employment circumstances. Yet in recent years there has been considerable criticism[2] both of the methods of wage negotiation in the industry and of the complex wage structures resulting. It seems generally held that the very complexity of the wage system has created considerable anomalies, and that institutional and procedural factors have a very important influence on the internal wage structure. For example, it has been suggested that the earnings of pieceworkers are subject to progressive increases because the institutional framework enables them to benefit directly from technical change, from changes in work methods and from errors in fixing piecework times and standards.[3] This is a source of 'primary' wage drift,[4] the earnings of pieceworkers at plant level rising more quickly than the minima established through national bargaining. If this is so

[1] See pp. 66–8 below.

[2] See, for example, National Board for Prices and Incomes, *Report No. 49, Pay and Conditions of Service of Engineering Workers*, Cmnd. 3495, 1967 and National Incomes Commission, *Report No. 4 (Final), Agreement of November–December, 1963 in the Engineering and Shipbuilding Industries*, Cmnd. 2583, 1965.

[3] S. W. Lerner and J. Marquand, 'Workshop Bargaining, Wage Drift and Productivity in the British Engineering Industry', *Manchester School of Economic and Social Studies*, Vol. 30, 1962.

[4] E. H. Phelps Brown, 'Wage Drift', *Economica*, N.S., Vol. 29, 1962. Labour shortages are held to be the second major source of primary drift.

there will be a tendency for the earnings of pieceworkers to outstrip those of timeworkers, thus creating pressure for further adjustments in timeworkers' earnings which will give rise to 'secondary' drift.

There are, then, two main areas of enquiry with respect to the internal wage structure which will be considered in Chapter 5. The first is to describe the various methods of wage payment adopted by plants and the extent to which different characteristics are discernible for plants in separate labour market areas. Second, the analysis considers the factors which have shaped developments in the internal wage structure over time to highlight the problems which arise from wage payment systems as presently applied. The most important issue is whether the wage settlements which arise from the bargaining framework are consistent with the predictions of economic theory, or, alternatively, whether the institutional forces underlying these settlements frustrate the evolution of a rational system of wage differentials.

4. LABOUR TURNOVER

Labour turnover consists of two flows, an inflow to and an outflow from an employing unit. Each flow in a given period can be expressed as a percentage of the appropriate stock,[1] but from many points of view the outflow or separation rate is more important than the recruitment or accession rate. Much recruitment simply involves replacement of labour which has left the plant, so that the recruitment rate is largely a function of the separation rate. Further, a study of the separation rate permits an investigation of the influence of economic variables in a manner not possible for the recruitment rate. This is partly a consequence of the methodology of labour market studies, but also results from other more general considerations. The methodological point is that the characteristics of the appropriate stock (those employed by a plant), to which separations are compared, can be ascertained. For example, the age and length of service of the stock can be compared to the characteristics of those who leave the plant in any given time period. A similar study of the factors influencing the recruitment rate is not possible, as the appropriate stock concept is all persons who *might* have joined the plant, and the characteristics of this stock cannot be established.

A more important reason for analysing separation rates in greater

[1] Turnover can be measured in many different ways (see F. T. Pearce, *Financial Effects of Labour Turnover*, University of Birmingham, Studies in Economics and Society, Monograph A4, 1954), but for our purposes the simple, 'crude' turnover rate is most appropriate.

detail is that labour may not be able to join a plant even if it wishes to do so, whereas it can always leave a plant. The influence of relative earnings and of other factors may then be concealed where only the recruitment rate or net changes in employment are considered. For example, if a plant with relatively high earnings has a low recruitment rate or a below-average increase in employment this need not indicate that labour is irresponsive to wage differentials. It may simply reflect the fact that few job openings are available in the high wage plant. Conversely there is no restriction on the separation rate, so that if relative wages affect labour mobility, we would expect high wage plants to have low separation rates and vice versa.

The factors influencing the total separation rate may conveniently be considered under two broad headings—personal characteristics, such as age, skill and length of service, and environmental influences, such as employment conditions and wages. The influence of personal characteristics on separation rates has been examined fairly extensively,[1] and there is no need at this stage to recap on the findings of previous studies. It is appropriate, however, to observe that certain facets of separation rates are less clear and can be usefully re-examined. Both age and length of service affect separation rates, but most studies have found it difficult to say which has the greater effect. Similarly, although the evidence suggests that unskilled manual occupations tend to have higher separation rates this may reflect force of circumstance rather than choice, i.e. they may have higher voluntary quit rates or they may be more liable to be declared redundant.

Little is known about the extent to which differences in economic conditions modify the influence of personal characteristics on separation rates. Previous studies appear to indicate that the profile of separation rates is largely independent of employment conditions;[2] whatever the employment conditions, young workers have higher separation rates than older workers, and short-service workers are more mobile than long-service workers. These studies have, however, dealt with comparatively small groups of workers, and it is difficult to judge whether the separation rates of certain groups are particularly affected by differences in employment conditions. Once again, it is necessary to be able to study separation rates in detail and to

[1] For a summary of the main conclusions, see D. T. Bryant, 'A Survey of the Development of Manpower Planning Policies', and A. Young, 'Models for Planning Recruitment and Promotion of Staff', *British Journal of Industrial Relations*, Vol. 3, 1965.

[2] For example, M. Jefferys, *Mobility in the Labour Market* (1954), p. 11, and G. L. Palmer, *Labor Mobility in Six Cities* (1954), p. 125.

relate them to employment conditions in different labour market areas.

The discussion so far has been concerned with the behaviour of the total separation rate simply because investigations of labour turnover have usually been conducted along these lines. Practical considerations have imposed this rather unsatisfactory compromise on researchers, but as total separations consist of persons leaving plants for a wide variety of reasons, it is evidently desirable to distinguish these various categories of leavers and especially those leaving plants voluntarily. Fortunately, this is possible in the present study,[1] and henceforth we concentrate our attention on voluntary leavers whose behaviour, because they leave a plant of their own volition, is the critical test for economic theory.

We can illustrate this by considering the impact of environmental factors on the decision to leave a plant. It is well established that the total separation rate is inversely related to the current level of unemployment. However, changes in the total separation rate, in response, say, to a rise in unemployment, are a compound of two opposing movements—a fall in the voluntary quit rate and a rise in the number of redundancies.[2] To understand how employment conditions influence *choice* it is therefore necessary to deal with those who leave their current employment voluntarily. On this subject, little is known beyond the broad generalization that quits are likely to be inversely related to the level of unemployment. This is hardly very helpful. We do not know whether quits are a function of present unemployment, past unemployment or expected unemployment in some future period; whether they are affected by the size of plant or by the demand for labour; whether they are a response to wage differentials or to changes in wage differentials and so forth. Each of these possibilities has to be considered. Labour turnover is important in two respects. In the negative sense it is a cost which must be borne by industry, while in the positive sense it is an essential prerequisite for most types of labour mobility. It follows that it is important to have a greater understanding of the factors which shape and determine the level of turnover. This forms the subject matter of Chapters 6–8.

[1] We can distinguish between those who left a plant of their own accord (i.e. voluntarily), those declared redundant, those dismissed through unsuitability and misconduct, and those leaving for other reasons such as retirement, sickness or death.

[2] *Wages and Labour Mobility*, pp. 65–8 and H. Behrend, 'Absence and Labour Turnover in a Changing Economic Climate', *Occupational Psychology*, Vol. 27, 1953.

5. RECRUITMENT PATTERNS AND 'EXTERNAL' LABOUR MOBILITY

The recruitment policies of plants are likely to respond to changes in external labour market conditions which affect their ability to obtain labour. In a 'slack' market, where labour is in plentiful supply, we would expect the bulk of recruitment to consist of persons with the required skills and experience resident close to the establishment. Managements are likely to have a preference for such labour, as no special training has to be provided and those recruited do not have to be compensated for the costs involved in residential mobility or long journeys to work. As the market tightens, however, managements may have to adopt more aggressive and costly methods in their search for labour, widening the area of their search in spatial, industrial and occupational terms. The pattern of recruitment and the factors which shape it have seldom been subject to direct investigation, unlike labour mobility which is simply the recruitment process looked at from the standpoint of the worker. However, management attitudes and actions towards recruitment will help to shape labour mobility just as the preferences and behaviour of potential recruits will influence the sources from which management obtains labour. Therefore while the remainder of this section will concentrate on discussing labour mobility which has been the subject of much previous research, in Chapters 9 and 10 we shall discuss this process from two angles—firstly, the sources from which the case-study plants recruit their labour force and, secondly, the type of mobility undertaken by that labour.

Labour mobility involves a change in job status.[1] It therefore involves a change in job within one's present plant ('internal' mobility) or a change in plant ('external' mobility). While external mobility may, and normally will, involve a change in occupation, industry or area of work, internal mobility almost by definition excludes industrial or geographical mobility. Moreover, internal and external mobility are likely to be subject to different influences. For this reason they are discussed separately—internal mobility in the subsequent section, while the remainder of this section considers the nature of external mobility.

Studies of employer changes show an impressive amount of job changing in any given time period. It has been estimated that in the United States some one-quarter to one-third of the labour force

[1] Conceptually the definition should include changes in employment status, e.g. between employment and unemployment. This latter aspect cannot, however, be investigated in the present study.

change jobs in the course of a year. In Britain, job changing is also considerable, and the evidence suggests that the bulk of job shifts involve 'complex' mobility, i.e. a change in industry, area or occupation (or some combination of these) as well as a change in employer. For example, Jefferys[1] in her study of Dagenham and Battersea workers found that one-third of all job changes combined a change in area, in industry and in occupation; one-half involved a change in occupation, and three-quarters a change in industry.

Of these types of mobility—geographical, occupational and industrial—the first is least common. While the low rate of geographical mobility is confirmed by all labour market studies,[2] inter-industrial mobility is considerable, and it is difficult to find a systematic pattern in the movements observed. Industrial boundaries do not seem to constitute much impediment to inter-industrial mobility, but this does not preclude the possibility that there are industries which share like characteristics and between which movement is fairly common.[3] As far as this study is concerned the points at issue are the extent to which engineering workers are mobile industrially and, related to this, whether the engineering industry tends to recruit workers over the range of all industries or from certain specific industrial groups.

Occupational attachment appears to be largely a function of the length of training involved. Other things being equal, occupational mobility and skill are inversely related, occupational change becoming less frequent the higher the skill level of the occupation group considered.[4] The explanation for this is simple enough. The worker who has undertaken a long training is naturally reluctant to change occupations if his acquired skill cannot be utilized in the new job. The greater the length of training and the more specific the skill, the greater will tend to be the degree of occupational attachment. Amongst manual workers, therefore, one would predict that the skilled are likely to show least occupational mobility on changing employers, while the occupational attachment of the unskilled will be weakest.

[1] *Mobility in the Labour Market*, p. 59. See also, Palmer, *op. cit.*, pp. 74–5.

[2] See, for example, Government Social Survey, *Labour Mobility in Great Britain, 1953–63* (H.M.S.O., 1966); H. R. Kahn, *Repercussions of Redundancy* (1964) and D. Wedderburn, *Redundancy and the Railwaymen*, University of Cambridge, Department of Applied Economics, Occasional Papers, No. 4 (1965).

[3] For a summary of the evidence on inter-industrial mobility see L. C. Hunter and G. L. Reid, *Urban Worker Mobility* (O.E.C.D., 1968), pp. 66–73.

[4] See G. Bancroft and S. Garfinkle, 'Job Mobility in 1961', *Monthly Labor Review*, Vol. 86, 1963, p. 905.

What would the evidence of previous empirical research lead us to expect as far as our case-study plants are concerned? First, it seems likely that the bulk of manual labour will be drawn from areas adjacent to the plant, both because of employers' behaviour and because job changers will tend to look for work within the immediate neighbourhood. Second, recruits will be drawn from a number of industries although certain industries may have especially strong links with engineering. Third, those recruited from skilled occupations will show considerable occupational stability compared to semi-skilled and unskilled manual workers. These predictions may, however, require some modification to reflect the different circumstances of separate labour market areas. For example, the conclusion that geographical mobility is limited and that manual workers tend to seek employment within their immediate neighbourhood is derived from studies which have generally been based on much smaller labour market areas than Birmingham or Glasgow. Further, in these two areas and in the other labour markets examined the pattern of geographical, industrial and occupational mobility may vary according to differences in employment conditions and the recruitment policies adopted by the plants. The situation is analogous to that found for labour turnover. For example, in Birmingham greater difficulties in securing labour may cause changes in the recruitment pattern, and this may affect geographical, industrial or occupational mobility in different degrees or have a more marked impact on particular types of labour. Such hypotheses can only be satisfactorily discussed after an empirical investigation of mobility patterns. This is undertaken in Chapters 9 and 10.

6. INTERNAL LABOUR MOBILITY

One of the chief advantages of the present study is that it allows a detailed investigation of a much-neglected aspect of labour mobility, namely, job changing within a plant. Previous studies have been almost exclusively concerned with job changing accompanied by employer change—external labour mobility in our terminology. As a result, although a number of interesting hypotheses[1] relating to internal mobility have been developed, few of these have been subject to empirical verification.[2]

[1] See C. Kerr, 'The Balkanization of Labor Markets', in E. W. Bakke (ed.), *Labor Mobility and Economic Opportunity* (1954), and P. B. Doeringer, 'Determinants of the Structure of Industrial Type Internal Labor Markets', *Industrial and Labor Relations Review*, Vol. 20, 1966–7.

[2] A notable exception is H. M. Gitelman, 'Occupational Mobility within the Firm', *Industrial and Labor Relations Review*, Vol. 20, 1966–7.

The labour market internal to a plant can take two extreme forms. A *closed* or structured market is one in which present employees get favoured treatment. New workers are recruited only at the lowest grade and all other vacancies which arise are filled by internal promotion. In contrast, an *open* or unstructured market is one in which there is a port of entry for every grade of work so that a vacancy is filled by external recruitment rather than by internal mobility. Neither of the above extremes is likely to be met in practice, but the concepts are important as points of reference to which we can compare actual behaviour.

The degree to which the employer discriminates between 'ins' and 'outs'[1] when filling vacancies, and the direction, nature and volume of movement between jobs within the plant, will depend on a number of factors. These include the type of unionism in the plant, the nature of technology, and the employment conditions prevailing in the external market and in the plant itself. On the first point, craft unions may reserve certain jobs for workers with particular qualifications so that employers may be forced to recruit additional labour in the external market. On the other hand, the prevalent form of mobility where the labour force is organized in an industrial union may be mobility within the plant. The union may regulate such mobility by control over ports of entry and by laying down the criteria (e.g. seniority) which determine how vacancies are allocated between employees. The nature of the production process may encourage internal mobility because experience on one job can be transferred to another. Alternatively, it may discourage such mobility if there are substantial discontinuities in the division of labour such that there are large differences in the skill and experience required from different groups of workers.

The policy adopted by management may also be influenced by employment conditions. Where the labour market is tight the plant may be forced back on its own resources as a means of filling vacancies, and this process may be reinforced if the plant's own labour force is expanding. It is also possible that institutional barriers to mobility will be less important when employment conditions are favourable and when employment opportunities in the plant are increasing. Pressure on the plant's productive capacity may encourage a review of customary practices with a view to making more efficient use of the present labour force. These and other possibilities are considered in Chapter 11.

[1] 'Ins' are those persons already employed by the plant; 'outs' are non-employees.

7. PERSONNEL AND MANPOWER POLICIES

The last main subject area of this enquiry is concerned with the process of decision-making in a number of areas covered by the omnibus heading of personnel and manpower policies—specifically, manpower forecasting, apprentice training, recruitment, personnel selection and redundancy. Manpower forecasting is an attempt to forecast future developments in the plant's labour force so as to assist in the formulation of current policy. The extent to which manpower forecasting provides such guidance will depend in great degree on the techniques involved—on the time span, detail and method by which forecasts are made. In recent years the subject of manpower forecasting has received considerable attention,[1] and many areas of government planning rely on the estimates made by plants of future labour needs.[2] Yet little is known as to the usefulness or accuracy of manpower forecasting as presently carried through by British plants, and without such knowledge it is not certain that present techniques form a sound basis for private or public action.

Management action with respect to apprentice training raises a number of similar problems. The ability of the economy to adapt to changes in skill requirements brought about by changes in technology and in the pattern of demand will to a considerable extent depend on the manner in which individual plants determine their training programmes. Such programmes may be based on hunch or on sophisticated forecasts of future labour requirements, or they may be based on rules of thumb such as the present size of the labour force or the training facilities currently available. The method chosen will have important implications, for if it yields good results it will assist the plant and the economy to adjust to changing labour requirements. Alternatively, if no attempt is made to anticipate the future, the plant may find it impossible to secure the labour necessary to meet production targets, or may be forced to seek such labour in the external market, thus increasing the competitive pressures on other establishments.

The above issues are discussed in Chapter 12, while Chapter 13 deals with the methods of labour recruitment and personnel selection. The methods by which plants recruit labour are of importance both for public policy and for the general operation of local labour

[1] See R. A. Lester, *Manpower Planning in a Free Society* (1966); Department of Employment and Productivity, *Company Manpower Planning: Manpower Papers No. 1 (New Series)* (H.M.S.O., 1968); and Ministry of Labour, *The Metal Industries, Manpower Studies No. 2* (H.M.S.O., 1965).

[2] The outstanding example is *The National Plan* (1965), Cmnd. 2764.

markets. A labour market, or indeed any market, cannot operate effectively without an adequate information system. Of crucial significance, therefore, is the manner in which a plant communicates its demand for labour to the external market. American studies[1] have suggested that informal methods of labour recruitment are especially important in the case of manual labour and that a relatively minor role is played by the public employment services. These studies, however, have often related to areas where unemployment was high by British standards, and it is possible that methods of labour recruitment will change with employment conditions. For example, it seems likely that employers will adopt less costly methods of recruitment in slack labour markets, and that methods of job-search may differ between large and small communities.

While this argument seems straightforward enough, it is not easy from first principles to ascertain how the operation of the public employment services will be affected by external market conditions. For example, in a tight market the plant may be forced to adopt more active methods of seeking labour so that a higher proportion of its recruits are obtained through the exchanges. Equally well, however, the plant could find that the employment exchanges cannot provide the labour it requires. In this case it may not notify vacancies to the exchanges. If this happens the statistics of unfilled vacancies, which are often used as an indicator of the unsatisfied demand for labour, may not provide an accurate measure of labour market conditions, and the exchanges may play only a minor role in placing labour in new employment.

Chapter 13 pays especial attention to relationships between the case-study plants and the exchanges and attempts to integrate the analysis with a consideration of the alternative methods of recruitment open to the plants. The chapter also considers the organizational framework within which personnel selection takes place and the criteria or 'hiring standards' which are used in selecting employees from amongst job applicants. It is important to know whether plants apply explicit hiring standards on the basis of a systematic analysis of job content, or whether hiring standards are implicit and based on subjective judgments. In the former instance, plants will be able to make a rational choice; in the latter, certain categories of labour may find it difficult to find work even though there are no objective reasons which militate against their employment.

[1] F. T. Malm, 'Recruiting Patterns and the Functioning of Labor Markets', *Industrial and Labor Relations Review*, Vol. 7, 1953–54; and J. C. Ullman and D. P. Taylor, 'The Information System in Changing Labor Markets', *Proceedings of the Industrial Relations Research Association*, Vol. 18, 1965.

The procedures operated in a redundancy situation are examined in Chapter 14. A number of studies have investigated the difficulties experienced by redundant workers in seeking new jobs,[1] but relatively little work has been undertaken on the mechanisms through which managements regulate redundancies.[2] Our main task is, then, to describe and analyse the policies adopted by managements in redundancy situations; for example, the extent to which redundancies are avoided or reduced by manpower forecasting; the extent to which redundancy procedures are regulated by formal agreements, and the methods by which those to be declared redundant are chosen. In addition, we look at the type of labour discharged in redundancy situations in Chapter 8, and the analysis is taken a stage further in Chapter 14 which discusses whether employers discriminate between different occupational groups. This raises the vexed question of 'labour hoarding' and the extent to which employers retain labour, especially skilled labour, in circumstances where the labour force is greater than that required to meet present production needs. The discussion also includes an analysis of the size and frequency of redundancies and attempts to ascertain whether these variables are related to the size of the plant.

8. SUMMARY

The above areas of investigation are considered, as it were, in isolation because this is necessary for the purposes of exposition, but, of course, they are interrelated, and the distinctions drawn between the subject areas are somewhat artificial. This must be borne in mind when considering the broader implications of the study. The primary aim is to investigate the labour market policies of plants in the engineering industry and to consider how these policies are modified by the labour market environment in which the plants operate. The kinds of question we shall consider include the following. Do managements react to a labour shortage by raising wages? In a tight labour market, are competitive forces so strong that each plant has to pay *the* market rate for a particular grade of labour? Are imperfections introduced by lack of knowledge, by non-economic motives and by collusion between employers, so that labour market behaviour bears little relationship to that presupposed by assumptions such as economic rationality and perfect knowledge? Can recruitment diffi-

[1] See Kahn, *op. cit.*, and Wedderburn, *op. cit.*

[2] The best summaries of the present position in Britain are: Ministry of Labour, *Dismissal Procedures, A Report* (H.M.S.O., 1968); and Acton Society Trust, *Redundancy, A Survey of Problems and Practices* (1958).

culties be met, not by raising wage levels, but by lowering hiring standards, by intensified recruitment campaigns, by greater internal mobility, by expanded training programmes or by some combination of these and other measures? The important point is that management has a choice between a set of alternative policies in all the fields considered, whether it be wage-setting, labour recruitment or redundancy situations. It is only by investigating behaviour in a particular labour market context that we can discover the policies adopted and the factors which mould these policies.

CHAPTER 2

METHOD OF ENQUIRY

1. INTRODUCTION

The labour market areas on which this study is based were chosen to provide a 'mix' of different circumstances which would assist analysis and interpretation of the relationship between a plant's labour market policies and the external labour market environment. The same considerations dictated the choice of period, 1959–66, which covers virtually two cycles in each labour market area.[1] Hence it is possible to contrast behaviour between areas and also within areas as labour market conditions changed.

The present enquiry differs from many previous labour market studies in that its main purpose is to describe and explain the actions of employers rather than the labour market behaviour of employees. The main source data were the personnel records of engineering plants. It appears that only one previous study[2] has attempted to obtain such information on a similar scale. In that study, however, the purpose was to identify and explain the mobility of employees *between* different employers, a subject which is, in fact, not easily investigated by this method.[3] No attempt was made to cover the range of topics considered in the present investigation.

The distinction between the actions of employers and employees is in a sense misleading. One would expect considerable interaction between these two groups in that each has to take into account the behaviour patterns of the other. Thus, the present study contains a considerable amount of information on the process of labour turnover and labour mobility just as previous studies primarily concerned with employees have provided valuable insights on the labour

[1] It was not possible to begin the study in 1958 because of the break in the *Standard Industrial Classification* which occurred in that year.

[2] C. A. Myers and W. P. Maclaurin, *The Movement of Factory Workers* (1943).

[3] In the above study based on two Massachusetts towns, most of the larger plants co-operated. Despite this, it proved possible to observe only a small proportion of all employer changes.

market policies adopted by plants. Nonetheless, our emphasis on employer action is worth stressing. Much less is known about the labour market policies adopted by plants than is known about the actions and attitudes of the labour force, and, given this, our understanding of the processes at work is incomplete and imperfect.

The methodology of this enquiry is therefore different to that commonly employed in labour market surveys. Most surveys have been based on a relatively small group of employees seldom numbering more than a few hundred, and have studied job histories through questionnaire techniques. The strength of this approach has been its ability to study mobility over time and to probe employee attitudes, aspirations and motives in some depth. Its weakness has been its high cost, which has severely restricted the size of the sample. It is often difficult to isolate the various factors influencing mobility and it is seldom possible to give quantitative expression to the relationship between variables such as earnings, unemployment and turnover. Further, the data collected have usually related to labour mobility as evidenced by changes in employer so that little is known about labour mobility within the plant.

A different method of analysis based on employer records obviously means that we have less data on job changes, and can at best compare only single job transactions. Thus, for those case-study plants whose personnel records provide the necessary details, we can examine whether the employee on entering the plant changed his industry, occupation or area of work. This analysis of labour mobility cannot, however, be carried any further back to examine any previous changes in jobs. Nor can it be carried further forward, as no details of the individual's work history are available once he leaves a case-study plant. Yet the present study does allow a detailed analysis of labour turnover, a process important in its own right and one which is related to mobility. An understanding of the factors which determine turnover may assist in understanding the economic, institutional and personal considerations which shape and influence labour mobility. Here the present study has the advantage of being able to examine a large number of cases and to relate these to employment conditions, earnings and other variables in a manner not possible where the original sample is small and is investigated through survey techniques.

The sources from which employers recruit labour, and the process of labour mobility when individuals take up employment with the case-study plant, are examined in the following chapters. In addition, it is possible to investigate the channels through which employees seek new work and the part played in this process by the public employment service. The main emphasis, however, is placed on the labour market policies of plants, this term being defined to include such topics as wage policies, upgrading and downgrading within plants, redundancy procedures, manpower forecasting, hiring standards, recruitment and selection procedures and so forth. At all points the discussion is linked to the employment conditions and other characteristics of the labour market areas in which the plants operate, in an attempt to establish how policies are modified by underlying economic forces. It follows from this that the choice of labour market areas was crucial, for to understand the relationship between economic conditions and plant policies it was necessary to select plants which were faced with different market environments. Section 2 explains the reasons for the choice of the labour market areas included in this enquiry. The selection of case-study plants within these labour market areas is described in Section 3, while Section 4 outlines the statistical information at our disposal.

2. SELECTION OF THE LABOUR MARKET AREAS

Most of the major labour market studies have been based on American conditions, and the results obtained may require substantial modification before being applicable to local labour markets in Britain. The institutional framework differs between the two countries and, probably more important, American research has been based on local labour areas where unemployment has been high by comparison with the levels experienced in post-war Britain. This is certainly true of the 'first generation' of labour market studies in the 1930s and early 1940s but the same reservation applies, although in a more modified form, to those conducted in the post-war period. Since 1945 the level of unemployment in the United States has generally been substantially above the British average,[1] and there has been no study of labour market behaviour in a market as 'tight' as that in Birmingham over 1959–66. In Britain, greatest attention has been paid to behaviour in redundancy situations,[2] which, although important, is not typical of normal market conditions in a full employment economy.

In view of these considerations, the choice of Birmingham[3] as one

[1] This statement remains valid even when allowances are made for differences in the definitions used to measure unemployment. See J. S. Zeisel, 'Comparison of British and U.S. Unemployment Rates', *Monthly Labor Review*, Vol. 85, 1962 and President's Committee to Appraise Employment and Unemployment Statistics, *Measuring Employment and Unemployment* (1962), p. 255.

[2] The exception is Jefferys, *op. cit.*

[3] The Birmingham labour market is taken to cover the employment exchange

of the labour market areas to be examined was logical enough. Of the conurbations in Britain outside London it was the one in which unemployment was lowest[1] and labour in shortest supply over 1959–66. For males, unemployment rarely exceeded 3 per cent and fell as low as 0·5 per cent. Female unemployment never reached 2 per cent and was most often less than 1 per cent of the insured labour force. As a contrast to Birmingham, Glasgow was chosen as the second labour market area.[2] Like Birmingham, it is a large urban market with a long-established engineering industry, but the general level of employment was contracting, as was employment in engineering, and unemployment was substantially in excess of the British average throughout 1959–66.[3]

Similarly, the three other labour market areas in Scotland, North Lanarkshire, New Town and Small Town,[4] each offered the possibility of studying the functioning of labour markets under a different set of conditions, although each had in common the establishment of new engineering plants in their respective localities. Most of these plants had been located in the 1950s or 1960s and were building up their labour forces over the period of the study. This expansion in engineering employment was taking place against very different backgrounds. North Lanarkshire is an area of intermediate size, insured employees numbering 173,000 in 1966 compared to 549,000 in Glasgow and 695,000 in Birmingham. It is an industrialized area with a number of medium-sized towns, where coal-mining was until recently an important industry and with a long tradition of engineering employment. However, the plants which are native to the area are mainly engaged in 'heavy' engineering and differ in important respects from the case-study plants most of which were located in the region during the post-war period. The remaining areas, New Town and Small Town, differed from Birmingham, Glasgow and North Lanarkshire in that their labour force was smaller and manufacturing industry was relatively unimportant. The economy of New Town had been based on coal-mining in which employment fell sharply over

areas of Aston, Birmingham, Handsworth, Selly Oak, Small Heath and Washwood Heath.

[1] From the *Employment and Productivity Gazette* fifteen observations were obtained of average unemployment rates in the major conurbations over 1959–66. With the exception of London, the unemployment rate in Birmingham was more often below than above the unemployment rate in each of the other conurbations.

[2] The Glasgow labour market comprises ten employment exchange areas in the County of the City of Glasgow plus the adjoining burgh of Clydebank.

[3] See Figure 3.1, p. 56 below.

[4] North Lanarkshire consists of twelve employment exchange areas, New Town of four and Small Town of one.

1959–66. The insured labour force in the latter year numbered 65,000 compared to 8,400 in Small Town, which is a small town situated in a remoter area of Scotland, heavily dependent on primary industry.

The nature of production technology also differed substantially between the areas. Most of the engineering plants in Glasgow specialized in one-off, or small-batch, custom-built production of producer durable goods. Large-batch production of consumer durables was much more typical of the engineering industry in Birmingham and of the case-study plants in the Other Scottish Areas. These differences in technology may have influenced the types of labour required and the use made of that labour, so that an examination of these contrasting situations might go some way to increasing our understanding of the operation of local labour markets.

3. SELECTION OF THE CASE-STUDY PLANTS

The decision to base this study on the engineering industry was made for a number of reasons. The scope of the subjects which the enquiry set out to investigate required a detailed examination of individual establishments, and the problems to be investigated were inter-related. For example, labour turnover is likely to be influenced by wage-setting and by personnel policy with regard to selection, training and dismissal. To hold such a discussion together would have been difficult, or even self-defeating, had an attempt been made to cover all types of employing unit. There were more positive reasons for confining the investigation to the engineering industry. It is the most important single sector of manufacturing industry, the source of much technological progress and the major exporting industry in Britain. Viewed simply on labour market considerations, engineering is important for a number of reasons, not least the size and skill content of the labour force engaged, the complex system of wage determination and the tendency of the industry to be a 'wage leader'.[1]

The enquiry was confined to manual workers, mainly because the method of investigation, which was suitable for a study of manual labour, did not lend itself to a similar examination of staff employees at clerical, technical and managerial levels. It is known that most manual workers seek employment within a fairly narrowly defined

[1] More accurately the Vehicles sector is a wage leader. At the end of our period (October 1966) adult male weekly earnings in Vehicles averaged 439s compared to 406s for all industries. Weekly earnings were higher only in Paper, Printing and Publishing. *Statistics on Incomes, Prices, Employment and Production*, No. 21, June 1967 (H.M.S.O.), pp. 26–7.

geographical area. This being so, it is possible to analyse the actions of plants and of manual workers by reference to employment conditions within the locality.[1] On the other hand, technical and, particularly, managerial staff are likely to be more mobile geographically, and hence an examination of their labour market behaviour requires a different context to that provided in this study. For this reason the only staff employees included in this enquiry were those who had been promoted from the manual labour force during the study period.

In the Other Scottish Areas no attempt was made to obtain a representative sample of the population of engineering plants. Instead the investigation was deliberately focused mainly on engineering units which had recently been located in these areas and had been expanding their labour force over the period 1959–66. The two case-study plants in Small Town were the only engineering establishments in the town. Five plants were taken from North Lanarkshire and a further five from New Town. In Birmingham and Glasgow, on which we base the bulk of our analysis, a sample of plants was selected to represent some of the main characteristics of the engineering industry in these areas, the sample comprising 25 plants in Birmingham[2] and 27 in Glasgow. The sampling frame was based on the size of units and on the product manufactured as shown by the Main Order Headings of the 1958 Standard Industrial Classification.[3] It should be made clear that the case-study plants in Glasgow and Birmingham form a 'representative' or 'indicative' rather than a random sample of engineering units. While for statistical reasons a random sample is generally to be preferred, practical difficulties rendered such an approach impossible in the

[1] This raises the question how one defines the terms 'locality' and 'local labour market'. Here a local labour market is taken as that area in which most manual employees normally seek work and from which employers recruit the bulk of the labour force. The concept is therefore difficult to define operationally and may vary between occupational groups and from one plant to the next. For empirical purposes, however, some imprecision has to be accepted. Whether this imprecision introduces serious errors depends on the patterns of geographical mobility and recruitment in those localities taken as labour market areas in this study. This is examined in Chapter 9.

[2] This represents the number whose personnel records were processed and analysed. A further two plants in Birmingham were interviewed and did provide wage data. This information was utilized when it was appropriate.

[3] The 1958 *Standard Industrial Classification* contained 24 Main Order Headings which were further divided to give 153 Minimum List Headings. The engineering industry as defined in this study consists of three Main Order Headings (VI, VIII and IX) plus the Minimum List Heading of Marine Engineering.

circumstances of this present enquiry. First, a handful of plants originally selected for inclusion in this study were subsequently rejected because their personnel records were not sufficiently comprehensive or accurate for the purposes of the enquiry. Second, and more important, it was found, as expected, that the 'response rate' achieved varied significantly according to the size of the establishment considered. In particular, the response rate was low for small units with 200–500 employees, and especially so in Birmingham. The procedure adopted was to replace any plant refusing to cooperate by a plant with like characteristics in terms of size and product division. This being so, the sample could not be generated by a random process, but the alternative would have been a sample dominated by the larger units.

One further qualification should be borne in mind. There was an extremely large number of small engineering plants in both conurbations, especially in Birmingham, about which little was known except that they employed less than 200 or 250 employees. Of some 1,300 engineering units in the latter area almost 1,200 had fewer than 250 employees. In Glasgow, 93 in a total of 156 units employed fewer than 200 persons. It was not possible to provide adequate representation for these smaller units although the sample drawn, on the basis of employment in 1966, did include four units with fewer than 250 employees in Birmingham and two units with fewer than 200 employees in Glasgow.[1] The samples of case-study plants in Birmingham and Glasgow were therefore intended to be representative only of those establishments above the 'cut-off points' of 250 and 200 employees respectively, but the great bulk of the engineering labour force was employed in such units. The extent to which this intention was realized can be judged by expressing the number of case-study plants by size range and by product division as a proportion of the total number of case-study plants in Birmingham and Glasgow above these cut-off points. This can then be compared to the characteristics of the total population of engineering units with more than 250 and 200 employees respectively.

The Glasgow sample contains no units from Marine Engineering or Metal Goods Not Elsewhere Specified, so that there is a bias in the industrial distribution of the sample in favour of Engineering and Electrical Goods, and Vehicles. In other respects the samples in both Birmingham and Glasgow show characteristics very similar to the engineering industries in these areas when the smaller units are

[1] The information used to construct the sampling frame was not always accurate. In particular it seemed on subsequent examination to exaggerate the number of employees in small to medium-sized units.

excluded. An examination of the behaviour of the case-study plants should, therefore, provide a good indication of the manner in which a plant's labour market policies are influenced by the external labour market environment.

Percentage of plants by size range[1]

No. of employees	Birmingham		Glasgow	
	Sample	Population	Sample	Population
250–499[2]	42·9	44·5	48·0	46·7
500–1,499	42·9	39·1	36·0	38·3
1,500 plus	14·3	16·4	16·0	15·0

Percentage of plants by product division[1]

Product division	Birmingham		Glasgow	
	Sample	Population	Sample	Population
Engineering and Electrical goods	36·0	37·3	88·0	79·3
Marine Engineering	0·0	0·0	0·0	10·3
Vehicles	28·0	24·5	12·0	6·9
Metal Goods not elsewhere specified	36·0	38·2	0·0	3·4

Note: 1. Individual items may not add to 100·0 because of rounding.
2. In Glasgow, 200–499 employees.

4. SOURCES OF DATA

As has been indicated, this enquiry was primarily concerned with the plant as a decision-making unit and in studying the reactions of plant management to differences in labour market conditions. It was therefore necessary to obtain a detailed record of the level and composition of employment in the case-study plants over 1959–66, so as to analyse through these data the impact of market forces. In all, four main sources of statistical data were utilized, supplemented by discussions with management representatives responsible for labour, manpower and personnel policy in the case-study plants.

Plant personnel records were the basic source material of the enquiry. When an individual is first engaged by a plant it is common practice to complete an employee record card showing details such as date engaged, job taken, date of birth, marital status and home address. This record card is then retained and updated where necessary to show details of changes in job, department, rates of pay, date of leaving, etc. Although the amount and the form of documentation varied considerably between plants, an initial survey revealed that certain items of information were recorded by almost all establishments. In the light of this, a standard form, the 'data sheet', was devised to record from employee record cards such information as was considered relevant.

The data sheet contained four sections. The first dealt with the personal characteristics of the employee such as sex, marital status, date of birth, home address and training. Second, there was a section which recorded details of entry to and exit from the plant—dates of entry and exit, job on joining and reason for leaving. The third section was designed to record work experience with the plant, so providing details of internal mobility. Fourth, space was provided to record information relating to the employer, industry and occupation from which the employee had been recruited. The above information was coded and entered on punch cards and magnetic tapes for subsequent analysis. The design of the data sheet and the coding procedures adopted are described in detail in the Appendix.[1]

It was not practicable to obtain for large units a complete record of all those employed as manual workers over 1959–66. In these cases, therefore, a sample of employees was taken on the following basis: a 1 in 2 sample for plants with 1,000 to 3,500 employees; 1 in 4 for plants with 3,500 to 5,000 employees, and 1 in 6 for plants with more than 5,000 employees.[2] In the following analysis the data sheets for such plants are 'grossed up' by the appropriate factor so as not to overweight the importance of plants with less than 1,000 employees.[3] In most plants, however, a data sheet was completed for all manual employees, no matter how short the period of employment over 1959–66. Thus a 100 per cent sample was taken for 15 of 25 plants in Birmingham, for 20 of 27 units in Glasgow and for 8 of the 12 establishments in the Other Scottish Areas.[4] This process yielded

[1] See p. 418 below. The reader interested in the details of the methodology is advised to consult the Appendix before proceeding further.

[2] As plant records were usually held in alphabetical order, the sample was drawn by taking every second, fourth or sixth record card for individuals employed by the plant at any time over 1959–66. Initial tests showed that this gave similar results to those obtained by a random number technique, and the sampling errors resulting were small.

[3] For reasons which are explained in Chapters 9 and 10 and the Appendix, this procedure was not followed in certain instances.

[4] In Birmingham, a 1 in 2 sample was sometimes applied to those who had *left* the plant over 1959–66 even where the plant employed less than 1,000 persons. This occurred when the turnover rate was extremely high. A 100 per cent sample

46,000 data sheets for Birmingham plants, 33,000 for Glasgow and 18,000 for the Other Scottish Areas.

Apart from sampling procedure, data collection had to allow for variations in terminology between plants and for the fact that certain items of information were not available for all employees. Hence certain reservations or qualifications may apply to particular items of information obtained from personnel records, and these will be noted at the appropriate points in the ensuing discussion. For the moment, it is sufficient to observe that the data collected from personnel records allow a close insight into the problems and difficulties encountered by case-study plants over 1959–66.

The second source of statistical data was the wage returns supplied by the Department of Employment and Productivity and by the Engineering Employers' Federation. A number of case-study plants in each of the labour market areas had supplied wage returns to the D.E.P. showing average plant earnings for males and females over 1959–66 and average occupational earnings for males over the period 1963–6.[1] Both types of D.E.P. earnings returns related to earnings in a particular week and were available on a biannual basis. The E.E.F. obtained annually from its member units a return of average male occupational earnings in a particular week. As with the occupational earnings collected by the D.E.P., these returns showed by occupational group total weekly earnings, overtime premium payments, total and overtime hours worked, and the number of manual workers employed. The Federation returns were available for Glasgow and for the Other Scottish Areas for the years 1959–62 and 1964–6, but were not, unfortunately, available for any of the units in Birmingham. The case study plants were asked for permission to consult these documents which, like the D.E.P. wage returns, provided wage information on a standardized basis and were of considerable value in investigating wage-fixing policies, wage structures and related problems such as turnover.

Third, background information relating to employment conditions

was taken of those employed at the end of 1966 and the appropriate weighting factor applied to 'past employees'.

[1] The former D.E.P. returns, which were collected under voluntary arrangements, were confidential, and each plant concerned gave the survey team written permission to draw on the material in respect of the specific reference dates. Information about occupational earnings, and also the data on 'L returns' (see p. 46), were collected under the Statistics of Trade Act, 1947. As a result, no information derived from these returns about an individual undertaking could be disclosed without the prior consent in writing of the undertaking concerned. Such authority was obtained by the survey team from every plant surveyed and in respect of each reference date used.

and to engineering plants in the labour market areas was supplied by the D.E.P., the sources being unpublished returns of insured employees, unemployment, etc., and 'L returns'. The former were exceptionally important as they permitted analysis of the influence of labour market conditions as represented by employment, unemployment and unfilled vacancies. In addition, it was possible to examine the industrial and occupational distribution of employment and unemployment. The L returns, although less extensively used, provided data on industrial and occupational employment, sex distribution and overtime and short-time working *by plant*. As a result, the characteristics of the case-study plants could be compared to those of other engineering establishments in the labour market areas.[1]

The fourth type of empirical data was recruitment information obtained through the case-study plants and the public employment exchanges. Thirty of the 64 case-study plants in the labour market areas were able to collect details showing the recruit's occupation, date of entry and method of recruitment—how the recruit had first heard about the job vacancy. In addition, employment exchanges in the labour market areas were asked to provide records of vacancies notified by and placements made with the case-study plants during 1966. The latter records were made available for Glasgow and for the Other Scottish Areas. A comparison with recruitment by the case-study plants (obtained from the data sheets) revealed the extent to which plants notified vacancies to the exchanges and the role played by the latter in meeting recruitment needs. From these two sources it was possible to examine the channels of information through which employers and employees made contact and the extent to which the use made of different media varied between labour market areas.

In addition to the four sources of statistical data, management representatives of each plant were interviewed by members of the research team. Wherever possible two or more managers were interviewed separately, and the interviews were usually conducted with personnel managers, works managers, managing directors or accountants. They were designed to allow an unstructured discussion, but, to maintain consistency, an attempt was made to obtain answers from all plants to a set of predetermined questions. The areas covered in these interviews dealt with manpower forecasting and

[1] As explained above (p. 45, n. 2), the appropriate authority to use L return data was obtained from each plant concerned (case-study and non-case-study) in respect of all relevant reference dates. Comparison of L returns for case-study and non-case-study establishments in Birmingham and Glasgow showed that in each area both groups had like characteristics in terms of overtime and short-time working between 1959 and 1966, and in terms of the occupational structure and the distribution of employment by sex.

planning, redundancy, recruitment methods, selection procedures, internal mobility, training and wage structure. These interviews gave a valuable insight into a number of topics which could not be approached solely through statistical data and form the basis for the discussion of management's views and opinions contained in the following chapters.

CHAPTER 3

THE LABOUR MARKET AREAS

1. INTRODUCTION

The labour market policies adopted by a plant are likely to be influenced by the relative ease or difficulty with which it can hire labour of the requisite quality. Normally the plant will prefer to hire manual labour from within a fairly restricted geographical area in order to minimize recruitment costs. In other words, we assume that impediments to geographical mobility exist[1] and that the behaviour of management will be conditioned by employment conditions within the relevant labour market area. This, in turn, depends upon the balance between the demand and supply of labour, on employment trends and on the existing distribution of employment between industries and occupations. It is, therefore, important to examine the characteristics of the labour market areas included in this study in order to establish the different circumstances faced by the case-study plants in the separate localities.

We have already noted some of the more significant contrasts between the labour market areas selected for examination. Birmingham and Glasgow are both major conurbations with established engineering industries. They differ in that employment increased in Birmingham but fell in Glasgow over the period 1959–66. In Glasgow, unemployment was substantially above the national average, while Birmingham is best regarded as a labour market which experienced only temporary departures from full employment. North Lanarkshire is a region embracing a number of medium-sized towns with an established industrial tradition. Over 1959–66, employment rose substantially, most noticeably in engineering, so that unemployment tended to decline from the high levels experienced at the beginning of the period. The secular trend in unemployment was upwards in New Town where the increase in engineering employment, although extremely rapid, was insufficient to counteract the

[1] The concept of local labour markets itself depends on the existence of barriers to geographical mobility. Needless to say, this does not exclude *any* geographical mobility but merely presupposes that such mobility is restricted in various ways.

substantial decline of the coal-mining industry. In Small Town, engineering employment rose, but the overall level of employment showed little change over the period.

The remaining sections of this chapter add some substance to this skeleton description. The composition and distribution of the labour force and the employment trends in each labour market area are outlined in Section 2. Section 3 investigates the relationship between labour demand and supply by means of vacancy and unemployment data. Finally, Section 4 examines the experiences of the case-study plants in recruiting labour. This can then be set against the pattern which emerges from the empirical data to establish the extent and the nature of the recruitment problems facing plants in the different labour markets.

2. EMPLOYMENT

The most significant differences in the industrial structure of employment occur between Birmingham, Glasgow and North Lanarkshire, on the one hand, and New Town and Small Town on the other. In the former areas, manufacturing industry was well established at

TABLE 3.1

INDUSTRIAL DISTRIBUTION OF TOTAL EMPLOYMENT:[1] LABOUR MARKET AREAS, JUNE 1966

Industrial group	Percentage				
	Birmingham	Glasgow	North Lanarkshire	New Town	Small Town
Primary	0·1	0·1	0·5	2·5	17·4
Extractive	*	0·6	1·4	16·0	*
Engineering	38·5	15·7	27·6	9·2	12·6
Other Manufacturing	16·9	22·8	24·3	21·3	21·6
Construction and Services	44·5	60·7	46·3	51·0	48·4

Notes: 1. Insured employees less unemployment.
* Signifies less than 0·05 per cent.

the beginning of our period, with substantial numbers engaged in engineering production. Engineering employment is a relatively recent development in New Town and Small Town which, in the past, have been heavily dependent on primary and extractive industries. There was no large quantity of skilled or experienced labour on which new engineering establishments could draw. These differences were still in evidence near the end of our period, as Table 3.1 demonstrates.

TABLE 3.2

PERCENTAGE CHANGE IN EMPLOYMENT:[1] LABOUR MARKET AREAS AND GREAT BRITAIN, JUNE 1966 OVER JUNE 1959

Category	Birmingham	Glasgow	North Lanarkshire	New Town	Small Town	Great Britain
Males	+10·4	−10·7	+22·4	−19·1	−2·9	+ 5·4
Females	+9·3	+4·9	+37·3	+25·6	+13·2	+14·1
All employees	+10·0	−4·9	+27·1	−7·2	+2·4	+8·5

TABLE 3.3

PERCENTAGE CHANGE IN ENGINEERING EMPLOYMENT:[1] LABOUR MARKET AREAS AND GREAT BRITAIN, JUNE 1966 OVER JUNE 1959

Category	Birmingham	Glasgow	North Lanarkshire	New Town	Small Town	Great Britain
Males	+4·1	−19·3	+40·7	+99·9	+72·2	+11·5
Females	+1·7	−15·2	+73·1	+832·1	+105·5	+18·8
All employees	+3·4	−18·6	+48·8	+191·8	+76·5	+13·2

Note 1. Insured employees less unemployment.

In 1966, 16 per cent of the employed labour force in New Town was still engaged in extractive industries (effectively coal-mining) despite the fact that such employment had fallen from 22,500 to 10,000 since 1959. Primary industry accounted for a substantial proportion of the work force only in Small Town, although here too employment was declining both absolutely and relatively. In Birmingham, Glasgow and North Lanarkshire, primary and extractive employment was of little consequence and manufacturing industry correspondingly more important. This is most obvious in the case of Birmingham which is essentially a manufacturing centre dominated by the engineering industry.

Employment trends over the period 1959–66 are shown in Tables 3.2 and 3.3. Table 3.2 provides details of the percentage changes in total employment and Table 3.3 presents similar data for engineering.

The most rapid rise in total employment (males and females combined) occurred in North Lanarkshire, the rate of increase being three times the national average. In Birmingham, too, total employment rose more quickly than the national average although the growth of female employment was relatively slow. It is also evident that only in Birmingham was the trend in female employment less favourable than that for males. As the demand for female labour was extremely buoyant in Birmingham throughout 1959–66, this must reflect a low elasticity of supply of female labour arising from a high participation rate[1] and low unemployment relative to other areas at the beginning of our period. There was simply less 'slack' in the market which could be taken up as employment opportunities for females increased.

Male employment declined in Glasgow, New Town and Small Town over 1959–66, the decline being sufficiently rapid to offset the rise in female employment in the two former areas. Glasgow also experienced a rapid contraction in engineering employment for both males and females, contrary to the national experience and to that of the other four labour market areas. The most rapid expansion in engineering occurred in New Town but, because the base from which this expansion took place was small, it was not sufficient to offset declining employment in coal-mining. Similarly, total employment expanded very slowly in Small Town, as increased engineering employment was largely counterbalanced by the contraction of the primary industries. The dependence of New Town and Small Town on declining industries can be further illustrated by comparing employment trends to those in North Lanarkshire. The percentage increase in engineering employment in North Lanarkshire was, for both males and females, substantially below the increases in either New Town or Small Town. Despite this, total employment expanded very rapidly in North Lanarkshire while declining or remaining stagnant in New Town and Small Town. This was due to the greater importance of the engineering industries in North Lanarkshire at the beginning of the period and to the relative absence of major declining industries in that area.

The contrasts between the two conurbations are just as striking. Total male employment fell in Glasgow at the same rate as it rose in Birmingham, and female employment increased at only half the Birmingham rate. While engineering employment showed some tendency to increase in Birmingham over 1959–66, Glasgow was the only area in which engineering employment was contracting. This has implications for the labour market policies of the Glasgow case-study plants, especially when we take into account the parallel

[1] The participation rate is given by the number of employed and unemployed (the 'economically active') as a percentage of all females aged 15 years and over.

decline of the shipbuilding industry in Glasgow. Although shipbuilding has been excluded from the engineering industry as defined in this study, it has amongst its labour force groups with skills similar to those employed in engineering establishments, such as welders, fitters, sheet metal workers and electricians. Over 1959–66, shipbuilding employment in Glasgow fell from 31,500 to 16,500, thus providing to the case-study plants a source of labour with skills relevant to engineering employment.

The remaining feature of the industrial structure to which attention should be drawn is the distribution of employment between sectors of the engineering industry. This varies from area to area—a fact

TABLE 3.4

SECTORAL DISTRIBUTION OF TOTAL ENGINEERING EMPLOYMENT: LABOUR MARKET AREAS, JUNE 1966

Engineering Sector	Percentage				
	Birmingham	Glasgow	North Lanarkshire	New Town	Small Town
Engineering and Electrical	33·4	66·7	77·1	88·6	90·9
Vehicles	34·8	18·3	12·8	4·1	0·4
Other Metal Goods	31·8	7·6	10·1	6·9	1·8
Marine Engineering	0·0	7·5	0·0	0·3	7·0

which is of some importance in explaining observed differences in labour market policies. Table 3.4 shows the distribution of engineering employment in each labour market area at June 1966.

It is apparent that Vehicles and Other Metal Goods assume greater importance in Birmingham than in the remaining areas where the predominant sector is Engineering and Electrical Goods. Indeed, the dependence of Birmingham on the Vehicles sector, based largely on motor-car production, is greater than the above figures imply because of the linkages between the car assembly plants and their component suppliers in the area. One result of this is that male unemployment, and particularly male unemployment in engineering, is subject to sharp cyclical swings in Birmingham although any increase in unemployment was generally short-lived over 1959–66. In the other areas, the Vehicles sector is smaller and is based on the production of commercial vehicles, aircraft components or locomotives, where employment levels have proved more stable over the short run.

These differences between the areas in vehicle output underline the need to look behind the broad aggregates on which Table 3.4 is based. Indeed, when the Main Order Headings are broken down to the Minimum List Headings, differences in product 'mix' are still concealed by the Standard Industrial Classification. In Birmingham, the motor-car industry, its component suppliers and other consumer durable production predominate, so that most engineering units are based on large batch production of standardized units. Although products differ, production technology is similar in New Town and

TABLE 3.5
PERCENTAGE DISTRIBUTION OF MANUAL EMPLOYMENT BY OCCUPATIONAL GROUPS IN CASE-STUDY PLANTS: MALES, 31/12/66

Occupational group	Birmingham[1]	Glasgow	North Lanarkshire	New Town	Small Town
Apprentices	3·3	6·7	5·4	1·9	4·2
Skilled	16·0	29·4	21·0	11·6	17·8
Skill not specified[2]	9·6	1·1	1·9	1·9	3·2
Semi-skilled, production	51·0	45·2	57·4	75·2	62·5
Unskilled, production	8·2	8·9	6·1	4·5	4·0
Others[3]	11·9	8·7	8·2	4·9	8·3

Notes: 1. Data for two of the smaller case-study plants excluded because of a large 'not known' element.
2. The nature of the work indicated that the bulk of this group (welders, inspectors, etc.) was engaged on semi-skilled work.
3. Includes those whose job was 'not known'. The proportion of 'not knowns' is greatest in Birmingham and Glasgow, at 3·3 per cent and 0·4 per cent respectively.

Small Town and in the recently established engineering plants in North Lanarkshire included in this study.

In contrast, the 'typical' engineering unit in Glasgow specializes in 'one-off' or small batch, custom-built production of producer durable goods. Hence the technology of the Glasgow case-study plants differs substantially from that of the case-study plants in Birmingham and even from that of units in the Other Scottish Areas, despite the apparent similarity conveyed by the predominance of the Engineering and Electrical Goods sector. In addition, Glasgow is the only area, apart from Small Town, with a significant proportion of marine engineering employment, a sector closely tied to shipbuilding and one in which employment was halved over 1959–66.

This difference in production technology is reflected in the occupational composition of the labour force of the case-study plants. Table 3.5 shows the distribution of male manual employment by skill groups at the end of our period. The highest proportions of skilled men and apprentices are found in the Glasgow case-study plants, especially relative to New Town, but also when comparisons are made with Birmingham, North Lanarkshire and Small Town. The difference in production techniques is also reflected in the sex composition of the labour force. The percentage of female employees is lower in Glasgow than in any other area except Small Town. The figures are 15, 21, 33, 56 and 11 per cent for Glasgow, Birmingham, North Lanarkshire, New Town and Small Town respectively. Hence the specialization of Glasgow plants on 'one-off' or 'small batch' production demands a highly skilled labour force capable of working to fine tolerances and adapting to changes in the 'work mix'. In the other labour markets a greater proportion of the plants are engaged on large batch production of standardized items. The repetitive nature of the work enables the unit to employ large numbers of semi-skilled workers, usually accompanied by a high proportion of female labour.

3. LABOUR SUPPLY AND DEMAND

The balance between the demand and supply of labour is generally measured by comparing the number of unemployed registered with, and the number of unfilled vacancies notified to, the employment exchanges.[1] Unemployment is taken as the relevant measure on the supply side showing the unused reserves of labour available to employers, while the number of job vacancies notified to the exchanges and remaining unfilled is the appropriate measure on the demand side showing unsatisfied manpower requirements. Unfortunately neither series is a wholly satisfactory proxy for the variable it is taken to represent. The unemployed may not constitute a useful reserve of labour where they are predominantly persons who are unskilled, in the older age groups, or mentally or physically handicapped. On the other hand, where the financial incentive to register as unemployed is weak or absent (as is the case for many married women) unemployment statistics may underestimate the number of persons who would take up employment opportunities were they available. This is likely to be particularly important in areas where registered unemployment is already high.[2]

[1] Notified vacancies are all job openings registered with the exchanges. Unfilled vacancies are the number of notified vacancies which remain outstanding at a given date. Both may differ from the 'true' number of vacancies given by the number of job openings which employers are currently attempting to fill.

[2] See p. 60 below.

The unused reserves of labour on which plants can draw may, therefore, be greater than or less than the number registered as unemployed. Similarly, unfilled vacancies notified to the employment exchanges are unlikely to coincide with the total unsatisfied demand for labour. Employers may not notify vacancies, because they can obtain such labour from other sources or because they believe that the exchanges cannot satisfy their requirements in a 'tight' labour market. In the latter situation, it is also possible that employers might deliberately overstate their requirements in order to obtain a larger share of a restricted labour supply.[1] On *a priori* grounds we might suppose that at any point in time, notified vacancies are likely to understate the total demand for labour in a 'slack' labour market; in a 'tight' market, understatement, or overstatement, appears equally plausible. We shall see later that understatement appears the most likely outcome in most labour market areas, but there is no way of establishing the degree of understatement and this may still vary with employment conditions. For the moment, we must take the unemployment and vacancy data on trust. The reader might be consoled by the observation that the broad conclusions which emerge are consistent with the views expressed by the case-study plants.

(i) *Males*

Figures 3.1 to 3.4, p. 56, show percentage unemployment for all males and for males in engineering in each labour market area at quarterly intervals over 1959–66.[2] The figures relate to total unemployment including those 'temporarily stopped' although we shall see later that in Birmingham it is often more appropriate to exclude those temporarily stopped when using unemployment data to measure changes in labour supply.

Male unemployment in Birmingham was, in all periods save the last quarter of 1966, substantially below the level of unemployment in each of the remaining labour market areas. In Glasgow and Small Town, percentage male unemployment was in all quarters but one more than twice the national average. It seldom fell below 4 per cent in Glasgow and fluctuated between 6 and 12 per cent in Small Town. Unemployment, while not as high in North Lanarkshire and New Town, was always above the national average, showing a secular tendency to fall in North Lanarkshire, while the long trend was

[1] See J. C. R. Dow and L. A. Dicks-Mireaux, 'The Excess Demand for Labour. A Study of Conditions in Great Britain, 1946–56', *Oxford Economic Papers*, N.S. Vol. 10, 1958.

[2] No data are presented for engineering unemployment in Small Town. Because of the numbers involved a small absolute change in unemployment leads to violent fluctuations in the rate of unemployment.

upwards in New Town. In contrast, male unemployment in Birmingham was equal to or less than the national average in 25 of 32 quarters over 1959–66. In Birmingham, the number of unfilled

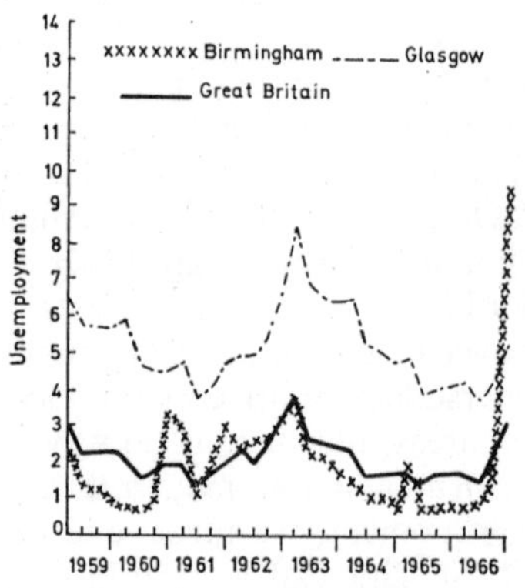

Figure 3.1. Percentage unemployment: Birmingham, Glasgow and Great Britain, males, 1959–66

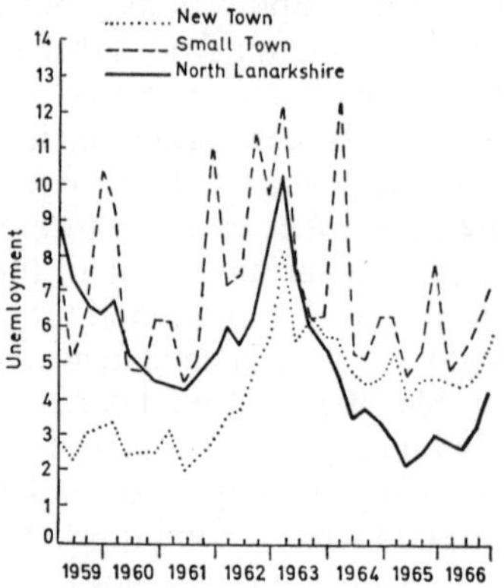

Figure 3.2. Percentage unemployment: North Lanarkshire, New Town and Small Town males, 1959–66

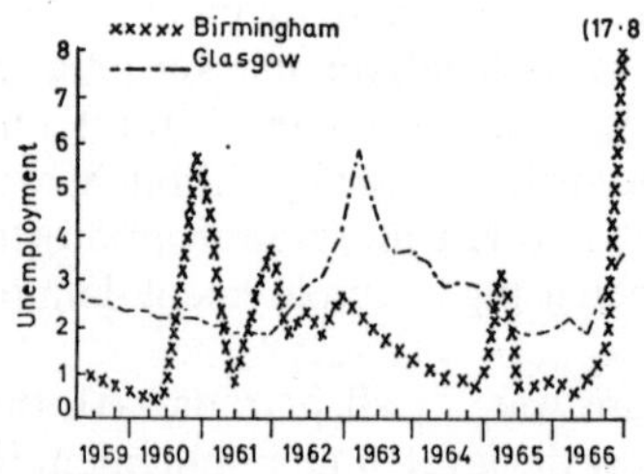

Figure 3.3. Percentage unemployment in engineering: Birmingham[1] and Glasgow males, 1959–66

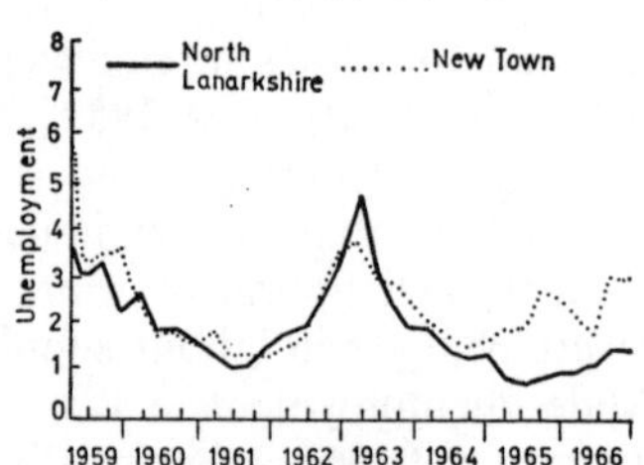

Figure 3.4. Percentage unemployment in engineering: North Lanarkshire and New Town males, 1959–66

Note: 1. In Birmingham no data of unemployment in engineering were available for March and September of 1959, 1960 and 1961. Missing values were estimated by extrapolation.

vacancies for males exceeded male unemployment in 18 of 32 quarters. On the other hand, in Glasgow, North Lanarkshire, New Town and Small Town, unemployment always exceeded unfilled vacancies by a substantial margin. In the *most favourable* employment situation in Glasgow the ratio of unemployed to unfilled vacancies was 4:1 and the ratios in North Lanarkshire, New Town and Small Town were 3:1, 5:1 and 6:1 respectively.

Male unemployment in the engineering industry in Glasgow and in North Lanarkshire and New Town was always below the corresponding percentage for all males. It was often as low as 2 per cent in North Lanarkshire and New Town and tended to decrease over 1959–66. In Birmingham, there was seldom much difference between the two series, where the only notable contrast was the greater cyclical instability of engineering unemployment. This arises because of the importance of the Vehicles sector which is subject to sudden changes

TABLE 3.6

RATIO OF UNEMPLOYMENT TO UNFILLED VACANCIES: BIRMINGHAM AND GLASGOW, MALE OCCUPATIONAL GROUPS, JUNE 1959–66

Year	Birmingham				Glasgow				
	Skilled	Semi-skilled	Unskilled	Tool-makers	Skilled	Semi-skilled	Unskilled	Turners	Fitters (precision)
1959	0·5	0·9	12·5	0·8	12·1	16·8	890·5	59·0	35·3
1960	0·1	0·2	1·8	0·1	3·7	22·2	244·2	2·6	5·6
1961	0·3	1·0	1·9	*	1·8	14·7	71·5	0·1	6·9
1962	1·0	3·3	82·8	1·2	10·1	42·9	150·4	1·6	2·0
1963	1·9	5·3	21·3	1·9	30·4	87·5	279·7	124·0	45·8
1964	0·2	0·3	3·9	*	4·4	17·9	522·0	1·3	14·1
1965	0·1	0·3	1·7	*	0·5	5·3	85·4	0·2	0·8
1966	0·1	0·6	2·0	*	1·1	3·4	85·1	0·2	2·6

Note: 1. The occupational groups are based upon the D.E.P's Classification of Occupations. The skilled and semi-skilled groups contain only occupations in engineering and allied industries. The unskilled group is composed of those classified as labourers who may be engaged in any industry. See Appendix p. 420–1.

* Signifies less than 0·05.

in output accompanied by temporary 'lay-offs' amongst the labour force. This category, known as the 'temporarily stopped', accounts for much the greater part of the occasional sharp jumps in engineering and total male unemployment shown for Birmingham in Figures 3.1 and 3.3. As the temporarily stopped almost by definition[1] are unlikely to experience long-term unemployment, the upward changes in Birmingham unemployment are often quickly reversed. Thus employment conditions were generally favourable. Unfilled male vacancies in engineering exceeded engineering unemployment in 14 of 32 quarters, a position which was never approached in Glasgow and only rarely in North Lanarkshire, New Town or Small Town.

The occupational characteristics of the unemployed show that in

[1] The temporarily stopped are 'Registered unemployed persons who, on the day of the [unemployment] count, are suspended from work by their employers on the understanding that they will shortly resume work. . . .' *Ministry of Labour Gazette*, Vol. 76. Jan. 1968, p. 98.

each area unemployment increased in inverse ratio to the skill of the group under consideration. Employment conditions were always most favourable for skilled engineering workers and least favourable for the unskilled. In consequence, unfilled vacancies could exceed unemployment amongst skilled workers even in areas like Glasgow where the general level of unemployment was always high. This is illustrated by Table 3.6 which shows the ratio of unemployment to unfilled vacancies by skill groups for Birmingham and Glasgow at June of each year 1959–66. Similar data for selected skilled occupations are also included for reasons which emerge subsequently.

In each area the ratio of unemployment to unfilled vacancies was always lower for skilled than for semi-skilled, while the ratio for the latter was always less than that for the unskilled. It is also apparent that employment conditions were always more favourable in Birmingham, and we would therefore expect to find that employers in Glasgow encountered less difficulty in recruiting labour. Labour shortages may have been very widespread in Birmingham, as the number of unfilled vacancies for semi-skilled males often exceeded the number unemployed, while for the unskilled the balance between unemployment and unfilled vacancies was much more even than in Glasgow. In other words, structural unemployment was less in evidence in Birmingham than in Glasgow so that in the former area recruitment difficulties may have emerged, at one time or other, for all types of labour.

(ii) *Females*

Female unemployment[1] in Birmingham never exceeded 1·3 per cent of the insured labour force throughout 1959–66. It was greater than the national average in only one quarter, and the number unemployed was less than the number of unfilled vacancies in all but four quarters. Much the same applies to female unemployment in engineering, which diverged very little from the overall female average. For both series female unemployment was always lower in Birmingham than in any of the other labour market areas. Female unemployment in Glasgow was substantially below the unemployment rate for males, and unfilled vacancies occasionally exceeded unemployment amongst females. In North Lanarkshire, New Town and Small Town, however, female unemployment was high, averaging 7·9 per cent in New Town and 5·9 per cent in North Lanarkshire and Small Town, compared to 1·8 per cent in Glasgow and 0·8 per cent in Birmingham.[2]

[1] As with males, unemployment is taken to include those 'temporarily stopped'.
[2] These figures are unweighted averages of the 32 quarterly observations.

The differences between the labour supply position in the separate markets are likely to have been even greater than the above figures suggest. Female unemployment in Birmingham had been extremely low for a long period prior to 1959, whereas it had exceeded the national average in each of the remaining labour market areas. Such differences in registered unemployment may conceal the true extent of under-employment, since participation rates are likely to be low in areas where registered unemployment is already high. Thus in Glasgow, and in North Lanarkshire, New Town and Small Town particularly, female participation rates could be expected to be below

TABLE 3.7

PARTICIPATION RATES BY AGE GROUPS: WEST MIDLANDS AND CENTRAL CLYDESIDE CONURBATIONS, MALES, FEMALES AND MARRIED FEMALES, 1966

Age group (years)	West Midlands			Central Clydeside		
	Males	Females	Married females	Males	Females	Married females
15–20	76·6	71·0	48·7	76·1	71·9	42·7
21–24	95·0	62·6	47·1	93·8	62·1	40·7
25–34	98·3	45·4	39·6	97·7	39·9	30·6
35–44	98·8	58·0	55·2	98·0	50·9	44·7
45–64	96·6	52·8	49·3	95·1	46·2	39·2

Source: Sample Census, 1966, Great Britain, Economic Activity Tables, Part 1.

those ruling in Birmingham. Although no information on participation rates is available for areas coincident with our labour markets, it does exist for the West Midlands and Central Clydeside conurbations. These conurbations are dominated by Birmingham and Glasgow as herein defined, so that any error introduced will not be substantial. Participation rates for males, all females and married females are shown by age groups in Table 3.7.

Save for females aged 15–20 years, participation rates are always higher in the West Midlands than in the Central Clydeside conurbation. This difference between the areas is, however, extremely small in the case of all male age groups. Males are 'primary workers' with a continuing commitment to labour force activity. They are, with the exception of certain minority groups, either in work, or looking for work and unemployed, when of 'working age'. In consequence, male participation rates do not vary greatly from area to area, and the number of unemployed males is a good indication of the supply of male labour available for work. The same is not necessarily true for females. As Table 3.7 demonstrates, participation rates are usually much higher for females in the West Midlands conurbation, the difference over Central Clydeside being especially marked for married females. Amongst females, and especially married females, there is a high proportion of 'secondary workers', persons with a primary commitment not to a labour force activity but to domestic responsibilities. These females may not register as unemployed when they cannot find paid employment, but may simply retire from the labour force to devote more time to their domestic responsibilities. If this is so, female participation rates will fall as unemployment rises, and unemployment will not be an accurate measure of the number of females who would seek work were more job opportunities available, since there will be a considerable amount of 'concealed' female unemployment in the form of low participation rates.[1]

The differences between labour market conditions for females in Glasgow and Birmingham were, therefore, greater than would be suggested by an analysis based purely on unemployment statistics, and the same argument applies when contrasting Birmingham with North Lanarkshire, New Town and Small Town. In Birmingham, female unemployment was insignificant and participation rates high, so that the supply of unutilized labour was less elastic than in the other areas. As a result, it was the only area where, over 1959–66, the increase in female employment was lower than that achieved for males. Again, while female employment did increase in Birmingham, the increase was much smaller than that achieved in North Lanarkshire and New Town despite the fact that unfilled job vacancies for females were always higher in the former area. Yet although female employment increased by more than one-third in North Lanarkshire and by one-quarter in New Town over 1959–66, female unemployment remained high in each area, reflecting in large part increased participation rates as more females sought paid employment. The conclusion which must be drawn is evident enough: engineering plants in Birmingham are likely to have had some difficulty in securing additional female labour, while the supply of female labour is likely to have been particularly elastic to plants in North Lanarkshire, New Town and Small Town.

[1] Most of the literature on this subject is based on American data (e.g. T. F. Dernburg and K. T. Strand, 'Hidden Unemployment, 1953–62', *American Economic Review*, Vol. 56, 1966 and 'Cyclical Variation in Civilian Labor Force Participation', *Review of Economics and Statistics*, Vol. 46, 1964) but the regional distribution of unemployment and participation rates does suggest similar labour market behaviour in Britain.

4. RECRUITMENT EXPERIENCE OF THE CASE-STUDY PLANTS

We have seen that unemployment and vacancy data are subject to certain deficiencies so that they cannot be taken as providing perfect measures for unused labour reserves and the unsatisfied demand for labour. Indeed, it has been suggested that these indicators may be positively misleading as measures of labour availability in different market areas.[1] To assess the position, we can check the view of employment conditions in the different areas as presented in Section 3 against the recruitment difficulties as reported by the case-study plants.

On any reasonable definition, Birmingham enjoyed full employment through most of 1959–66. Female unemployment never reached 2 per cent and male unemployment exceeded 3 per cent in only three of 32 quarters. Unfilled vacancies exceeded the number unemployed in 18 of 32 observations for males and in 24 of 32 observations for females. Hence, even if no allowance is made for the probability that unfilled vacancies understate the true demand for labour,[2] departures from full employment were of a temporary nature. If one accepts the proposition that full employment characterized the British economy through most of 1959–66, then the 'tightness' of the Birmingham labour market is more easily appreciated. Male unemployment in Birmingham was generally below the national average, and female unemployment exceeded the national rate in only one of 32 quarterly observations.

Therefore, on the evidence of the unemployment and vacancy data, we would expect Birmingham plants to experience widespread recruitment difficulties. They would be greatest for skilled males and for females, but in certain periods even semi-skilled and unskilled male labour may have been difficult to obtain. In Glasgow, on the other hand, engineering plants should have had few problems in securing an adequate supply of semi-skilled and unskilled labour, and even with skilled labour, recruitment difficulties might be expected to be less pronounced than those evidenced in Birmingham.

The position in North Lanarkshire, New Town and Small Town is less easy to evaluate. In all areas the general level of unemployment was high both for males and females. However, as can be seen from Table 3.3, engineering employment was expanding in each area over 1959–66. The pace of this expansion was extremely rapid, so that, even in North Lanarkshire with its tradition of engineering employment, plants may have found it difficult to recruit labour with

[1] See G. Davies, 'Regional Unemployment, Labour Availability, and Redeployment', *Oxford Economic Papers*, N.S., Vol. 19, 1967.

[2] See pp. 352–4 below.

experience of the engineering industry. New Town and Small Town had no large pool of engineering labour on which expanding plants could draw. It may, then, be inappropriate to focus attention on the general level of unemployment. The supply of labour would only be adequate if it proved itself adaptable, and if plants could supply the training facilities necessary to provide recruits with the required skills.

Despite the rapid expansion of engineering employment and the restricted supply of trained engineering labour in North Lanarkshire, New Town and Small Town, the experience of the case-study plants in hiring new labour had more in common with the position in Glasgow than that in Birmingham. In the former areas, and in Glasgow, plants did occasionally find difficulty in obtaining an adequate supply of skilled labour, the most notable shortages occurring in the case of turners and skilled toolroom workers. However, none of the 27 plants in Glasgow or the 12 establishments in North Lanarkshire, New Town and Small Town reported a general shortage of labour. Only eight indicated that skilled labour shortages had adversely affected output, and these had generally been of marginal importance and of a temporary nature. Thus, only one plant in Glasgow classified the loss of output as 'severe'. Shortages of skilled labour, therefore, appear to have been a minor irritant rather than a serious obstacle to productive efficiency.

In Birmingham, only two of the 27 plants interviewed reported no recruitment problems.[1] Of the remainder, no less than 21 reported a specific or a general shortage of skilled labour and emphasized the difficulties of obtaining and retaining labour in a 'tight' labour market. Another contrast between Birmingham and the other labour markets is that in Birmingham a significant number of plants had difficulty in obtaining female labour and male semi-skilled and unskilled labour. For example, despite the impression conveyed by unemployment and vacancy data for the unskilled, seven plants in Birmingham found the supply of unskilled labour inadequate for their needs. No plant in Glasgow or in North Lanarkshire, New Town and Small Town reported any difficulty on this score.

The tightness of the labour market in Birmingham is reflected in recruitment policies adopted by the plants. Recruitment activities in advance of production need may occur because the new recruits require training, or because some time must elapse before the plant can contact and secure additional labour. In Birmingham the 'tight' labour market situation made it difficult to acquire labour with the necessary skills. Hence recruitment activities were almost invariably

[1] The two exceptions were wage leaders engaged in motor-car production.

begun in advance of production needs, and the advance period was generally greater than that required for training purposes, to allow sufficient time to hire the additional labour.

The situation in Glasgow is quite different. In the latter area, most establishments did not commence recruitment activities until the need for additional labour arose. Only in the case of skilled labour was the market 'tight' enough to warrant the commencement of recruitment activities in advance of production need, and even this was confined to a minority of units. The explanation is simply that few plants had difficulty in securing semi-skilled labour with appropriate industrial experience. In these circumstances, less prior training was required and, given slack labour market conditions, additional manpower could be obtained with little need for advance recruitment activity.

The reported experience of the case-study plants is, therefore, consistent with that which would be expected from a study of unemployment and vacancy data. Over the period 1959–66 most units in Birmingham had difficulty in satisfying their demands for skilled labour through external recruitment. Indeed, at some period or another over 1959–66, labour shortages in this sense have occurred for almost all occupational groups. In Glasgow, the supply of semi-skilled and unskilled labour was always adequate and skilled labour shortages were of a less pressing and more temporary nature. The general level of unemployment was high in North Lanarkshire, New Town and Small Town, but engineering employment was increasing rapidly and the available reserves of labour lacked engineering skills, particularly in the two latter areas.

We might expect these differences in labour market conditions to influence the labour market policies adopted by the case-study plants. For example, an inability to meet labour needs through external recruitment might facilitate labour mobility external to, and within, the plant, promote greater flexibility in the use made of labour and encourage the provision of additional training by the plant. Again, a 'tight' labour market might affect the policies adopted in a redundancy situation, stimulate the development of manpower forecasting and influence the prevailing pattern of wage differentials. Such issues are considered in the following chapters.

PART II: WAGES

CHAPTER 4[1]

THE INTER-PLANT WAGE STRUCTURE

1. INTRODUCTION

As we have seen, traditional labour market theory predicts that competition between employers for the labour available, and economic rationality on the part of employees, will set in train forces which tend to equate the net advantages offered by alternative employments. Employees compare the separate elements of different jobs and from this comparison they pick that job which *on balance* is most advantageous. Movements of labour therefore occur in response to differences in *net advantages* which may or may not be reflected in the existing pattern of wage settlements. Differences in plant wage levels could, then, denote 'real' differences in net advantages, or simply the fact that high wage plants suffered some non-monetary disadvantage (more dangerous, dirty or monotonous work, greater job insecurity, etc.) which offset any pull which the higher wage might have exerted. Hence, we cannot demonstrate that labour markets do not attain, or even approach, equilibrium by simply observing that wage earnings differ from one unit to the next. Such behaviour may be quite consistent with the competitive model, so that further evidence has to be assembled before any final conclusion can be reached.

This is the nub of the difficulty which faces all applied research in labour economics. Labour market theory is so versatile, it admits of so many possible influences which may be important in job choice, that it is possible to rationalize almost any set of observed events. The researcher in the field cannot easily measure the non-pecuniary factors which influence job choice, and his own subjective estimate of their importance may not coincide with those of the employee. Such factors are extremely numerous, and their importance will vary from job to job, from individual to individual and, even, from one period

[1] Some of the material in this and the subsequent chapter was first published by D. I. MacKay in *Local Labour Markets and Wage Structures* (D. Robinson, ed., Gower Press, 1970).

to the next. For practical purposes, therefore, empirical investigations have concentrated on that element—wages—which is most amenable to quantification, and is in any case probably the most important single influence on job choice. This enquiry is no exception. The following analysis is largely concerned with measuring inter-plant wage differentials and changes in such differentials over time. Initially, no attention is paid to those factors influencing job choice which are not captured by the wage variable. We have ultimately to consider these non-wage factors, but at this stage the analysis becomes more hesitant. It is simply not possible to give quantitative expression to the many other influences apart from wages which might affect the behaviour of employers and employees. The discussion of these other elements is, therefore, always unsatisfactory in greater or lesser degree. Nonetheless, some important pointers to wage behaviour do emerge. The insights gained into the behaviour of wage differentials cast some doubt on a number of hypotheses which have sought to explain the existence of wage differentials within local labour markets, and certain alternative explanations do emerge which might suggest more fruitful avenues for future research.

The plan of this chapter is as follows. In Section 2 we set the scene by examining the institutional framework within which wage negotiation in the British engineering industry is conducted.[1] The nature of the wage data available is discussed in Section 3, while the core of the chapter is contained in Sections 4 and 5, which describe and analyse the pattern of inter-plant wage differentials and changes in that pattern over time. Section 6 attempts to widen the analysis by considering alternative explanations for the large wage differentials found to exist in practice, and a summary of the conclusions reached is provided in Section 7.

2. WAGE NEGOTIATION

In practice, it may be misleading to talk of the wage 'structure' in the engineering industry. There is no structure in the commonly accepted sense of the word; there is no system of wage payments which is deliberately and formally constructed according to a mutually consistent set of principles. Rather there is a series of outcomes resulting from wage settlements—with certain observable differences between regions and occupations, between timeworkers and pieceworkers, between different sectors of the industry, between engineering estab-

[1] The reader familiar with this subject can proceed to Section 3.

lishments of different size[1]—which arise from complex *ad hoc* bargaining at national, district and plant levels. The term 'structure' must, therefore, be interpreted as a shorthand expression relating to the result of a series of wage settlements.

The main features of the wage negotiating machinery are fairly well known[2] and need only be outlined briefly. There are two important levels of bargaining, national and plant, with a third level, district negotiations which can largely be ignored as an historical anomaly of dwindling importance. The national agreement concluded between the Engineering Employers' Federation and the Confederation of Shipbuilding and Engineering Unions centres around two key wage rates for fitters and labourers. It establishes national time rates, known as consolidated time rates (C.T.R.), for these two groups which determine the minimum level of earnings for a standard working week,[3] and act as the basis on which overtime and shift premia are usually calculated. The C.T.R. for fitters and labourers also acts as the datum to which the minimum time rates for other groups are related. No standard formula is applied. For some occupations, rates are negotiated at the national level between the E.E.F. and individual unions;[4] for others, adjustments are made to minimum time rates at district or plant level on the basis of a 'recognized' or 'accepted' relationship. The rates established for fitters and labourers have, therefore, very wide repercussions through the industry.

The C.T.R. acts as a 'fall back' wage for pieceworkers as well as for timeworkers. In addition, the agreement establishes a piecework supplement[5] for fitters and labourers, and lays down that piecework times and bonuses should be established by each plant so that the worker 'of average ability' can earn at least 45 per cent more than the 'basic rate'.[6] The piecework supplement and the basic rate plus 45

[1] See, K. G. J. C. Knowles and T. P. Hill, 'The Structure of Engineering Earnings', *Bulletin of the Oxford University Institute of Statistics*, Vol. 16, 1954.

[2] See references above, p. 24, n. 1 and n. 2; also National Board for Prices and Incomes, *Report No. 65 (and Supplement), Payment by Results Systems*, Cmnd. 3627, 1968; and *Report No. 104, Pay and Conditions of Engineering Workers*, Cmnd. 3931, 1969.

[3] We ignore the minimum earnings level guarantee which has now been established because it falls outside the period with which we are concerned.

[4] A. Marsh, *Industrial Relations in Engineering* (1965), pp. 147–8.

[5] This is the mechanism through which national wage increases are incorporated in pieceworkers' earnings.

[6] In 1968, a new piecework standard and an altered basic rate were introduced to the national agreement. These changes have been ignored in the text, along with the minimum earnings levels provisions, as they fall outside our period 1959–66, and do not introduce any radical departure from the situation prevailing over those years.

per cent together make up the minimum piecework standard for the 'average' fitter or labourer on piecework. The national agreement merely establishes *national minima* for fitters and labourers, whether timeworkers or pieceworkers. The federated employer (and, of course, the non-federated) can, and usually does, pay above the national minima, which bear little relation to actual earnings even in the case of fitters or labourers. For example, at the end of our period, in 1966, the C.T.R. for fitters yielded a wage for a standard week of £10 16s 8d while the national average weekly earnings for fitters on timework was £22 7s 6d.[1] The E.E.F. estimated that in 1962 only 0·4 per cent of fitters and 6 per cent of labourers were earning no more than their appropriate C.T.R.[2] It is, therefore, apparent that the national agreement fixes a floor for earnings and not a ceiling, and that negotiations below the national level are extremely important in determining actual earnings.

The second crucial level of bargaining occurs at the plant level. In a sense, the term 'plant bargaining' is a misnomer, for it obscures the very fragmented nature of the bargaining process. In the case of pieceworkers, bargaining only rarely takes place on a plant or even a group basis. Most often the prices and times fixed for pieceworkers emerge from haggling between individual employers and ratefixers, with the shop steward and the foreman acting as the first court of appeal. Bargaining on a wider basis is common only for timeworkers. The upshot of a system in which plant bargaining assumes such importance, and in which the bargain often affects only small groups or individuals within the plant, is an extremely complex pattern of wage settlements. National bargaining does not, then, impose *a* solution. The main effect of such bargaining is that 'a national agreement lifts the floor above which plant negotiations take place'.[3] The scope for variations in earnings levels is, therefore, considerable, and the outcome in practical terms, as this and the subsequent chapter will show, is a system of wage[4] differentials of considerable complexity.

3. THE WAGE DATA

The wage data at our disposal have already been outlined briefly,[5]

[1] See National Board for Prices and Incomes, *Report No. 49*, *op. cit.*, p. 51 and *Statistics on Incomes, Prices, Employment and Production*, No. 19, Dec. 1966 (H.M.S.O.), p. 30.

[2] Marsh, *op. cit.*, p. 182.

[3] National Board for Prices and Incomes, *ibid.*, p. 9.

[4] Throughout the remainder of this and the two following chapters it is convenient to use wages and earnings as interchangeable terms. Where wage *rates* are referred to this will be made explicit in the text.

[5] See p. 45 above.

but it is necessary to add certain further comments before proceeding directly with our analysis. Two types of earnings returns are collected by the D.E.P. The first, the W.E. returns, shows the total wage bill for the wage earners employed by each plant and the number of wage earners distinguished by sex. The returns relate to a given working week in April and October of each year and are available over the entire period 1959–66. No distinction is made by occupational groups, and it is only possible to calculate average male and female earnings *including* overtime payments.[1]

While the W.E. returns are useful because they cover the whole survey period and are available for a substantial number of plants in each local labour market[2] they do suffer from two disadvantages. First, average earnings across plants may differ solely as a result of different skill mixes, even if the forces of competition are such as to establish a going or market rate for a given type of labour in the whole market. Second, W.E. returns only allow the calculation of gross weekly earnings (i.e. total earnings *including* all overtime payments). This may, indeed, be the appropriate variable to represent 'wages'. Overtime working is widely accepted in Britain as a means of increasing the relative attraction of employment in a particular plant. If, then, the 'wage' as seen by the labour force includes overtime earnings we cannot exclude such earnings in our measure of wages, for then we would be failing to consider an important factor in the choice between alternative employment opportunities.

On the other hand, as all our earnings returns relate to a particular week, and as the overtime worked by a plant may show substantial variation over the short run, the measure of gross weekly earnings may not be representative of the longer-run wage position of the establishment under examination. A change in wages as measured by gross earnings may, therefore, simply disclose a transient situation caused by short-run variations in overtime earnings. Given these difficulties, we have provided where possible a measure of standard weekly earnings (i.e. earnings for a standard working week *excluding* all payments made in respect of overtime working), as well as of gross weekly earnings.

The deficiencies of the W.E. returns are made good by the occupational earnings returns compiled by plants for the D.E.P. and the E.E.F. The E.E.F. returns are made once yearly (except for 1963) and

[1] Details are provided for men (aged 21 years and over), and for women (aged 18 years and over)—full-time and part-time separately in each case—and for youths and boys, and for girls. In the following text all the calculations from the W.E. returns relate to full-time male and women workers.

[2] W.E. data are available for 14 units in Birmingham and 18 units in Glasgow.

show for male[1] occupational groups the numbers employed, total and premium earnings and total and overtime hours worked. Average gross weekly earnings can then be calculated for each occupational group, as can average standard weekly earnings. These returns were not available for Birmingham plants, but the D.E.P. occupational earnings returns provide similar information to that obtainable from the E.E.F. data and are, indeed, modelled on the latter. Both sets of occupational earnings returns distinguish between skilled[2] and unskilled males, and those of the D.E.P. provide additional information for male, semi-skilled employees.

Certain difficulties remain, the most important of which is that we are forced to use average plant or average occupational earnings and are unable to describe and measure the spread of earnings within the plant or occupational group. The distribution of earnings, like all income distributions, tends to be positively skewed and high-earning employees tend to be those with long service.[3] This may have an important effect on labour turnover, and therefore on the response of management, which can be obscured by the use of plant and occupational averages. For example, the turnover rate, say for fitters, could differ between plants with the same average level of fitters' earnings according to the spread of fitters' earnings around that average. Thus the turnover rate for fitters might be higher in a unit where the distribution of fitters' earnings was markedly skewed and where high earnings were enjoyed only by long-service employees. We cannot consider this possibility with the data at our disposal, but the information to hand allows us to study inter-plant differentials in some detail, and to examine changes in such differentials over time in a manner not usually possible in earlier studies.

4. INTER-PLANT WAGE DIFFERENTIALS

We begin our investigation of the inter-plant wage structure with Tables 4.1 and 4.2. below. Occupational earnings returns provided to the E.E.F. by 14 Glasgow plants form the basis of Table 4.1. D.E.P. earnings returns by 13 Birmingham units were used to construct Table 4.2. Because different earnings returns are used, the dating differs between the tables, and the choice of occupational groups was

[1] No occupational data are available for females, who are mostly semi-skilled. The bulk of the following analysis therefore relates to males although W.E. returns for females are used for Birmingham establishments.

[2] For skilled males, details are also provided for fitters, turners, toolroom workers, etc.

[3] See Robinson, *Wage Drift, Fringe Benefits and Manpower Distribution*, pp. 74–5; and *Local Labour Markets and Wage Structures* pp. 245–53.

also influenced by this factor. In the latter instance, however, other considerations were more important. Fitters and labourers are included in the Glasgow data because these are the two key groups in

TABLE 4.1

INTER-PLANT EARNINGS DIFFERENTIALS: GLASGOW MALES, JUNE 1959 AND OCTOBER 1966

A. Standard weekly earnings

Weekly earnings	Fitters	Turners	Labourers	All workers
1959				
Range, lowest to highest (£)	9·3 to 15·6	9·4 to 15·9	7·8 to 12·9	9·1 to 15·3
Interquartile range (£) (Q_3–Q_1)	2·3	1·6	2·2	2·1
Coefficient of variation	14·4	14·2	15·4	14·6
1966				
Range, lowest to highest (£)	12·7 to 22·8	14·6 to 22·5	10·2 to 15·3	13·8 to 22·0
Interquartile range (£) (Q_3–Q_1)	2·5	4·7	1·7	3·0
Coefficient of variation	14·1	13·2	12·2	13·8

B. Gross weekly earnings

Weekly earnings	Fitters	Turners	Labourers	All workers
1959				
Range, lowest to highest (£)	9·3 to 18·7	10·3 to 17·3	8·0 to 14·8	10·4 to 16·5
Interquartile range (£) (Q_3–Q_1)	2·8	3·5	1·9	3·0
Coefficient of variation	18·3	15·4	16·5	13·8
1966				
Range, lowest to highest (£)	16·3 to 26·4	19·1 to 26·6	12·5 to 20·2	17·2 to 24·7
Interquartile range (£) (Q_3–Q_1)	3·6	3·1	3·3	4·3
Coefficient of variation	13·2	11·1	14·3	12·1

national negotiations, and turners because they were the type of skilled labour most difficult to recruit. Toolroom workers were in shortest supply in Birmingham,[1] and labourers were again included

[1] See Table 3.6, p. 57 above.

because of their importance in national bargaining. Turners and fitters do not appear in the Birmingham results because these types of

TABLE 4.2

INTER-PLANT EARNINGS DIFFERENTIALS: BIRMINGHAM MALES, JUNE 1963 AND JUNE 1966

A. Standard weekly earnings

Weekly earnings	Toolroom	Semi-skilled	Labourers	All workers
1963				
Range, lowest to highest (£)	13·0 to 24·1	11·6 to 20·7	7·7 to 13·6	11·9 to 20·3
Interquartile range (£) (Q_3–Q_1)	7·6	5·3	1·4	5·1
Coefficient of variation	20·1	17·8	13·0	16·3
1966				
Range, lowest to highest (£)	14·1 to 31·5	14·3 to 25·4	9·9 to 17·4	15·1 to 25·7
Interquartile range (£) (Q_3–Q_1)	9·3	7·0	4·2	6·5
Coefficient of variation	22·7	19·9	16·0	17·5

B. Gross weekly earnings

Weekly earnings	Toolroom	Semi-skilled	Labourers	All workers
1963				
Range, lowest to highest (£)	14·1 to 29·4	14·1 to 23·6	9·4 to 19·9	13·7 to 24·0
Interquartile range (£) (Q_3–Q_1)	6·8	5·7	3·2	4·4
Coefficient of variation	19·7	16·7	18·2	14·5
1966				
Range, lowest to highest (£)	19·9 to 36·0	16·2 to 28·5	14·0 to 22·0	19·5 to 28·8
Interquartile range (£) (Q_3–Q_1)	7·7	5·9	2·4	6·4
Coefficient of variation	17·6	15·8	12·1	12·8

labour are employed by relatively few plants in that area. This reflects the production technology of Birmingham units, and the importance of semi-skilled employees to that technology dictated their inclusion in Table 4.2. In all other respects the results of Tables 4.1 and 4.2 were obtained by identical methods. They show various measures of dispersion calculated for average standard and average gross weekly earnings.

Three methods of measuring the inter-plant earnings spread are used in Tables 4.1 and 4.2. Taken singly, they may not tell us much about the structure of earnings; as a whole, they add up to a fairly complete picture. The range of occupational earnings between the lowest and the highest wage-paying plant is a measure of the absolute spread of earnings. It is of interest for obvious reasons but suffers from the disadvantage that it may be seriously distorted by extreme observations. To overcome this difficulty, a second measure of spread, the interquartile range, is used, which excludes plants above and below the upper and lower quartiles respectively. Finally, the most 'sophisticated' measure, the coefficient of variation which takes all observations into account, is a measure of *relative* dispersion and therefore provides the most appropriate measure of spread when comparing earnings distributions with different means.

It is important to note that there is no unique measure of dispersion. For example, taking Glasgow plants and dealing with standard weekly earnings in 1966, we find the interquartile range was greatest for turners, while the range from lowest to highest, and the coefficient of variation, was greatest for fitters. Yet the difficulty of measuring dispersion should not be allowed to obscure the self-evident, which is that whichever measure is used the spread of plant earnings was extremely large. There is nothing which resembled a market wage for a particular occupational group; only a range of earnings, with earnings in high-wage units much above earnings in low-wage units. To take just one example, we find that the average gross weekly earnings of fitters in Glasgow in 1959, in the highest wage-paying plant, were more than twice as high as earnings in the lowest wage-paying plant.

Apart from this major conclusion, what else can we learn from the above tables? First, let us consider whether the extent of earnings differentials was greater in Glasgow or Birmingham in 1966 when earnings returns are available for both areas. Ignoring the heterogeneous semi-skilled group, we find that for standard and gross weekly earnings each measure of dispersion yields a greater spread of earnings for skilled toolroom workers in Birmingham than for either skilled group—fitters or turners—in Glasgow. For labourers, the evidence is more ambiguous. The spread of standard weekly earnings was greater for labourers in Birmingham than in Glasgow whatever measure of spread is used, but conflicting results emerge for gross weekly earnings. However, as standard weekly earnings are the best measure of the long-run position we can conclude that the dispersion of labourers' earnings was usually wider in Birmingham.[1] There is, then, fairly strong evidence which suggests that the spread of earnings was greater in the tighter Birmingham labour market. Even if we put it at its lowest, we can say that there is simply no suggestion that earnings differentials are narrower when employment conditions in the local market are more favourable. This is an important conclusion, for it is a severe blow to the hypothesis that inter-plant differentials are due to under-employment, and to the related view that wage differentials narrow in a 'tight' labour market. Inter-plant differentials appear to be a characteristic of all labour market situations[2] and do not seem to be significantly affected by local employment conditions.

In this regard it is interesting to note that in Birmingham wage differentials were *greater* for skilled toolroom workers than for either semi-skilled workers or labourers. This applies whichever measure of spread is adopted and whether the wage variable is measured by gross or standard weekly earnings.[3] Again, although in this instance the relationship is less systematic, plant wage differentials in Birmingham were generally greater in absolute and relative terms for the semi-skilled as opposed to labourers. The tendency for wage differentials to widen as skill increases is rather surprising given the fact that recruitment difficulties were greatest for skilled workers and least in evidence for the unskilled. In view of this, one might have expected competition between plants for scarce skilled labour to have led to a narrowing of plant differentials for such employees. It is true that in Glasgow the various measures of spread and of earnings give rather conflicting results, so that there is no indication that plant wage differentials were widest for the skilled; but, equally, there is nothing which suggests they were narrower. Hence, although competitive pressures were greatest for skilled employees, they did not reduce the spread of plant earnings relative to the spread for other groups.

The remaining issue is whether the inter-plant differentials observed are a permanent and stable feature of labour market be-

[1] See also Table 4.3, p. 76 below.

[2] A special wage analysis for North Lanarkshire indicated that extremely large differentials can emerge in labour markets much smaller than those of the two conurbations. To take only one example, standard weekly earnings for labourers ranged from £10 2s to £17 1s and gross weekly earnings from £12 4s to £19 8s.

[3] It might be objected that this conclusion largely rests on comparing coefficients of variation as the difference in earnings between the lowest and highest wage plant, and the interquartile range might be expected to be greater for toolroom workers because of their high earnings levels. However, if one expresses earnings in the highest wage plant as a *percentage* of earnings in the lowest wage unit, the resultant figure is always highest for toolmakers.

haviour or merely a transient phenomenon which disappears as the market adjusts to demand/supply relationships. It is not wholly satisfactory to rely on Tables 4.1 and 4.2 for the necessary evidence, as they provide a cross-sectional analysis for only two points in time with evident dangers that the results for one or both observations may be affected by the operation of special factors. Nonetheless, we can take these tables as our starting-point and introduce further data when this is appropriate. Given the tendency of wage levels to rise over time, the relevant measure is the coefficient of variation. Using this, we can see that in Glasgow low-wage plants had, comparing 1966 to 1959, improved their relative position for all occupational groups in terms of both standard and gross weekly earnings. However, the improvement was a slight one, as a glance at Table 4.1 will demonstrate. We cannot even say this in Birmingham. There was some contraction over 1963–6 in the spread of gross weekly earnings, while plant differentials as measured by standard weekly earnings, which are the best indication of the long-run position, tended to widen. Probably the only safe conclusion to be drawn is that plant wage differentials, however measured, were considerable in both markets, and there was no fundamental change in the extent of these differentials between the two periods examined.

It is important to carry this discussion forward, for it implies a conclusion, that substantial plant differentials persist over time, which, if true, could have important consequences for labour market analysis. We do this with the aid of Table 4.3 which was constructed in the following fashion. The W.E. returns which are available biannually over 1959–66 were used to obtain average male gross weekly earnings by plant. In addition the available occupational earnings returns[1] were used to calculate standard weekly earnings for the groups included in Table 4.3. From these figures, the coefficient of variation of plant earnings was calculated.[2] The results are shown in Table 4.3 together with the percentage of the male insured labour force wholly unemployed in the relevant quarter.[3]

Table 4.3 clearly demonstrates that substantial wage differentials do, indeed, persist over long periods of time for all the groups selected for examination. There are some differences in behaviour, but these are differences of detail rather than of degree. For example, the

[1] See p. 45 above.

[2] For occupational earnings, the calculations in all periods are based on the 14 Glasgow plants and the 13 Birmingham plants used to obtain the results of Tables 4.1 and 4.2 above. The W.E. returns in Glasgow are taken from 18 plants for all periods shown, save two, and from 14 units in Birmingham.

[3] It is convenient to divide each year into four quarters giving 32 quarters in all over 1959–66. See p. 133 below.

coefficient of variation of gross weekly earnings for males shows greater fluctuations over the short run than the same measure applied to standard weekly earnings for occupational groups. For all males in Glasgow, it varies between 10·7 and 20·2 compared to a range of 13·2 to 19·9 for turners' standard weekly earnings. The same contrast can

TABLE 4.3

COEFFICIENT OF VARIATION OF PLANT EARNINGS: GLASGOW AND BIRMINGHAM (GROSS WEEKLY EARNINGS) AND OCCUPATIONAL GROUPS (STANDARD WEEKLY EARNINGS); AND MALE UNEMPLOYMENT RATES, 1959–66

	Glasgow				Birmingham			
Quarter	Coefficient of variation				Coefficient of variation			
	All males	Turners	Labourers	U%	All males	Toolroom	Labourers	U%
2	16·9	14·2	15·4	5·7	15·6	—	—	0·7
4	14·2	—	—	5·6	17·9	—	—	0·5
6	18·7	15·6	18·2	4·7	17·0	—	—	0·3
8	17·5	—	—	4·4	15·9	—	—	0·7
10	15·0	14·9	18·1	3·8	16·3	—	—	0·7
12	15·0	—	—	4·7	18·4	—	—	1·0
14	15·9	15·0	14·5	4·9	17·9	—	—	1·3
16	18·2	—	—	6·6	17·3	—	—	1·7
18	16·7	—	—	6·9	22·6	—	—	1·5
20	18·5	—	—	6·4	21·2	20·1	13·0	1·1
22	10·7	19·9	15·9	5·2	14·9	20·9	16·9	0·7
24	15·1	—	—	4·7	14·9	20·2	13·0	0·5
26	15·1	17·2	12·5	3·9	14·9	19·3	13·6	0·5
28	15·2	—	—	4·1	14·6	18·7	14·4	0·5
30	20·2	13·2	12·2	3·8	16·7	21·7	17·8	0·4
32	13·2	—	—	5·2	13·1	22·7	16·0	2·0

be observed in Birmingham. This is not altogether surprising. Gross weekly earnings include overtime payments, and it appears reasonable to suppose that overtime payments are one of the least stable elements of the pay packet.[1] Hence the coefficient of variation of gross weekly earnings may show substantial changes from one period to the next which are not reflected to the same degree in standard weekly earnings. While this is so, there is no indication of any marked *long-term* trend in plant wage differentials when this is measured by gross or standard weekly earnings. Calculations similar

[1] We include in this the possibility of short-time working.

to those of Table 4.3, but based on gross weekly earnings for turners, toolroom workers and labourers, confirm this result.

Given that no long-term trend emerges, we have still to consider whether short-run changes in plant wage differentials as measured by the coefficient of variation are associated with changes in the employment conditions ruling in the market. If we take the percentage rate of male unemployment to represent employment conditions, we might suppose that the spread of plant earnings would narrow when unemployment falls and would widen as unemployment rises. The argument is that recruitment difficulties will force low-wage plants to bid up wages when they have difficulty in securing an adequate supply of labour (i.e. when unemployment is low), whereas competitive pressures will be less severe, and wage differentials wider, when labour is more easily obtained (i.e. when unemployment is high). This suggests that the coefficient of variation of plant earnings will be positively related to the level of unemployment. Unfortunately Table 4.3 provides little support for this hypothesis. Inspection will indicate no apparent relationship between the coefficient of variation of plant earnings and unemployment, an impression which is confirmed by more detailed examination. Correlating the coefficient of variation with the male unemployment rate, we obtain, reading left to right across Table 4.3, the following coefficients: +0·09, +0·22, +0·25, +0·22, +0·63 and +0·07. If we re-calculate the data for occupational groups to obtain the coefficient of variation for gross weekly earnings and correlate this with unemployment, then the coefficients obtained, again reading from left to right, are +0·36, +0·53, +0·02 and −0·61 for the four occupational groups included in Table 4.3.[1] It will be observed that the coefficients have the expected sign in 9 of the 10 cases, but they are extremely weak and do not approach the 5 per cent significance level. There is no evidence, therefore, to support the proposition that the spread of plant earnings in a market is affected by short-run changes in the level of unemployment.

There is no need to labour the point any further. In each market examined, substantial plant wage differentials existed for all the groups of manual employees considered. Moreover, there is no evidence to suggest that wage differentials were narrower in a 'tight' labour market such as Birmingham, or that the spread of differentials within a market was responsive to short-run changes in employment conditions. However, one important question has yet to be considered. At any point in time there must have been a hierarchy of

[1] Much the same results are obtained if we use the male unemployment rate in engineering and if unemployment is defined to include or exclude the temporarily stopped.

high-wage to low-wage plants, but we do not know whether this hierarchy was stable or unstable over time; whether some plants were always high-wage units, and some low-wage units, or whether high-wage plants in one period were low-wage units in the next and vice-versa. This issue is taken up in the following section.

5. THE WAGE HIERARCHY

Plant earnings can be arranged in a kind of league table with the high-wage units at the top of the league and the low-wage units at the bottom. If this is done for a number of periods we can investigate the permanency of plant positions in the league table through measuring rank correlation coefficients. We begin with Tables 4.4 and 4.5 which are based on gross weekly earnings. In Table 4.4, the 18 Glasgow plants for whom W.E. returns were available were ranked according to their position in the inter-plant wage structure in selected periods.

TABLE 4.4
RANKING OF PLANTS BY GROSS WEEKLY EARNINGS: GLASGOW MALES, SELECTED PERIODS

Plant Number	Oct. 1959	Oct. 1961	Oct. 1963	Apr. 1966	Oct. 1966
G1	1	5	12	10	10
G2	2	2	5	3	6
G3	3	3	1	6	1
G4	4	5	2	2	11
G5	5	9	9	11	12
G6	6	7	17	15	5
G7	7	1	3	4	4
G8	8	3	4	1	7
G9	9	16	7	17	18
G10	10	11	6	5	2
G11	11	13	10	7	8
G12	12	8	15	8	8
G13	13	11	13	11	17
G14	14	10	14	13	15
G15	15	13	8	9	3
G16	16	17	16	14	14
G17	17	15	11	16	13
G18	18	18	18	18	16
Rank correlation		+0·80	+0·57	+0·55	+0·42

TABLE 4.5

RANKING OF PLANTS BY GROSS WEEKLY EARNINGS: BIRMINGHAM MALES AND FEMALES, SELECTED PERIODS

Males						Females					
Plant Number	Oct. 1959	Oct. 1961	Oct. 1963	Apr. 1966	Oct. 1966	Plant Number	Oct. 1959	Oct. 1961	Oct. 1963	Apr. 1966	Oct. 1966
B1	1	1	2	2	10	B2	1	2	1	2	2
B2	2	2	1	1	2	B1	2	1	2	3	7
B3	3	3	4	3	14	B3	3	3	4	5	12
B4	4	8	6	7	1	B5	4	4	3	4	1
B5	5	6	3	6	4	B7	5	5	5	1	3
B6	6	5	8	4	6	B13	6	6	7	6	4
B7	7	11	7	5	3	B14	7	12	8	10	10
B8	8	4	11	14	5	B8	8	7	10	11	6
B9	9	9	5	9	8	B11	9	11	12	12	11
B10	10	13	14	13	13	B9	10	8	6	9	8
B11	10	12	9	11	9	B6	11	9	8	8	5
B12	12	10	10	12	12	B12	12	10	11	7	9
B13	13	7	12	8	7						
B14	14	14	13	10	11						
Rank correlation		+0·77	+0·86	+0·77	+0·36			+0·85	+0·85	+0·71	+0·34

The ranking of the plants in each period was then compared to their ranking in October 1959 by calculating the Spearman rank correlation coefficient. The results for Glasgow are based on average male gross weekly earnings by plants. In Birmingham the same process was repeated for 14 plants in the case of males and 12 in the case of females.[1] For ease of reference, Glasgow and Birmingham plants are given numbers prefixed by G and B respectively.[2]

All the rank correlation coefficients obtained are positive. Moreover, for each group considered the first three coefficients are significant at the 1 per cent level, and only the coefficients for Birmingham males and females in October 1966 fail to be significant at the 5 per cent level. In other words, the ranking of plants in terms of average gross weekly earnings appears to exhibit considerable stability over long periods of time. Yet superimposed on this may be sharp changes in plant rankings in the short run. If we take the case of Birmingham males, we find that in April 1966 the position of plants in the league table had changed relatively little as compared to October 1959. Although more than six years had elapsed, the rank correlation coefficient was, at +0·77, significant at the 1 per cent level and the stability of plant rankings was particularly evident for high-wage and low-wage establishments. Only in the middle ranges had there been much switching of positions. However, a mere six months later plant rankings had changed markedly. In October 1966, the rank correlation coefficient had fallen to only +0·36 and was not significant at the 5 per cent level. Much the same features can be observed for female gross weekly earnings in Birmingham units, and it is therefore necessary to seek some reason for these sudden shifts in ranking after a long period of stability.

The most plausible explanation runs along the following lines. Most engineering plants in Britain operate a working week in which some overtime working is regarded as 'normal' or 'systematic'. While the amount of overtime regarded as 'normal' will vary from unit to unit, each plant will take up a position in the wage structure, as measured by gross weekly earnings, which will be much the same in one period compared to the next. In abnormal situations, however, where the plant is faced with recruitment difficulties, and most particularly when its work load falls, it will react in the short run by

[1] Few Glasgow plants employed substantial numbers of female workers. See p. 135 below.

[2] A reference number is unique to a plant to facilitate comparisons the reader may wish to make between tables in this and the following chapter. Thus we can see from Table 4.5 that in October 1959 plant B1 in Birmingham was ranked first for average male gross weekly earnings and second for average female gross weekly earnings.

varying overtime hours and even by adopting short-time working. The overtime element in the wage packet will, then, on occasion be subject to sharp changes and with it the ranking of the inter-plant gross weekly earnings structure.

If we now re-examine Table 4.5 we can see that the substantial shift in plant rankings for male gross weekly earnings between April and October of 1966 was mainly due to the fall in the relative earnings

TABLE 4.6
RANKING OF PLANTS BY STANDARD WEEKLY EARNINGS: GLASGOW MALES, JUNE 1959 AND OCTOBER 1966

Plant Number	Plant ranking							
	Fitters		Turners		Labourers		All Workers	
	1959	1966	1959	1966	1959	1966	1959	1966
G4	1	3	1	4	1	2	1	3
G7	2	5	3	7	3	5	3	6
G3	3	1	2	1	2	1	2	1
G19	4	8	7	10	12	13	4	8
G20	5	6	5	5	5	14	5	5
G14	6	7	8	8	7	8	7	7
G15	7	4	9	3	11	4	8	4
G10	8	2	4	2	6	6	6	2
G12	9	13	11	12	4	7	9	10
G9	10	10	6	6	9	11	10	9
G11	11	11	10	13	10	10	11	11
G16	12	14	13	11	8	12	12	14
G6	13	12	14	9	14	3	13	12
G18	14	9	12	14	13	9	14	13
Rank correlation	+0·72 (+0·39)		+0·74 (+0·34)		+0·33 (+0·33)		+0·85 (+0·48)	

of two wage leaders—plants B1 and B3. Both were engaged in motor-car production and were normally high-wage units, but in October 1966 there were widespread redundancies in this sector of the Birmingham engineering industry accompanied by short-time working. As a result, plant B1 ranked second for males in April 1966, fell to tenth in October 1966, and plant B3 fell from third to last. Similar movements in female rankings are observable, with the fall in relative gross earnings for plants B1 and B3 again prominent.

What this suggests is that plant rankings will be more stable, particularly in the short run, if measured by standard weekly earnings rather than by gross weekly earnings. Unfortunately we do not have data on occupational earnings for periods which match those in Tables 4.4 and 4.5 and, hence, we cannot demonstrate this point conclusively for Birmingham. However, the data for Glasgow plants do corroborate the above argument, and Tables 4.6 and 4.7 below do allow us to standardize for occupational groups. Plants were ranked according to their position in the inter-plant wage structure

TABLE 4.7

RANKING OF PLANTS BY STANDARD WEEKLY EARNINGS: BIRMINGHAM MALES, JUNE 1963 AND JUNE 1966

Plant Number	Plant ranking							
	Toolroom		Semi-skilled		Labourers		All Workers	
	1963	1966	1963	1966	1963	1966	1963	1966
B1	1	2	3	2	1	1	2	1
B2	2	1	1	1	2	2	1	2
B15	3	5	2	3	11	3	5	6
B16	4	4	5	5	8	7	4	5
B3	5	3	4	4	4	5	3	3
B4	6	6	—	—	10	6	6	4
B5	7	7	6	6	9	8	7	7
B12	8	8	8	12	6	13	9	11
B7	9	9	7	7	5	4	8	8
B9	10	12	10	8	3	10	10	10
B11	11	11	12	11	12	11	11	12
B6	12	10	11	9	7	12	12	9
B14	13	13	9	10	13	9	13	13
Rank correlation	+0·95 (+0·91)		+0·90 (+0·94)		+0·38 (+0·44)		+0·94 (+0·92)	

for the occupational groups and periods shown.[1] From these rankings, based on standard weekly earnings, the Spearman rank correlation coefficient was calculated for each group. The process was repeated for gross weekly earnings, and the resulting rank correlation coefficients are shown in parentheses at the bottom of each table.

In Birmingham, rank stability is very high for most groups, resulting in a high and positive rank correlation coefficient. For toolroom, semi-skilled and all workers, the coefficient is 0·90 or greater, whether or not the potentially volatile overtime earnings are excluded. In all of these cases, the correlations are significant at the 1

[1] The plants included in Tables 4.6 and 4.7 are the same as those used to compile Tables 4.1 and 4.2.

per cent level. In other words, those plants which were wage leaders in 1963 tended to remain so in 1966, while the remaining units also tended to maintain a very similar position in the wage hierarchy. Two additional points may be made from the Birmingham results, which are of some importance to our concluding argument. First, it is apparent that the ranking of plant earnings was much less stable for unskilled workers even where the wage variable is taken to be standard weekly earnings. Second, plants B1, B2, B3 and B16, which were all motor-car assembly plants, were clearly established as wage leaders amongst the case-study plants and consistently occupied high rankings in the inter-plant wage structure.

For the Glasgow plants, the wage structure is much more stable when comparison is made on the basis of earnings for a standard working week than when gross weekly earnings are used as the unit of comparison. When overtime earnings are excluded, the rank correlations are positive and significant at the 1 per cent level for fitters, turners and all workers. None of the rank correlations are significant at the 5 per cent level when overtime payments are included. The low coefficients obtained for gross weekly earnings are due to major shifts in the position of two plants, both of which were normally high wage-payers. Thus, plant G4 which was always a high wage-payer in terms of standard weekly earnings[1] was, for all groups considered in Table 4.6, ranked first for gross weekly earnings in 1959. By 1966, its position in the inter-plant wage structure, as measured by gross weekly earnings, had fallen to tenth, seventh, ninth and tenth for fitters, turners, labourers and all workers respectively. These substantial shifts were, in large measure, due to a reduction in overtime working between the two periods considered. They reflected a temporary rather than a permanent shift in the plant's relative wage position.

These results are consistent with our previous suggestion that over the long run, where the appropriate wage variable is standard weekly earnings, major shifts in the inter-plant wage structure are relatively rare. This does not preclude any shift in rankings, but it is possible to discern certain units which usually are high-wage units and other establishments which generally are low wage-payers. The plant wage structure is less stable when overtime earnings are included, but this is only to be expected. However, major changes in the wage hierarchy as measured by gross weekly earnings are usually reversed within a fairly short period, so that stability is also likely to emerge when gross weekly earnings are considered over a long period.

[1] This remark is based on an examination of rank stability for seven periods over 1959–66.

We may conclude this section with two further observations. First, in Glasgow and Birmingham the inter-plant wage structure was least stable for labourers, and this applied to both standard and gross weekly earnings. Second, it is interesting to note that in a plant where average earnings for one group of workers was high the earnings for each group of workers also tended to be high and vice-versa. If, from Table 4.6, Glasgow plants are ranked according to standard weekly earnings for fitters in 1966 and this is then correlated in turn with the plants' ranking, in 1966, for turners, labourers and all workers, rank correlation coefficients are obtained of +0·83, +0·49 and +0·92 respectively. Repeating the process in Birmingham, but this time comparing plant rankings in 1966 for semi-skilled, labourers and all workers to that for toolroom workers, we obtain coefficients of +0·84, +0·74 and +0·93. All the coefficients are positive and significant at the 1 per cent level. Similar results are obtained for gross weekly earnings.[1] Further, a study of Tables 4·6 and 4·7 will show that, when a plant's relative wage position changed, this tended to affect the relative wage position of all workers employed by that plant.[2] This suggests that changes in plant earnings take place 'across the board,' all groups experiencing an improvement or worsening of their relative positions. We shall return to this finding when we discuss the internal wage structure in Chapter 6.

6. CAUSES OF INTER-PLANT WAGE DIFFERENTIALS

We have shown that the substantial inter-plant differentials observed are not transitory; that over long periods of time there is no tendency for wage differentials to disappear; and that there are plants which in all or most periods are high or low wage-payers. We have now to consider the factors which may give rise to and perpetuate these differentials. Let us start by investigating explanations which have been put forward as alternatives to the competitive model.

Empirical studies have in general accepted that the wage differentials found to exist in practice are in some sense 'real', and are not simply or even mainly due to offsetting non-pecuniary considerations or to differences in hiring standards. Although the authorities differ on the importance ascribed to the various factors, they would appear to agree that the following elements are important in giving rise to wage differentials: imperfect knowledge on the part of employers and

[1] The relevant rank correlation coefficients are +0·76, +0·71 and +0·92 in Glasgow and +0·83, +0·65 and +0·86 in Birmingham.

[2] See, for example, plants G7 and G15 in Glasgow, Table 4.6.

employees; inertia on the part of employees; collusion between employers in the form of 'anti-pirating' agreements; the application of seniority rules in filling job vacancies which discriminate in favour of employees and against non-employees, and the existence of underemployment in the labour market.[1]

While these factors may give rise to plant wage differentials, neither singly nor collectively are they the chief cause of the wage differentials observed in our labour market areas. Let us examine each in turn. There appear to be severe limits to the job knowledge possessed by the labour force. The job-seeker generally acquires information through a restricted circle of friends or relatives; he is often not able to evaluate correctly the wage levels offered by different jobs, and still less to judge precisely the importance of non-wage considerations. Some part of the labour mobility which takes place in practice must, then, be wasteful and haphazard. The effect of imperfect knowledge will be compounded if it extends to employers as well as employees, so that managers are unable to assess their plant's position in the plant wage structure or to judge changes in relative positions over time.

In our view, imperfect knowledge is indeed an important feature of labour market behaviour and will give rise to wage differentials and differences in net advantages in the short run. The 'short run' might even extend to months and years rather than to days and weeks, simply because imperfect knowledge prevents employers and employees from adjusting quickly to any change in labour market conditions. The usefulness of competitive theory does not, however, rest on the assumption that labour markets adjust instantaneously to any change in employment conditions. The existence of imperfect knowledge has always been recognized, but this does not necessarily imply that 'the greater part of labour turnover consists of a continuous and fruitless inter-change of workers between firms which involves the waste of national and private resources',[2] or that employers are so ignorant of the policies of their competitors that competitive forces have little influence on the inter-plant wage structure.

In the instance of our case-study plants, the wage surveys conducted by the local associations of the E.E.F. and by the plants themselves, and the information obtained through informal channels, provided managers with some knowledge of the standing of their plant in the inter-plant wage structure. For example, managements could

[1] See L. G. Reynolds, *The Structure of Labor Markets* (1951); and Lester; Myers and Maclaurin; and Robinson, *op. cit.*

[2] J. R. Long, *Labour Turnover under Full Employment*, University of Birmingham Studies in Economics and Society (1951), p. 12.

usually identify 'wage leaders'[1] and had a fairly clear notion whether their establishment was, in general, a 'high', 'average' or 'low' wage-payer. Very occasionally, some plants could go further. For example, each of the two case-study plants in Small Town, being the only two engineering units in the area, had considerable detailed knowledge of their rival's earnings levels. This was, however, exceptional and is likely to be repeated only rarely outside small localized labour market areas. In most cases the manager has only an imperfect understanding of the level of wages ruling in rival plants. He is like a traveller faced with a dense mist rather than with an impenetrable fog. He may not have lost his way but he is seldom certain precisely where he is. The behaviour of his competitors, which provides signposts for possible courses of action, is seen rather indistinctly. Errors of judgment therefore occur, and these can have serious consequences. For example, each of the five plants established in New Town during our period had to construct a system of wage payments *de novo*. However, although all these units had attempted to gather wage information from other engineering establishments on a systematic basis, none found this information satisfactory in an operating environment, so that substantial alterations had to be made to the structure and level of earnings.

The plants had, then, little *detailed* knowledge of the level of earnings ruling in other establishments in their local market area. This being so, it is quite possible that they might react slowly to changes in the wage policies adopted by other units, or that they might take action on false premises as did the plants recently established in New Town. An element of uncertainty or indeterminacy is therefore introduced which will cause differences in plant wage levels, and this will be reinforced if the labour force, too, is unable to discern accurately the position of plants in the wage structure. However, managers (and employees) *can* identify wage leaders and they can detect *major* shifts in relative earnings. Ignorance is not so pervasive that it can alone account for the substantial and persistent wage differentials found to exist in practice. Given a sufficient lapse of time, information *will* be disseminated through the market which will enable employers and employees to identify high-wage and low-wage plants.

A number of points might be made to buttress the above conclusion. The first, and most obvious, is that if wage differentials arose primarily because of imperfect knowledge there would be no tendency for a stable plant wage structure to emerge. Instead, over time,

[1] This was confirmed by relating the remarks of the plants to our detailed knowledge of actual earnings in the local labour markets.

plants would change their rankings at random. This is far from being the case. Second, plants do have some knowledge of their own relative position in the wage structure and can identify high- and low-wage units within their own labour market area. They may not, in fact do not, know the precise extent of the wage differentials which exist, but they *do* know that plant A is near the top of the league table, that plant B is near the bottom and so forth. For example, in Small Town there are only two units each of which knows in considerable detail the earnings levels prevailing in its rival's establishment. Despite this, and despite the fact that this knowledge is shared with the labour force in the area, differences in earnings of some 10 per cent for most occupations have persisted throughout the period 1959–66. It might be argued that these differentials are relatively small compared to those prevailing in Glasgow and Birmingham, and that the existence of extremely large differentials could then be due to inadequate information. Even this appears unlikely. In Birmingham, it is part of the accepted wisdom that motor-car plants offer high levels of earnings. When asked to identify high-wage units, managers almost invariably singled out motor-car plants, and our empirical evidence provides ample justification for their choice.[1]

Wage differentials may arise through a combination of factors which limit labour mobility and give each employer, in effect, a 'captive' labour force. These factors are inertia, particularly on the part of those in employment, many of whom are not 'in' the labour market but are satisfied with their current job;[2] fringe benefits which are related to seniority; discrimination in favour of employees in filling job vacancies; the application of seniority rules in determining promotion and demotion within the plant; and collusion between employers through anti-pirating agreements.

The first two points can be dealt with quickly. Conceptually, it is extremely difficult to distinguish inertia, which may make employees willing to put up with a lower wage than they would receive elsewhere, from the principle of net advantage. If employees are 'satisfied' with their current jobs and not therefore 'active' job-seekers, we must surely infer that other factors, such as attachment to known routines, habit, or a fear of the unknown, are sufficiently important to offset the low relative wage. In any event, we shall see that there appears to be a considerable amount of labour turnover, especially within the Birmingham and Glasgow conurbations.[3] Again, while the extent of the fringe benefits offered to manual employees does

[1] See p. 83 above.

[2] Reynolds, *op. cit.*, pp. 101–3.

[3] It is also worth noting that of those workers Reynolds found to be 'satisfied'

vary from unit to unit, and within each unit according to length of service,[1] their value, even for long-service employees in the engineering industry, does not usually exceed a few shillings per week.[2] Hence, it is difficult to believe that fringe benefits have more than a marginal effect in restricting mobility.[3]

Greater attention must be paid to the possibility that anti-pirating agreements and the operation of internal labour markets might create and perpetuate wage differences between establishments. American studies have placed especial emphasis on those aspects of labour market behaviour which are held to inhibit mobility between plants. thus creating, in effect, a series of sub-markets based on each employing unit. We can find little support for these suggestions and, while we have to take into account the labour market and institutional framework in which the present study is set, it would seem that collusive practices between employers and the existence of internal labour markets are not the main factors promoting plant wage differentials.

Almost all the studies based on American data have concluded that anti-pirating agreements are widespread,[4] and that an unwritten code is observed by most managements, which prohibits aggressive recruitment tactics and even the employment of those persons currently employed by other units. Hence the employee must, to find new employment, quit his present job or obtain the acquiescence of his current employer. The latter behaviour amongst the case-study plants included in this study was rare to the point of being nonexistent.[5] No examples were found of plants in Birmingham or Glasgow refusing or having any moral scruples about hiring labour currently employed by other establishments. Only with the two plants

with their present job in 1947, 26 per cent had changed employers by 1948 (*op. cit.*, p. 103). Although some of this movement will have been involuntary, it does indicate a fairly high degree of mobility.

[1] Employees often have to complete some minimum period of service before qualifying for fringe benefits such as pension schemes, sickness benefits etc., and the size of these benefits usually increases with length of service after this minimum period is completed.

[2] In 1964, fringe benefits (private social welfare payments, payments in kind and subsidized services) amounted to 4·0 per cent of total labour costs in British manufacturing industry. Department of Employment and Productivity, *Labour Costs in Great Britain in 1964* (H.M.S.O., 1968), p. 5.

[3] This was certainly the view of the case-study plants, a view which is supported by other investigators. See Hunter and Reid, *op. cit.*, pp. 133–7.

[4] See, for example, Myers and Maclaurin, *op. cit.*, pp. 40–1; R. A. Lester, *Adjustments to Labor Shortages* (1955), pp. 46–9, and *Hiring Practices and Labor Competition*, pp. 62–5; Reynolds, *op. cit.*, pp. 51–2.

[5] See pp. 348–9 below.

in Small Town, the sole major engineering employers in that area, was there any indication of an anti-pirating agreement, and even here it was not always operated. Again, while most units tended to avoid certain types of advertising in recruiting labour, it was not uncommon to find plants which took the initiative in attempting to secure labour from other establishments. This applied in the slacker Glasgow market as well as in Birmingham where recruitment difficulties were often intense. We might conclude, therefore, that anti-pirating agreements have less force in British labour markets, and that 'labour enticement' in which the plant plays an active role in hiring labour away from rival units is more common than appears to be the case in the U.S.A.[1] Nonetheless, the lack of collusion between employers seems to have had little effect on wage differentials, which remain substantial in all our labour market areas.

We must also reject the view that wage differentials arise primarily because of the operation of policies which inhibit labour mobility by 'tying' the labour force to the employing unit. Lester, writing in 1954, concluded that 'developments during recent decades, especially seniority rules and in-plant training and promotion, have enhanced the costs of inter-company movement for labor and have increased the independence between labor mobility and wage determination'.[2] The argument runs as follows. Plants often have a preference for filling jobs through internal promotion so that the new recruit to the establishment will have to begin at the bottom of the occupational ladder. His movement up that ladder is then largely determined by the acquisition of seniority which also reduces the risk of his being declared redundant. The process is reinforced by union pressure which attempts to institutionalize the procedures which discriminate in favour of long-service employees relative to short-service employees and in favour of both groups relative to non-employees. An employee who quits his present job must, therefore, sacrifice any benefits secured by seniority and begin with a new plant at the bottom of the occupational ladder. This deters mobility, particularly on the part of long-service employees.

There is no doubt that the concept of internal labour markets is a valuable addition to labour market theory which throws new light on a number of aspects of labour market behaviour. Yet there is very

[1] As the above discussion suggests, this conclusion may require modification according to the type of labour market investigated. American studies have usually been based on labour markets much smaller than the Birmingham and Glasgow conurbations and in these smaller markets the social pressures on employers to avoid pirating may be greater.

[2] *Hiring Practices and Labor Competition*, p. 29.

little empirical evidence on the extent of internal mobility or on the effect of the rules used to govern such mobility. However, anticipating the results of Chapter 11, we can reach the following conclusions. None of the case-study plants followed a policy in which there was only one port of entry at the foot of the occupational ladder. On the contrary, there was usually a port of entry for each major type of skill required by the unit. Some plants did follow fairly active policies of internal promotion,[1] but such promotion was never based on seniority alone. The latter factor was usually decisive only when the ability, acquired experience and training of prospective candidates were considered equal, a condition which can seldom be met in practice.[2] Nor was there any evidence that union officials or rank and file union members had been active in pressing policies aimed at encouraging internal mobility or governing the rules which determined such mobility. Formal rules and agreements were rare even in redundancy situations, and although the principle of seniority was important in this instance, it was often narrowly applied.[3]

It may be that the British engineering industry is atypical. Certainly, there are industries in Britain which place more emphasis on internal mobility,[4] and managements and unions in the U.S.A. seem to have paid greater attention to promoting and regulating this process. However, there is little 'hard' evidence on the extent of internal mobility, and we must be sceptical of any view that internal labour market policies have been developed sufficiently to explain the wage differentials found between our case-study plants. Some discrimination in favour of current employees is likely to exist and long-service employees are more likely to avoid redundancies than short-service employees,[5] but these relatively minor qualifications, while they may take us some part of the way in explaining wage differentials, still leave us with much of the road to travel.

Nor do we find under-employment in the labour market a satisfactory explanation of wage differentials. It is true that under-

[1] Particularly, be it noted, the plants under American management.

[2] Nonetheless, it is probable that internal mobility will be greater the longer the length of service of the employee (see pp. 309–10 below). This does not reflect the application of seniority rules as much as the fact that long-service employees have acquired skills and experience which assist upgrading.

[3] See pp. 380–1 below.

[4] Internal mobility appears to be greater in the chemical and the soap or candles industries than in engineering (see S. W. Lerner, J. R. Cable and S. Gupta, *Workshop Wage Determination* (1969)) and is also considerable in steel production (see L. C. Hunter, G. L. Reid and D. Boddy, *Labour Problems of Technological Change* (1970), p. 132).

[5] As we shall see (pp. 210–1 below), this is *not* the same thing as saying that a low proportion of redundant employees are long-service employees.

employment has been regarded not as the initial cause of plant wage differentials but as a factor which perpetuates wage differentials which have arisen for other reasons. Hence, under-employment is 'only a *necessary* and not a *sufficient* condition of differences in job attractiveness. It permits them to appear and continue but does not necessarily bring them into existence.'[1] Despite this qualification, the argument, on our evidence at least, appears unconvincing. Under-employment is not even a necessary condition. No conceivable definition of under-employment could be applied to the Birmingham labour market over 1959–66 and the extremely 'tight' labour market conditions ruling in that area had persisted for a long period prior to 1959. Yet plant wage differentials in Birmingham were as large as, or even larger than, those prevailing in the 'slacker' Glasgow market. Again, there was no evidence of any appreciable narrowing of wage differentials over time in Birmingham as would be expected if under-employment was a major factor in perpetuating differentials. Last, differences in plant wage levels were as substantial for skilled employees as for other groups, which is again inconsistent with the under-employment hypothesis, given the more favourable labour market conditions for skilled workers and the chronic tendency for the demand for skilled labour to outstrip the available supply.

Some of the suggestions put forward by previous empirical research to account for the existence of substantial plant differentials provide, at best, only a partial explanation. We must therefore provide some alternative view which fits the observed facts. Let us first consider whether we can reconcile our findings with the competitive model. The model predicts that within a local labour market there will be a tendency for labour of the same type and quality to obtain the same wage. If there is no tendency towards equalization, as we have suggested, there are only two explanations of this behaviour consistent with the predictions of traditional theory. First, that differences in wages are offset by non-pecuniary factors so that net advantages are equalized although wages are not. Second, that our occupational groups, even if defined in terms of 'turners', 'fitters', 'toolroom workers', etc., are not, in fact, homogeneous, so that we are not observing plant wage differentials for groups of the same skill, experience and productivity.

The most obvious point to be made is that if non-wage factors are the chief cause of variations in plant earnings then they must be extremely important in job choice, and heavily biased in favour of low-wage plants, to offset the large wage differentials which exist in a local market. This is by no means impossible, for most studies of

[1] Reynolds, *op. cit.*, p. 247. Italics in the original.

labour market behaviour clearly indicate that job choices are not simply based on considerations of relative wage levels.[1] Unfortunately, we were not able in this study to undertake the massive, and probably impossible, task of quantifying the importance of the host of non-wage factors which might conceivably condition job choice. We can only report our 'impressions' gained from visits to the case-study plants, which included an inspection of shop-floor conditions, and supplement this by putting together certain scraps of factual information which might be useful.

Fringe benefits such as pension, life assurance and sickness schemes were the only element not entering directly into the wage packet which proved susceptible to measurement. We have already seen that the value of such benefits was generally low. In any event, fringe benefits were usually of greatest importance in those units where wage earnings were high, so that one cannot explain plant differentials on these grounds. Of the twenty plants in Glasgow for which details were available, eight plants in the top half of the wage distribution had pension schemes for manual employees to which the plant contributed, six had sickness benefit schemes, and three had life insurance schemes. In the case of the ten plants in the bottom half of the wage distribution, only four had pension schemes, one a sickness benefit scheme, and three had life insurance schemes.

Visits to the case-study plants did not provide any indication that low-wage units enjoyed any advantage over high-wage units with respect to non-pecuniary factors. Indeed, physical conditions of work, social amenities and like conditions were generally more favourable in high-wage units. This is hardly conclusive evidence, for the assessment which is relevant is not that of the observer but that of the employee. Yet it is worth while noting that our impressionistic conclusion is consistent with that of earlier studies, namely, that there is no indication that non-wage factors are particularly favourable in low-wage units, and there is some evidence which points in the opposite direction.[2] Our only qualification to this finding is that high-wage plants tend to be large units whose employees have relatively long travel-to-work journeys.[3] Hence, while more powerful factors are at work giving rise to wage differentials, high wages must to some extent be regarded as compensation for the opportunity cost of these longer journeys.

The notion of the worker being presented with a choice between

[1] See H. S. Parnes, *Research on Labor Mobility: An Appraisal of Research Findings in the United States* (1954), pp. 147–60.

[2] See, for example, Reynolds, *op. cit.*, pp. 202–3 and 221–2; and Myers and Maclaurin, *op. cit.*, p. 54.

[3] See pp. 251–2 below.

different combinations of wage and non-wage factors has a parallel with the employer choosing between certain combinations of wages and effort. Thus, the employee may choose a plant where wages are low but where other conditions of work are favourable; the employer may pay low wages and accept less skilful or less productive labour. Non-homogeneity of the labour force has two aspects—differences in skill between workers and changes in the level of skill of the individual worker. If skill differences exist between workers then employers can choose among different combinations of wages paid and hiring standards. They can elect, for example, to pay low wages in return for relatively low labour efficiency, so that efficiency wages may be equalized across plants although differences in money wages exist. However, even if the hiring standards applied are identical it may still be rational behaviour for an employer to offer high relative wages. High wages might reduce the selection, recruitment and training costs associated with labour turnover, and if the efficiency of the individual worker increases through training and acquired experience while he stays with the plant, the reasons for minimizing quit rates are reinforced.

Let us first take up the question of differences in skill and effort between workers. Since we are dealing with an industry in which a substantial proportion of the work force are paid by piecework, one might suppose that differences in wage levels must, to some extent, reflect differences in effort and productivity. In other words, high-wage plants may enjoy high levels of output so that efficiency wages are equalized across units. There is likely to be an element of truth in this argument, but it is by no means the whole story. Plant wage differentials are extremely wide even where timeworkers are concerned. Thus toolroom workers and labourers are timeworkers in almost all units in Birmingham, and Table 4.2 showed that for these groups differentials are as wide as, or even wider than, those found for other types of employees.

It is, of course, still possible that this labour force will differ in quality and in efficiency because of the application of different hiring standards. As hiring standards are almost invariably subjective, their effect is impossible to quantify[1] and one is usually thrown back on indirect observation or measurement. The most interesting example of the latter approach is a study by Weiss which, using various personal characteristics as proxies for labour efficiency, found that, 'in general employers who for any reason pay high salaries receive "superior" labor in the bargain'.[2] Our own observations are consis-

[1] Hiring standards are considered in greater detail on pp. 359–64 below.

[2] L. W. Weiss, 'Concentration and Labor Earnings', *American Economic*

tent with this view at least to the extent that where large wage differentials are found they are accompanied by the application of stricter hiring standards in the high-wage units. This being so, we can explain at least some part of the variation in earnings between very low and very high-wage plants. It is much more difficult to say how big 'some' actually is, but our impression is that few plants made any attempt to carry through a systematic analysis of job content which could be related to the attributes of potential recruits. Again, while experienced personnel officers are no doubt often able to reach accurate assessments on what appears to be a perfunctory appraisal, selection procedures are often extremely casual. The best summary of the position is provided by Lester who found that 'notable differences in quality of the work force were evident only for two or three firms at the top of the wage hierarchy and for two or three at the very bottom. In between these extremes there appeared to be little evident correlation between quality and relative wage position.'[1]

Thus, while wage differentials may reflect differences in hiring standards, it still seems unlikely that efficiency wages are equalized in this manner. There remains, then, the analogous argument that the efficiency and usefulness of the employee may increase with his length of service. This is an important point, although a seriously neglected one in labour market analysis. A high-wage policy may be justified because it reduces labour wastage and associated costs. Indeed, we shall find, in Chapter 6, that an inverse relationship between wage *levels* and labour wastage does exist. It should also be observed, however, that if accumulated experience and training with the plant is a valuable asset to the employer, an important element of monopsony is introduced on the supply side. The skills acquired through experience and on-the-job training will in greater or lesser degree be specific to the plant, and the more these skills are differentiated from those possessed by non-employees the greater will be the difficulty experienced by the employer in replacing labour wastage, and the greater the room for forcing wages upwards.

This opens up a number of possibilities which are not admitted in the competitive model. The employer may not be able, without in-

Review, Vol. 56, 1966, p. 116. We should note two reservations. First, Weiss introduces so many personal characteristics that it is hardly surprising that some emerge as 'significant'. Second, those found to be 'significant' are often unsatisfactory proxies for efficiency. For example, the positive relationship between age and earnings could reflect increased efficiency on the part of older workers (as Weiss appears to assume) or, simply, that workers who find well-paid jobs stay in them and grow older.

[1] *Hiring Practices and Labor Competition*, p. 74. See also Reynolds, *op. cit.*, pp. 218–20.

curring heavy costs, to replace employees by recruiting labour in the external market. *A fortiori*, this applies when he is faced with collective pressure from his work force, for his alternatives in this instance are not to accede to wage demands or replace labour at the margin, but to risk the costs associated with strikes and low morale or replace his entire labour force by external recruitment. The latter is seldom a practical policy, as non-employees are imperfect substitutes for employees, and especially for those employees with long service and a high level of acquired skills and experience.

Widening the discussion still further, it would seem that the conditions prevailing in the particular plant may be important in a much broader sense. The plant's labour force may enjoy an economic rent (a payment over and above the level of wages necessary to keep the labour force attached to the plant) where profit margins are wide, or where the plant is able to absorb higher wages because of favourable product market conditions, high-quality management, efficient methods of production, economies of scale and so forth.[1] A high-wage policy may bring a plant benefits through reduced wastage and higher morale and productivity but, over and above this, it may reflect non-economic motives such as the desire to be thought of as a 'fair employer', or the exercise of 'pure bargaining power' by the unions or the rank and file. In other words, high-wage plants will tend to be profitable units or units which find it easier to pass on higher wages in the form of higher prices.[2]

The above argument does not imply that competitive forces are absent from the labour market so that wages are simply administered prices. No plant can entirely ignore the wage levels paid by its competitors, or, if it does, it is likely to rue the consequences. The very existence of a fairly stable inter-plant wage structure does suggest that competition for labour forces plants to pay some attention to wage movements in other units. In support of this proposition, we may point to the fact that plants are able to identify their approximate position in the wage structure over time, and that they are likely to

[1] The reader should observe that at this point the discussion is particularly conjectural. Interviews with the case-study plants did, however, provide some supporting evidence for the arguments advanced, and we shall see in Chapter 5 that changes in wages do appear to be responsive to conditions in the plant rather than to conditions in the external labour market.

[2] See W. G. Bowen, *Wage Behaviour in the Postwar Period* (1960); D. G. Brown, 'Expected Ability to Pay and Interindustry Wage Structure in Manufacturing', *Industrial and Labor Relations Review*, Vol. 16, 1962–3; J. W. Garbarino, 'A Theory of Interindustry Wage Structure Variation', *Quarterly Journal of Economics*, Vol. 64, 1950; and S. H. Slichter, 'Notes on the Structure of Wages', *Review of Economics and Statistics*, Vol. 32, 1950.

be under pressure from their work force to match wage increases in other units. Hence, although the wage structure may show substantial short-run variations when measured in terms of gross weekly earnings, over the long run it exhibits considerable stability for both gross and standard weekly earnings.

Two further points might be made in support of the competitive hypothesis. First, in both Birmingham and Glasgow the inter-plant wage structure was more stable for skilled and semi-skilled employees than for labourers. We might explain this by supposing that, since recruitment difficulties were least apparent for the unskilled, the plants were not particularly concerned with movements in relative wages for labourers. On the other hand, difficulties in securing skilled and semi-skilled labour, which were most important in Birmingham, compelled plants to match wage increases for such employees in other units so that plant rankings changed very little over time. Second, we shall see in Chapter 6 that quit rates were inversely related to wage levels, which suggests that labour mobility did respond to wage differentials, and hence that competitive pressures must exist to some degree.

While competitive forces do appear to set some limits to management action, the limits set do not appear to be particularly severe. The cynic might remark that the stability of plant rankings only emerges because wage differentials were so wide in the first instance that they could accommodate, without much shift in rankings, very disparate wage changes.[1] He might observe, and again with some justification, that competitive pressures did not eliminate or reduce plant wage differentials over time; that these differentials seemed as wide for skilled as for semi-skilled labour in each market, and that they were as wide in the 'tight' Birmingham labour market as in Glasgow. Moreover, the existence of an inverse relationship between wage levels and quit rates must imply that the wage differentials observed in the market were in some sense 'real', for, if this were not so, no negative correlation would emerge. Yet, despite the fact that labour turnover, as theory would suggest, responded to differences in net advantages, there was little indication of a tendency towards equalization of net advantages across plants.

In sum, our view is as follows. While very large wage differentials will be associated with differences in hiring standards, these differences are not sufficient to produce equalization of efficiency wages. The view that wage differentials are due to offsetting non-pecuniary factors is even less compelling. It does not seem that low-wage plants

[1] The stability of plant rankings need not, then, imply equivalent wage changes as Tables 5.4 and 5.5, p. 121–2 below, will show.

offer particularly attractive conditions of employment in other respects, but we should recognize that large plants are likely to have to compensate employees for long travel-to-work journeys. Market forces set certain limits to plant wage differentials which, due to differences in hiring standards, are rather narrower than a superficial reading of the data suggests. Nonetheless, there is a significant range of indeterminacy. In part, this may be due to imperfect knowledge, the operation of internal labour markets and anti-pirating agreements. However, none of these factors appears to us to be as influential as is commonly supposed. More important is the skill acquired by the work force during its service with the employer, which introduces an important element of monopsony in the labour market. This provides a basis upon which the bargaining power of the union or work group, or the exercise of management discretion, allows the plant's labour force to enjoy high relative earnings, when profits are high, when product market conditions are favourable, or when the plant is particularly efficient. This focuses attention on the conditions prevailing in the employing unit. It is a theme that we shall return to in the following chapter.

7. CONCLUSIONS

(i) Measuring plant wage differentials by occupational groups reveals a wide spread of earnings in Birmingham and in Glasgow. Although differences in detail exist according to the specific measure of spread and the definition of wages adopted, this holds as a general conclusion for each period, in both markets, and for each occupational group considered. The spread of earnings was *not* narrower in the 'tighter' Birmingham market; indeed, there is some evidence that plant wage differentials were more substantial in that area than in Glasgow. The failure of wage differentials to narrow in the face of severe recruitment difficulties is also illustrated by the fact that plant wage differentials did not appear to vary according to the skill of the group considered in Glasgow, and were wider for toolroom workers than for labourers in Birmingham. Large differences in plant wage levels were not confined to the two conurbations but were also apparent in North Lanarkshire.

(ii) At any point of time there is a considerable spread of plant earnings. There was no tendency for differentials to narrow over the long run for any group of workers considered, nor was there any indication that the spread of earnings was systematically related to short-run changes in market conditions. This conclusion, and conclusion (i) above, is unaffected by the measure of earnings adopted—

standard weekly earnings (excluding overtime payments) or gross weekly earnings (including overtime payments).

(iii) If plants are arranged in descending order according to their wage levels in different periods, then the ranking which emerges shows considerable stability in all periods examined where standard weekly earnings are adopted as the unit of measurement. Plant rankings also exhibited a fairly high degree of stability when gross weekly earnings are taken as the appropriate wage concept, but in this case rankings did, on occasion, change appreciably. Such changes were, however, usually reversed fairly quickly. Thus, while the overtime element in the wage packet, and hence gross weekly earnings, can show sharp fluctuations over the short run, it is possible to identify plants which are high-wage or low-wage units in most periods. A plant which is a high-wage payer for one group of worker tends to be a high-wage payer for all groups of workers and vice-versa. It follows from this that, where a plant's relative wage position changes, it tends to affect the relative wage position of all workers employed by the plant.

(iv) Previous labour market studies have suggested that large wage differentials between plants are due to imperfect knowledge, inertia on the part of employees, collusion between employers through anti-pirating agreements, discrimination in favour of employees against non-employees, and under-employment in the labour market. While each of these factors may be present in some labour market situations, and while some may be *a* cause of plant wage differentials in all situations, the evidence available does not suggest that they were the *main* cause of the large differentials observed in this study.

(v) Despite conclusion (iv) above, the behaviour of the inter-plant wage structure is only partly explicable in terms of competitive forces. Non-wage conditions of employment did not appear to be particularly favourable in low-wage plants, and although it seems likely that extremely large wage differentials were associated with differences in hiring standards, this is not a sufficient explanation of the differentials which arise in practice and which showed considerable stability over the long run. The limits set to wage differentials by market forces are not, then, very narrow. This is because the work force of a plant is differentiated from non-employees by acquired experience and skill and, probably more important, because wages are high in those plants where profits are high or where higher wages can be passed on through higher prices due to favourable product market conditions. The circumstances of the individual plant are, therefore, an important determinant of wage levels, a possibility which is excluded in the competitive model.

CHAPTER 5

THE INTERNAL WAGE STRUCTURE

1. INTRODUCTION

The discussion of Chapter 4 was directed towards explaining the behaviour of the *inter*-plant wage structure, that is, the relationships between the earnings levels of different plants located in the same labour market area. Here we are concerned with the intra-plant, or internal, wage structure—the relationships between the earnings of different occupational groups engaged by the *same* unit. These two aspects of wage behaviour are, of course, related. Just as labour market theory predicts that the inter-plant wage structure will be responsive to competitive pressures in the market, so it also predicts that wage relativities in the plant will be moulded by the labour market environment in which the plant operates. Changes in the wages paid by rival units may cause managements to reconsider not only the level of earnings in their unit but also the relationships between the earnings of different groups of employees. However, while we must ultimately relate our observations on the internal wage structure to external factors, the concept is a useful one analytically and provides us with a convenient starting-point for the following discussion.

Little is known about the characteristics and the behaviour of internal wage structures in Britain. There has been 'an almost complete absence of empirical investigation into plant-level wage structures'[1] despite the fact that it is an area of considerable theoretical and practical importance. Our first task is, then, essentially descriptive. It is simply to record certain features of wage-payment methods and of the internal wage structure resulting, so that we can establish the nature of the problem which confronts us. Once the main features of internal wage structures have been established they can be related back to the elements of labour market theory.

[1] Lerner, Cable and Gupta, *op. cit.*, p. 9. The honourable exceptions to this statement are D. J. Robertson, *Factory Wage Structures and National Agreements* (1960), and Robinson, *op. cit.*

The function of a plant wage structure is to reflect the balance, and changes in the balance, of economic forces determined by the demand for and supply of labour. On the supply side the primary reason for internal wage differentials is to induce a sufficiently large number of people to undertake the costs of training or to accept the more taxing content of skilled work. Skill differentials arising for the former reason can then be regarded as providing a stream of returns to the initial 'investment' in training. In the latter case they are a straightforward compensatory payment. As long as one of these two elements is present, as they are in engineering, and as long as they are not offset by other factors, we can expect skilled manual workers to have higher average earnings than the unskilled. But this is not the end of the matter, for we also want to know how the wage structure alters over time in response to demand conditions in the market. If plants want to recruit more skilled employees, for example, and are finding such labour difficult to obtain, economic theory suggests that the earnings of skilled employees will be bid up relative to those of other groups. But does the wage structure behave in this manner? Is it as responsive to the demand and supply conditions for different types of labour as the above statement suggests, or is the internal wage structure inflexible, so that wage increases for one group of employees are reflected in similar increases for other groups?

Of course, the simple competitive model does not provide a realistic explanation of factor pricing in the real world, and wage determination is not simply the result of economic calculus. Notions of equity and justice, of 'equitable comparisons', may provide the yardstick by which the 'fairness' of a wage increase is judged.[1] Such notions are derived from the existing order. 'Change—always, everywhere, in everything—requires justification: the strength of conservatism is that it is held to justify itself. It is not, therefore, surprising that the maintenance of standards, absolute or comparative, should be woven as warp and woof into the texture of wage discussions. . . .'[2]

If we accept the view that social pressures can affect wage settlements, then it seems that their influence is likely to be particularly powerful at the local, or plant, level. After all, employees may be expected to take as their point of reference the earnings and status of other groups with whom they have regular contact. Hence the pressure to maintain 'fair relativities' may be particularly intense within the plant. If this is so, the internal wage structure may be inflexible,

[1] A. M. Ross, 'The Dynamics of Wage Determination under Collective Bargaining', *American Economic Review*, Vol. 37, 1947.

[2] B. Wootton, *The Social Foundations of Wage Policy* (1955), p. 162.

and may not prove particularly sensitive to changes in the demand and supply for different types of labour.

We begin our analysis in Section 2 below by briefly outlining the current state of the argument on 'wage drift'. This is useful as a background to Section 3 where we consider the methods of wage payments adopted by the case-study plants and, in particular, the problems which arise from the widespread use of pieceworking in the British engineering industry. In Section 4, we look at the outcome of the system of wage payment adopted by means of a cross-sectional analysis of internal plant wage structures at a given moment of time. The objective of this section is to establish whether the pattern of occupational relativities is the same for all plants or, alternatively, whether it varies substantially from unit to unit. Subsequently, in Section 5, the static analysis of Section 4 is supplemented by an examination of changes in plant occupational earnings over time. The changes in occupational earnings which have occurred are related to the employment conditions prevailing in the respective labour market areas, in an attempt to ascertain whether the internal wage structure responds to economic pressures. A brief summary of the conclusions reached is provided in Section 6.

2. WAGE DRIFT

To discuss the controversy over wage drift, which has often centred around the engineering industry, is rather like picking one's way through a minefield, and it is politic to skirt the area quickly rather than plough through the middle. As we have seen, earnings levels in the British engineering industry are substantially above the minima established by national agreements due to the importance of bargaining at the plant level. The extent of this difference between national minima and actual earnings gives the 'earnings gap',[1] and 'wage drift' is the rate of increase of the earnings gap when overtime payments have been excluded.

It is useful to consider two types of wage drift. Primary wage drift arises when a group of workers is able, due to labour shortages, bargaining power, etc. to obtain a rise in earnings which disturbs 'customary' differentials. Secondary wage drift occurs when an attempt is made to restore customary relativities.[2] Particular stress has been placed by most authorities on the role of the pieceworking

[1] Strictly speaking, the earnings gap is given by the difference between earnings and *district* rates, but in the interests of simplification we shall ignore this complication.

[2] Phelps Brown, *op. cit.*

system in promoting drift, but while it is widely accepted that most wage increases originate with pieceworkers and then spread to time-workers, there is no unanimous view of the *causes* of the primary drift.

Empirical studies have failed to establish any strong relationship between wage drift and the level of demand for labour,[1] and so alternative explanations have been sought in productivity increases[2] and in the institutional framework within which wage bargaining takes place.[3] None of these 'explanations' necessarily excludes the others. Increases in productivity and a high level of employment create an environment within which wage drift occurs, but within this environment increases in earnings owe little to the demand conditions for, and the productivity increases of, the particular group considered. Hence, while earnings and productivity increases seem to be related for the engineering industry as a whole, the relationship becomes much weaker, or disappears altogether, for individual sectors of engineering.[4] Nor is there any evidence of a strong association between increased effort and changes in earnings at the level of the individual employee. Even if we made the heroic assumption that increased productivity reflected increased effort rather than additional capital investment, changes in production scheduling, etc., we would still leave much unexplained. Thus one of the rare studies of wage drift in the engineering industry found in a 'well controlled piecework system operating in near ideal circumstances . . . a wholly unproductive wage "drift" of about 1 per cent per year. It is a strong argument that such wholly inflationary "drift" is unavoidable in any piecework system.'[5]

It is also a strong argument for the view that institutional factors, particularly the fragmented method of piecework bargaining, combined with discontinuities in the production process, allow individ-

[1] See, for example, J. Marquand, 'Wage Drift: Origins, Measurement and Behaviour', *Woolwich Economic Papers*, No. 14 (1967).

[2] H. A. Turner, 'Wages: Industry Rates, Workplace Rates and the Wage Drift'; 'Wages, Productivity and the Level of Employment'; and 'The Disappearing Drift', in *Manchester School of Economic and Social Studies*, Vol. 24, 1956; Vol. 28, 1960 and Vol. 32, 1964, respectively.

[3] Lerner and Marquand, *op. cit.*, and 'Regional Variations in Earnings, Demand for Labour and Shop Stewards Combine Committees in the British Engineering Industry', *Manchester School of Economics and Social Studies*, Vol. 31, 1963.

[4] Lerner, Cable and Gupta, *op. cit.*, pp. 32–5; and R. J. Nicholson and S. Gupta, 'Output and Productivity Changes in British Manufacturing Industry, 1948–54', *Journal of the Royal Statistical Society*, Vol. 123, Series A (General), 1960.

[5] National Board for Prices and Incomes, *Report No. 65* (*Supplement*), p. 50.

uals or particular groups to obtain increases in earnings almost irrespective of demand conditions and changes in effort in their particular case. Primary drift arising with pieceworkers disturbs customary relativities, creates pressure on the wage structure and generates secondary drift as timeworkers attempt to restore the initial position. Yet while we might accept this conclusion, its very generality leaves many questions unanswered. Further enlightenment must depend on more detailed examination of the wage-payment methods adopted by individual plants. We take up this subject in the following section.

3. WAGE-PAYMENT METHODS

Using occupational earnings returns for 1966, we show for male manual employees[1] in Glasgow, Birmingham and the Other Scottish Areas, and for those case-study engineering plants for whom the information is available, the number of plants using timework, piece-work or some combination of both these methods. The data for Glasgow and Other Scottish Area plants are drawn mainly from E.E.F. occupational earnings returns, although where these were not available the D.E.P. occupational earnings returns were utilized.[2] This procedure yields information for 23 of the 27 case-study plants in Glasgow and for all 12 units in the Other Scottish Areas. In Birmingham, E.E.F. occupational earnings returns could not be obtained. The coverage is, therefore, restricted to those 14 plants for which D.E.P. occupational earnings returns were available.

In Table 5.1 the methods of wage payment adopted by Glasgow and Other Scottish Area plants are shown for fitters, turners, labourers and 'all workers'. The latter group comprises all manual employees covered by the E.E.F. occupational earnings returns, but excludes the bulk of semi-skilled employees for whom little earnings data are provided by this source. The skill composition of the 'all workers' group in Birmingham is substantially different, for D.E.P. occupational earnings returns do disclose details on semi-skilled workers. As the results for Birmingham units show, a high proportion of semi-skilled workers are paid through piecework, so that comparison of the 'all workers' group will understate the true extent of pieceworking in Glasgow relative to Birmingham.

[1] The following text is exclusively based on earnings data for male manual employees, for no occupational details are available for female employees. The bulk of female employees are semi-skilled and are paid by piecework.

[2] In Table 5.1 below we have departed from the practice of Chapter 4 and combined results from E.E.F. and D.E.P. occupational earnings returns although the returns relate to different dates in 1966. This is permissible because major shifts in methods of wage payment are unlikely to occur frequently.

TABLE 5.1
METHODS OF PAYMENT: MALES, 1966

Occupational group	Number of plants: Timework only	Piecework only	Timework and piecework	Not applicable	Total	Percentage paid by: Timework	Piecework
A. Glasgow plants							
Fitters	6	8	5	4	23	24·0	76·0
Turners	5	12	2	4	23	8·3	91·7
Labourers	18	3	2	0	23	80·0	20·0
All workers	4	3	16	0	23	36·8	63·2
B. Other Scottish Area plants							
Fitters	7	2	0	3	12	46·5	53·5
Turners	6	3	0	3	12	31·8	68·2
Labourers	11	0	1	0	12	70·9	29·1
All workers	9	0	3	0	12	71·1	28·9
C. Birmingham plants							
Fitters	1	1	4	8	14	42·4	57·6
Turners	1	2	3	8	14	25·2	74·8
Semi-skilled	2	0	11	1	14	40·5	59·5
Labourers	13	0	1	0	14	99·2	0·8
All workers	2	0	12	0	14	51·8	48·2

We shall begin by concentrating our attention on wage-payment methods in Glasgow and Birmingham as these seem fairly 'typical' of the British engineering industry as a whole. The evidence relating to the Other Scottish Area plants will be introduced later, as this offers an interesting contrast to behaviour in the two conurbations. The majority of plants in both Glasgow and Birmingham employed males on both timework and piecework, and it was quite common to find these different payment methods applied to employees in the same occupational group. Of the 23 Glasgow plants included in Table 5.1, only four paid all manual workers by timework,[1] and in only three units were all males on piecework. In Birmingham, none of the plants were exclusively pieceworking establishments and only two employed all males on timework. Pieceworking was more extensive in Glasgow than in Birmingham in the case of all three groups where direct comparison is possible—fitters, turners and labourers. This was even true of the 'all workers' category, although for reasons already stated the comparison will understate the extent of pieceworking in Glasgow relative to Birmingham.

There are three reasons for the greater use which was made of male pieceworkers by Glasgow plants. First, female employees accounted for a greater proportion of the labour force in Birmingham,[2] and women, who undertake many of the routine production tasks and are paid largely by piecework, are excluded from the above table. Second, skilled males in Birmingham were more often engaged in toolroom, or in development or prototype, work, than was the case in Glasgow. In such 'service' departments, where the speed of working is not set by the production line, timeworking was the common rule. The third reason for the more restricted use of pieceworking in Birmingham was also related to technological factors. In Birmingham, the engineering industry is largely concentrated on large-batch or mass production of standardized goods. On the other hand, Glasgow engineering plants tend to specialize in 'one-off' or small-batch, custom-built production of producer durable goods. The pace of work of manual employees can therefore be more easily controlled by management in Birmingham by regulating the flow of work through the shop. In Glasgow, changes in product mix provide an acute problem of supervising employee effort. The solution widely adopted has been to use the incentive element in a piecework system as a 'silent supervisor'.

Although pieceworkers accounted for a smaller proportion of male

[1] Even this probably overstates the position, as semi-skilled employees who are largely on piecework are poorly represented for Glasgow plants in Table 5.1.

[2] See p. 54 above.

manual employees in Birmingham, most plants employed *some* pieceworkers. Therefore, if pieceworkers' earnings increased through 'demand pull', productivity changes or institutional mechanisms, such primary wage drift was likely to create friction within most units and lead to secondary drift as timeworkers attempted to match increases in pieceworkers' earnings. The difficulty of maintaining wage relativities between pieceworkers and timeworkers was emphasized by almost all the case-study plants employing both types of workers in Glasgow and Birmingham. Increases in pieceworkers' earnings were seen by plant managers as the major source of primary drift, and considerable stress was placed on the difficulty of controlling secondary drift as timeworkers attempted to restore former relativities. The problem of controlling pieceworkers' earnings, and hence unit labour costs, arose in rather different forms in Glasgow and Birmingham, but in both cases the end result was the same—an upward pressure on earnings levels due to a tendency for pieceworkers' earnings to increase over time.

In Birmingham, the traditional method of piecework payment was still commonly employed, namely, a price 'per piece' (or unit of output). This rather unsophisticated form of pieceworking proved practical because of the more standardized nature of production, but even in Birmingham changes in product mix or in methods were sufficiently frequent to provide ample scope for haggling over new piecework prices. Many of the plants had no proper system of work measurement, so that ratefixers relied to a considerable extent on past practice, experience or guesswork. The errors which inevitably arose were difficult to locate and even more difficult to eradicate.

The Glasgow plants made much greater use of work-measurement. In almost all cases, a standard time was set for each job as the result of work-measurement, and piecework earnings depended on the difference between the standard time and the actual time taken to complete the task. Work-measurement was more extensively employed in an attempt to find a more 'scientific' basis for matching effort to reward, because a system of pieceworking on a price-per-piece basis would be extremely difficult to control given the existence of a small-batch technology. Nonetheless plant managers stressed the difficulty of preventing wage drift for much the same reasons as those described by Lerner and Marquand.[1] Nor was there any indication that wage drift was primarily due to 'demand pull' factors. Earnings appear to have increased at much the same rate in Glasgow and Birmingham despite substantial differences in employment conditions, and the problems of controlling wage drift within a piece-

[1] *Op. cit.*

working system were of as much concern to plant managers in both areas.

The tendency of pieceworkers' earnings to increase over time has forced plant management to adopt some method of adjusting time-workers' earnings to maintain relativities or, at least, to prevent too great a difference in the rate of increase of earnings for these separate groups. This has not been achieved through national agreements, which have provided for similar increases in the minimum standards for timeworkers and pieceworkers and which are, in any event, an inappropriate method of compensating for wage drift in piece-workers' earnings occurring at the plant level. Increases in piece-workers' earnings over any period of time vary substantially from one unit to the next and create different pressures which cannot be dealt with through a uniform increase for all timeworkers. It is pieceworkers in their own plants with whom timeworkers will make comparisons. It is, therefore, logical that plant managements should adjust timeworkers' earnings through domestic bargaining. Such adjustments can be achieved by a variety of methods: through establishing consolidated time rates above the agreed or recognized district minimum; through lieu bonuses or merit rates; through special payments of various kinds which sometimes include payments loosely related to output; or through systematic overtime working.

The payment of a consolidated time rate above the district minimum was widely used as a method of preserving or adjusting earnings differentials between timeworkers and pieceworkers. The C.T.R. is the most important constituent of the timeworker's wage packet, but with pieceworkers is only relevant in providing a floor below which his earnings cannot fall. Two-thirds of the plants in Glasgow and all but three of the 27 Birmingham units had established a C.T.R. above the district minimum. In fact, none of the time-workers employed by the case-study plants failed to achieve earnings above the minimum levels set through national and district negotiations. Thus, all the plants in Glasgow and Birmingham which incorporated the district minimum C.T.R. in the wage structure paid timeworkers a lieu bonus in addition to the C.T.R., and it was also quite common to find plants who added lieu bonuses to a C.T.R. which was already above the district minimum.

While a lieu bonus, as its name implies, is a compensatory payment to timeworkers, it may be paid through a number of different methods. More than half the Glasgow case-study plants made lieu bonus payments to some group of timeworkers (or to all timeworkers), and half of these paid a lieu bonus at a *fixed* rate per hour. In Birmingham, lieu bonuses were more regularly related to pieceworkers' earnings. The rationale behind such a system was to provide timeworkers with an incentive to greater output; in practice, this aim was seldom realized. In some cases, lieu bonuses were paid to all timeworkers, and when this occurred the lieu bonus is best regarded as having the same effect as a C.T.R. fixed above the district minimum. Even where lieu bonuses were restricted to certain groups of timeworkers, and pressure to generalize such payments to other groups was successfully resisted, it was relatively rare for the size of the bonus to be related to the earnings of pieceworkers with whom the timeworkers on bonus were associated. In these instances, the size of the lieu bonus was usually determined by reference to piecework earnings in the plant as a whole, so that the incentive element in lieu bonus payments was relatively weak.

Merit payments were also used frequently as a method of adjusting timeworkers' earnings. Such payments were made by about half of the case-study plants in Glasgow and in Birmingham. In one Birmingham unit, merit payments were made to all timeworkers, although at varying rates, but it was more common to find merit payments restricted to skilled timeworkers. One might expect that merit payments would show substantial variations between individuals with the same occupation, for, in theory, they should be payments to reward additional responsibility, special skills or other attributes not adequately reflected in alternative methods of wage payment. In fact, it was usual to find that most workers in the same occupational group received the same merit rate or that the spread of merit rates was comparatively narrow.[1] Plant managements stressed the difficulty of resisting pressure to narrow the band of merit rates for workers of the same skill level or to extend merit payments to other groups. In some cases the original purpose of merit payments had been lost and such payments were used, in the words of one Glasgow manager, as 'a method of adjusting the wage structure in favour of skilled timeworkers'. One reason for this is that managers find the existing system of merit payments difficult to justify. Only one Birmingham plant had made a serious attempt to evolve a more objective method of assessing merit rates. In other cases where any discretionary element existed, assessment was usually the responsibility of foremen or works superintendents, who were seldom given any guidance on how the awards should be determined. The subjective element was, therefore, strong and the system was difficult to defend in the face of charges of 'favouritism'.

[1] See also K. G. J. C. Knowles and D. Robinson, 'Wage Movements in Coventry', *Bulletin of the Oxford University Institute of Economics and Statistics*, Vol. 31, 1969.

Because of the arbitrary nature of many merit rates and the difficulty of preventing them spreading to other groups, five of the twelve Glasgow plants making merit payments at the beginning of our period, 1959–66, had abandoned such payments by 1966. In such cases, the merit rates had been consolidated in the C.T.R. There was little evidence of a similar trend in Birmingham, where only one of thirteen plants had abandoned merit awards over 1959–66.

Systematic overtime working was another method of adjusting the relationship between timeworkers' and pieceworkers' earnings. Average weekly overtime hours worked by timeworkers and pieceworkers were calculated for Glasgow and Birmingham plants from the E.E.F. and D.E.P. occupational earnings returns respectively. The results are shown in Table 5.2. Overtime working was more important for timeworkers than pieceworkers in all periods shown,

TABLE 5.2

AVERAGE OVERTIME HOURS WORKED BY TIMEWORKERS AND PIECEWORKERS: GLASGOW AND BIRMINGHAM MALES

A. Glasgow Plants

	June 1959	June 1960	June 1961	June 1962	1963[1]	June 1964	Oct. 1965	Oct. 1966
Timeworkers	3·8	4·5	6·3	4·8		6·8	9·1	6·8
Pieceworkers	2·5	4·5	4·4	3·0		3·0	4·9	5·1

B. Birmingham Plants

	June 1963	Jan. 1964	June 1964	Jan. 1965	June 1965	Jan. 1966	June 1966
Timeworkers	6·4	6·7	6·7	6·0	6·8	5·8	6·6
Pieceworkers	2·6	2·9	2·9	2·7	3·1	1·8	2·7

Note: 1. No E.E.F. earnings returns were available for 1963.

save one.[1] It is clear, therefore, that the systematic use of additional overtime working was widely used to adjust the earnings structure in favour of timeworkers.[2] This tendency was especially marked in Birmingham where in every period shown timeworkers averaged at

[1] This result is not affected to any important degree if one excludes maintenance workers when calculating overtime working by timeworkers. In the Other Scottish Areas also, timeworkers worked longer hours than piece-workers in plants employing both methods of wage payment.

[2] It may also have the effect of increasing the relative earnings of those in less skilled occupations. See Robertson, *op. cit.*

least twice as many overtime hours as pieceworkers. In the former case, overtime working usually averaged six hours per week, while for pieceworkers average overtime worked only once exceeded three hours.

The above discussion suggests the following conclusions. Two broad systems of payment for timeworkers were in general use. First, timeworkers were paid a C.T.R. substantially above the district minimum.[1] This was the simplest method of payment although lieu bonuses, merit awards or other forms of payment could be added to the C.T.R. Second, the C.T.R. was equal to or close to the district minimum. Here there was always, amongst the case-study plants, some form of supplementation to the C.T.R. through lieu bonuses and, often, through merit awards. In principle, the latter system was supposed to provide timeworkers with an incentive to greater output or to reward special responsibilities, skills, etc. In practice, the incentive element was seldom strong, for merit rates often did not vary substantially within an occupational group, and lieu bonuses were often fixed in amount or were tied to the earnings of pieceworkers, whose output was not closely linked with the effort of timeworkers. Timeworkers also worked longer overtime hours than pieceworkers, and there was a wide range of additional emoluments to specific groups, such as 'incentive bonuses' (some of which were in no way linked to output or effort), bonus payments on tonnage produced, bonuses linked to profitability, length of service awards,[2] 'dirty money', quality allowances and so forth.

The complex systems of wage payment which resulted can be illustrated by reference to one of the case-study plants in Birmingham, B1, which, in common with most other establishments, paid some employees by timework and some by piecework. The timeworkers were divided into two groups. Group A consisted of electricians and their mates, welding plant attendants, skilled and semi-skilled toolroom attendants and toolsetters. They received an hourly payment which was based on the average hourly earnings of pieceworkers minus a fixed amount the size of which depended upon the type of worker employed. For example, the deduction was 9d for skilled electricians and 3s 6½d for their mates. The same procedure was followed for toolmakers and toolsetters except that they had respectively a 'loading bonus' and a 'toolsetters' bonus' in addition. Finally, a 'corrective factor' of 3½d an hour was deducted from the toolsetters' earnings.

[1] In a few cases, this had been consolidated into a guaranteed weekly wage for a standard working week.

[2] These were relatively rare and usually accounted for only a small proportion of total earnings.

These procedures established the time rates for workers in Group A. However, to demonstrate that human ingenuity is inexhaustible, there was a further payment known as the 'general incentive bonus'. Each worker received a number of points equivalent to his hourly rate (calculated as above) expressed in pence. The value of each point was given by multiplying each unit of output produced in the factory by a predetermined money value and dividing the resulting figure by the total number of points 'owned' by the labour force. The general incentive bonus for the individual worker was then given by the bonus value per point times his number of points as determined by his hourly rate.

The timeworkers in Group B (which includes labourers, storemen, inspectors, etc.) were, fortunately, paid by a simpler system. Each group had a consolidated time rate to which was added a group-based merit payment and/or a lieu bonus. In addition, they participated in the general incentive bonus. Overtime payments were based on the C.T.R. plus any merit award but ignoring lieu bonus. Amongst pieceworkers there were again two systems of payment. For some, a standard time was set for each job and piecework earnings depended on the time saved. Alternatively, a price per unit of output was fixed, earnings in this case being determined directly by the number of units produced. In addition male pieceworkers in certain departments received a merit rate which was added to the basic rate used for piecework calculations, while females, together with some male workers, were eligible for certain length-of-service awards.

While the methods adopted by plant B1 were an extreme case of the intricacies to which payment systems in the British engineering industry sometimes aspire, they do illustrate certain features which are by no means uncommon. For example, the manner in which the earnings of timeworkers in Group A was tied to pieceworkers' earnings imparted a strong inflationary tendency to wage costs as pieceworkers' earnings increased over time. Again, we can see how the wage level may move upwards through time because of pressure to maintain relativities. Thus, the general incentive bonus was initially introduced for Group B, timeworkers only, to compensate that group for the fact that they did not benefit automatically from increased pieceworkers' earnings as did Group A. The general incentive bonus then became general in fact as well as in name, as it was claimed by, and granted to, Group A timeworkers. The payment of merit rates and lieu bonuses evolved in a similar fashion. Finally, the application of merit rates to certain groups of pieceworkers calls into question the efficacy of the incentive system in the plant.

It is apparent that many adjustments to the wage structure are not necessarily a response to the external pressures of the labour market but the result of a situation where an increase in earnings for one group of employees is likely to set off a series of further wage claims. Similarly, in a number of instances, plants which had experienced difficulties in recruiting specific types of labour had refrained from attempting to adjust the wage structure in favour of these groups for fear of provoking a general increase in earnings. There is a tendency, therefore, for the earnings of different groups to be interlinked so that a certain rigidity is imparted to the wage structure. From time to time, plants attempt and achieve an adjustment of occupational wage differentials in the light of the relative ease or difficulty with which they can obtain different types of labour. However, such adjustments are not lightly contemplated in view of the possible repercussions which the plant might be unable to control. Hence there may be a considerable time lag before a change in labour market conditions is reflected in the internal wage structure.

The other feature of wage behaviour which emerged from discussions with the case-study plants was that the wage structure develops through a series of *ad hoc* decisions in which one expedient is piled on top of another to raise the earnings of timeworkers. In this process the major factor is the need to maintain, or to prevent too great a fall in, timeworkers' earnings relative to those of pieceworkers. Such decisions are often made for particular groups with little thought of long-term consequences. This is due to the fragmented system of bargaining which places considerable power, in fixing pieceworkers' earnings, in the hands of the ratefixer and the foreman, and to the fact that, where the process of wage determination is centralized within one plant, power is vested in those responsible for day-to-day production decisions. The pressure on ratefixers, foremen and works managers alike is to 'keep the shop going'. In other words, they are more likely to accede to wage demands to prevent a disruption of production in the short run than to risk such unrest in order to conform to a consistent and agreed system of wage payment.

Such decisions, once made, are seldom reversible even if the original reasons for the decision no longer apply. With the passage of time, the wage structure becomes more complex and tends to develop anomalies just as the keel of a ship collects barnacles. It is not possible to understand and explain the wage structure which exists at a given moment of time unless one takes into account the historical process by which one form of wage payment was added to another to retain workers, to maintain relativities, to prevent unrest, or to satisfy other requirements. This does not indicate that economic forces are unimportant, but the upshot of the system of wage determination is a wage structure which does not at all points reflect that which would be struck by the purely economic forces of supply and demand.

It is worth while contrasting the wage payment methods of Glasgow and Birmingham plants with those of the twelve case-study plants located in the Other Scottish Areas. No less than nine of these units employed timeworkers exclusively, and much wider use was made of job evaluation and measured day work than in Glasgow and Birmingham.[1] In consequence, the methods of wage payment employed are much more easily understood. In part, this probably reflects the fact that all save one of the plants in these other areas had been established since 1945, the majority in the late 1950s and the 1960s. Their wage-payment systems had not, then, been exposed for a long period to the pressures and strains operating on a wage structure. This was not the only influence at work, for there was a clear cleavage in the methods adopted by plants under British and American management. The six plants under American control employed timeworkers only, and in three cases the wage structure consisted simply of an hourly rate based on job evaluation. In contrast, three of the six plants under British management employed mainly pieceworkers, and the payment systems adopted for timeworkers invariably contained some form of payment in addition to a C.T.R. above the district minimum.

British managements do, therefore, appear to react to particular situations by introducing additional wage payments under different guises. As a result, they offer more hostages to fortune, for the wage-payment system becomes more complicated with the passage of time and the internal wage structure more difficult to control. On the other hand, American managements are more concerned to evolve a 'rational' or 'systematic' wage structure and appear to be more reluctant to disturb the system arrived at. This supports the conclusion of one previous writer who has remarked, 'As a broad generalisation it is suggested that American employers are more concerned to maintain control over internal wage structures than are British . . . employers', and that 'In Britain . . . some firms have made their wage structure so flexible as to abdicate all effective control.'[2]

[1] Four Glasgow plants adopted systems of wage payment based on measured day work over 1959–66. No such trend was apparent in Birmingham.

[2] Robinson, *Wage Drift, Fringe Benefits and Manpower Distribution*, p. 69.

4. THE INTRA-PLANT EARNINGS STRUCTURE

The relationship between the earnings of different groups within the plant are, of course, extremely important. The structure of earnings should encourage and recognize effort and skill and should be flexible enough to reflect changes in the demand for and supply of different types of labour. Unfortunately, labour market theory does not provide any clear indication as to what pattern of relativities might be expected to arise at any given moment of time. The relationships between the earnings of different groups might be the same for all plants, or may show substantial differences from one unit to the next. Either observation can be reconciled with the economic model of labour market behaviour. Nonetheless, an examination of intra-plant earnings structures will at least establish which of the two extremes accords most with reality. Subsequently, in Section 5, we can obtain some idea how relativities change in response to the employment conditions prevailing in the market by examining changes in internal wage structures over time.

We begin our investigation with Table 5.3. The table is based on D.E.P. occupational earnings returns for June 1966[1] and includes all those plants for which the relevant D.E.P. returns were available. Standard weekly earnings have been used as the unit of measurement because, for reasons outlined previously, they provide the best indicator of the long-run position. In *each* plant, standard weekly earnings for the occupational groups shown have been expressed as a percentage of the standard weekly earnings of labourers on timework. The same process was repeated for gross weekly earnings, and the results of this analysis are referred to in the text when it is thought necessary.

The relationships between the earnings of different occupational groups correspond in a general sense with the static predictions of the competitive hypothesis as long as we treat timeworkers and pieceworkers as separate groups. Thus, in all plants in Glasgow and Birmingham the standard weekly earnings of skilled fitters, turners and toolroom employees on timework exceeded those of semi-skilled timeworkers who, in turn, earned more than unskilled timeworkers, a finding which conforms with our expectations, given differences in training requirements and the tendency for recruitment difficulties to become more acute as the skill level rises. Similarly, if we confine our attention to pieceworkers, a fairly consistent ranking of intra-plant

[1] In this case, D.E.P. occupational earnings returns have been used for Glasgow plants as well as for Birmingham plants. These are preferred to E.E.F. returns, for the latter virtually exclude semi-skilled workers.

occupational earnings emerges. Within each unit, fitters, turners and toolroom workers on piecework tended to receive higher standard weekly earnings than semi-skilled and unskilled pieceworkers. However, in Birmingham plants B2, B9 and B17 there were skilled pieceworkers with lower standard weekly earnings than semi-skilled pieceworkers, and in two plants (G5 in Glasgow and B15 in Birmingham) standard weekly earnings for unskilled pieceworkers exceeded those for semi-skilled pieceworkers.[1]

The complexities of internal wage structures become most apparent when we widen our comparison to include both timeworkers and pieceworkers, measuring wages by standard or gross weekly earnings. Taking standard weekly earnings as in Table 5.3, we find that, in each plant, semi-skilled and unskilled pieceworkers had higher standard weekly earnings than timeworkers of the *same* skill. This may have reflected greater effort on the part of pieceworkers, but amongst the skilled groups timeworkers could obtain earnings higher than pieceworkers at the same skill level. The payment method adopted had, on occasion, sufficient influence to submerge the effect of skill. In Glasgow, semi-skilled pieceworkers in plants G5, G16 and G18, and unskilled pieceworkers in plant G5, had higher weekly standard earnings than one or more of the groups of skilled timeworkers. In Birmingham, the same feature emerged for semi-skilled pieceworkers in plants B5 and B14 and for unskilled pieceworkers in plant B15. As only a limited number of comparisons are possible on the basis of the above table, it would appear that it is not unusual for pieceworkers to have higher standard weekly earnings than timeworkers with a greater level of skill.[2] Indeed, if a more detailed analysis of the rather heterogeneous semi-skilled group was possible, and if earnings data were available for individual employees, the reversal of skill differentials between pieceworkers and timeworkers would be even more frequent than suggested by our data. However, what we have already learnt about the systems of wage payment lends support to the conclusion that 'all too often . . . reversed differentials are not the result of considered and agreed policy but of a haphazard development arising from *ad hoc* decisions'.[3]

It is also apparent that the internal wage structure shows such considerable variation from one unit to the next that it is extremely

[1] These tendencies become more evident if gross weekly earnings are considered.

[2] When gross weekly earnings are considered, this tendency is somewhat less marked. The difference between the results based on standard and gross weekly earnings arises because timeworkers usually work more overtime hours than pieceworkers.

[3] National Board for Prices and Incomes, *Report No. 49*, p. 11.

TABLE 5.3

STANDARD WEEKLY EARNINGS OF OCCUPATIONAL GROUPS AS A PERCENTAGE OF THE STANDARD WEEKLY EARNINGS OF LABOURERS ON TIMEWORK: GLASGOW AND BIRMINGHAM MALES, JUNE 1966

A. Glasgow Plants

Occupational group	Plant Number								
	G1	G2	G4	G5	G11	G16	G18	G19	G21
Timeworkers									
Fitters	141		132	121	138	173	153		
Turners			135	141		149	140		
Toolroom			144	141		178	167	152	174
Semi-skilled	106	115	113	110	116	123	126	118	123
Pieceworkers									
Fitters		174			158			143	
Turners		173			147		171	152	153
Toolroom		161			163				
Semi-skilled		139		136		171	142	129	152
Labourers				144		129	121	111	

TABLE 5.3 *continued*

B. Birmingham Plants

Occupational group	Plant Number													
	B1	B2	B3	B4	B5	B6	B7	B9	B11	B12	B13	B14	B15	B17
Timeworkers														
Fitters		177	191		157						135		171	
Turners			206		157		130				130			
Toolroom	166	190	200	152	166	165	131		161	211	159	108	169	175
Semi-skilled	127	125	121		124	144	112		122	125	128	107	122	125
Pieceworkers														
Fitters		157	183	193							151		183	
Turners		181	187		189						149			148
Toolroom				193				133	172					
Semi-skilled	164	163	165		179		122	134	136	167	130	134	162	161
Labourers													173	

difficult to establish any general rules, even if we confine our attention to the standard weekly earnings of timeworkers, where short-term changes in intra-plant occupational differentials are likely to be least in evidence.[1] Taking Birmingham plants, we can see that the standard weekly earnings of fitters on timework ranged between 135 per cent and 191 per cent of the standard weekly earnings of unskilled timeworkers. The range for turners, toolroom and semi-skilled workers on timework was 130–206 per cent, 108–211 per cent and 107–144 per cent respectively. This lack of a consistent relationship between the earnings of different groups appears to apply, in greater or less degree, whatever comparison is made.

It is difficult to evaluate the results obtained in the light of theoretical considerations, for wage theory does not suggest that earnings relativities within plants must be the same for all units. Differences in the structure of intra-plant earnings could arise because of differences in training requirements or through the effort and 'quality' of those employed in the separate plants. Again, plants may not experience the same difficulties in recruiting a given type of labour, and the non-wage conditions attached to particular tasks are likely to vary between establishments. For these and other reasons, it is conceivable that differences in intra-plant earnings structures could result from rational responses to economic and other pressures. We shall discuss this in Section 5, but, for the moment, it must be observed that the outstanding characteristic of the earnings relativities which prevail in any plant is that they are unique; there appear to be as many internal wage structures as there are plants.[2] Whatever the force of demand and supply factors or of equitable comparisons, therefore, they do not give rise to similar occupational differentials in each establishment. There is no such thing as *a* skill differential determined by market forces, nor is there *an* 'accepted' relationship between the earnings of different groups which the employee carries with him as he moves from plant to plant.

5. CHANGES IN OCCUPATIONAL EARNINGS

The cross-sectional analysis of Section 4 has revealed a very complex system of occupational differentials in which each plant appears to be unique. Nonetheless, this form of analysis cannot provide an

[1] Rather surprisingly the spread of occupational differentials from one plant to the next, while still considerable, appears less substantial for pieceworkers than for timeworkers in Birmingham. Week to week changes in earnings are, however, likely to be much greater for the former group so that a cross-sectional analysis at a given moment of time may give misleading results.

[2] It is interesting to observe that, although the principles underlying wage

unambiguous answer to the question whether we can reconcile observed intra-plant differentials with wage theory. The problem has to be approached in a more indirect manner. Wage theory predicts that where particular types of employees are difficult to recruit, competition between employers will drive up the earnings of such employees relative to those obtained by other groups. Should such a tendency be observable, we could conclude that, to some extent at least, the wage structure was sensitive to economic pressures.

In Glasgow, the ratios of unemployment to unfilled vacancies for fitters and turners were substantially lower than those for semi-skilled workers, while employment conditions for the latter were always more favourable than those for the unskilled.[1] The impressions conveyed by unemployment and vacancy data are confirmed by conversations with plant managers. While a number of plants reported occasional difficulties in obtaining skilled labour, shortages of semi-skilled labour were much less acute, and none found any difficulty in securing an adequate supply of labourers. The greatest recruitment difficulties arose with turners. Thus the ratio of unemployment to unfilled vacancies was generally lower for turners than for fitters through 1959–66, and plant managers reported that labour shortages were more intense for turners than for any other single group. In Birmingham, both the evidence of unemployment/unfilled vacancies ratios and that of the case-study plants indicate that toolroom workers were the category of engineering labour in shortest supply. Again, although recruitment difficulties were more widespread and acute in Birmingham than in Glasgow, and although, in consequence, some plants found even unskilled labour difficult to obtain, the same broad pattern emerges. Labour shortages were most severe for the skilled occupations, particularly toolroom workers, and least marked for the unskilled, with the semi-skilled taking up an intermediate position.

The employment conditions ruling in the Glasgow and Birmingham markets would then lead us to expect that earnings would increase most for skilled workers and least for unskilled workers in each area, *so long as the internal wage structure was flexible and labour market pressures were met solely or mainly through an adjustment of occupational earnings*. We shall take up the first proviso, the flexibility of the internal wage structure, later on in this analysis.[2] Let us consider here whether the second proviso holds, or whether there are

payment methods in the Other Scottish Areas are easily grasped, the resulting patterns of intra-plant differentials seem as complex as those in Glasgow and Birmingham.

[1] See Table 3.6, p. 57 above.

[2] See p. 123 below.

other methods of meeting labour shortages for particular groups of workers which do not involve the competitive bidding-up of wages. One alternative which immediately suggests itself is that an employer, instead of bidding-up the wages of, say, turners who are in short supply, may reduce his hiring standards for turners. This possibility is best explained by Reder[1] in an ingenious argument which points out that the supply of a particular group of employees can be increased by lowering 'quality' requirements. In effect, the employer, instead of adjusting relative wages in favour of turners and moving along a given labour supply curve, attempts to shift the supply schedule to the right by accepting turners of lower quality. He can do this by lowering the hiring standards applied to those entering the plant as turners, by encouraging internal mobility in the plant, by improving his training programme, by altering his methods of production, etc. If this process of 'upgrading' semi-skilled and unskilled labour is carried far enough and if the demand for labour is high enough to absorb the 'labour reserve', then it is conceivable that labour shortages will become most acute for the unskilled. In this case the end result would be a narrowing of skill differentials, rather than the widening of skill differentials which would occur if labour shortages were met primarily through wage adjustments favourable to skilled employees.

The process which Reder describes is, of course, likely to be present. Faced with acute shortages of skilled labour, employers are likely to lower hiring standards and encourage internal mobility, but there are severe limits to the effectiveness of these policies, as we shall see subsequently.[2] A necessary condition for the transmission of acute shortages of skilled labour, through a reduction of hiring standards, to a shortage of unskilled labour and a narrowing of occupational differentials, is that the labour market is fully employed and the 'labour reserve' is exhausted. This certainly does not apply in Glasgow, and although it does in Birmingham, full employment is not a sufficient condition to achieve this result. It must be accompanied by a high degree of intra-factor substitutability so that skilled labour can easily be replaced by less skilled labour. This condition is not met in the British engineering industry where mobility into skilled jobs is limited by institutional restrictions and, more importantly, by the technology of the industry. With a few exceptions, it is difficult to transfer semi-skilled men to skilled jobs, so that entry to the skilled trades is, to a considerable extent, regulated by the

[1] M. W. Reder, 'The Theory of Occupational Wage Differentials', *American Economic Review*, Vol. 45, 1955.

[2] See particularly pp. 307–9 below.

apprenticeship system, involving a long period of training. Once we accept this, the evidence points unambiguously to the conclusion that labour shortages have been most acute for skilled labour and least acute for the unskilled. In these circumstances we would expect, on the basis of labour market theory, a markedly favourable shift in

TABLE 5.4

PERCENTAGE INCREASE IN OCCUPATIONAL EARNINGS BY PLANTS (STANDARD WEEKLY EARNINGS): GLASGOW MALES, OCTOBER 1966 OVER JUNE 1959

Plant number	Fitters	Turners	Labourers	All workers
G3	61·7	46·1	45·7	52·8
G4	25·0	23·3	11·6	20·3
G6	59·2	72·3	66·7	61·3
G7	26·0	27·0	19·0	25·2
G9	39·8	36·7	33·7	42·6
G10	79·7	59·0	35·6	72·3
G11	42·7	35·3	33·7	43·0
G12	30·5	36·5	19·6	34·2
G14	42·3	39·7	34·9	41·0
G15	52·0	72·7	58·8	55·6
G16	22·1	56·9	25·0	32·7
G18	82·8	37·7	43·0	53·8
G19	29·9	31·1	29·1	27·7
G20	37·7	43·8	14·6	41·7
Average increase (unweighted)	45·1	44·2	33·6	43·2

the earnings of skilled labour (turners and fitters in Glasgow and tool-room workers in Birmingham) relative to those of labourers. One might also expect that in Glasgow turners' earnings would show a more rapid rate of increase than fitters' earnings in most plants, while in Birmingham, the increase in earnings for the semi-skilled would be less than for toolroom workers but more than that achieved by labourers. Inspection of Tables 5.4 and 5.5 will indicate how far these expectations are borne out in practice. The former shows for Glasgow the percentage increase, over 1959–66, in standard weekly earnings by occupational group and plant. Percentage changes in plant standard weekly earnings over 1963–6 for occupational groups in Birmingham are shown in Table 5.5.[1] The results from similar

[1] The tables are based on the same case-study plants and occupational groups as those included in Tables 4.1 and 4.2, pp. 71–2 above.

calculations based on gross weekly earnings are discussed in the following text.

In Glasgow, standard weekly earnings for skilled turners increased more rapidly than standard weekly earnings for labourers in no less than 13 of 14 plants. The earnings of the latter group lagged

TABLE 5.5

PERCENTAGE INCREASE IN OCCUPATIONAL EARNINGS BY PLANTS (STANDARD WEEKLY EARNINGS): BIRMINGHAM MALES, JUNE 1966 OVER JUNE 1963

Plant number	Toolroom	Semi-skilled	Labourers	All workers
B1	20·1	34·8	27·9	31·1
B2	31·3	22·5	27·5	23·4
B3	24·6	19·5	19·5	20·1
B4	36·0	—	32·8	27·4
B5	21·4	26·8	26·0	27·6
B6	26·0	30·5	1·4	29·1
B7	26·7	21·3	38·1	22·1
B9	1·7	26·6	2·9	13·9
B11	20·5	26·0	19·4	22·4
B12	23·0	−3·7	−13·7	7·3
B14	7·9	9·6	70·2	27·8
B15	15·0	16·9	55·7	21·2
B16	18·3	20·3	28·3	19·0
Average increase (unweighted)	21·0	20·9	25·8	22·5

behind those of skilled fitters in 11 of 14 cases. If gross weekly earnings are used as the unit of measurement, labourers come out of the comparison rather better, but their relative position still worsens in the majority of establishments. This is consistent with the view that greater competition for skilled workers should bring about a shift in intra-plant earnings differentials in their favour. In other respects, there is less support for the prediction that adjustments in the intra-plant earnings structure will reflect the employment conditions ruling in the labour market. In half of the Glasgow plants, standard weekly earnings rose more quickly for fitters than turners, in spite of the greater difficulty experienced in recruiting turners.[1] It is even more difficult to reconcile the results for Birmingham plants with

[1] Increases in gross weekly earnings were greater for fitters in eight of the 14 plants included in Table 5.4 above.

labour market conditions in that area. Standard weekly earnings for labourers increased more rapidly than those for toolroom workers in seven of 13 cases and more rapidly than standard weekly earnings for semi-skilled workers in five of 12 cases. Increases in the standard weekly earnings of semi-skilled employees outstripped those for toolroom workers in eight of 12 establishments. For gross weekly earnings also, it is difficult to find any pattern at all in the Birmingham results. None of the three groups of employees improved its position relative to the others, although there were substantial contrasts in the degree of difficulty encountered in recruiting these different types of employees.

It would appear, therefore, that changes in intra-plant differentials may not be wholly, or even mainly, explicable in terms of local labour market conditions. At the very least, changes in such differentials are not acutely sensitive to employment conditions ruling in the labour market.[1] It may be that the internal wage structure responds only slowly to economic pressures, but if this is so the time-lag, on the evidence of the above tables, must be considerable. Yet there is some indication in Glasgow of the influence of economic pressures, and it could be that the failure of the intra-plant wage structure to adapt in the expected direction in Birmingham is due to the existence of secondary drift. Competition for labour which is particularly difficult to recruit might drive up earnings for such labour in each plant. Such primary wage drift will disturb customary relativities and create demands for compensatory wage increases from other employees. If these demands are met, the wage structure of each plant will move bodily upwards over time with no tendency for the earnings of any group to grow more quickly than any other.

A plant can adjust the earnings of different groups by a number of methods, but the two extreme types are a general wage adjustment for all employees, leaving intra-plant differentials unchanged, and an adjustment of the earnings for a particular group which alters intra-plant differentials. If external labour market conditions dictate changes in the intra-plant wage structure, then persons in those occupational groups where recruitment difficulties are most acute will have relatively large increases in earnings. If such primary drift did *not* give rise to secondary drift, then, reading down the columns of Tables 5.4 and 5.5, we would expect to find fairly similar increases in earnings while more substantial differences, related to supply and demand factors in the external market, would emerge across the rows. Now, suppose that increases in earnings for one group, for

[1] The same conclusion is reached from a study of changes in occupational earnings in North Lanarkshire and New Town.

whatever reason, *do* result in secondary wage drift which attempts to restore customary relativities. In the extreme case, earnings for all groups employed by the plant would rise by the same amount (in percentage terms).[1] It is unlikely that such inflexibility would ever be realized in practice, but where equitable comparisons were important, differences across the rows would be less substantial than differences down the columns of Tables 5.4 and 5.5.

To ascertain which of these views is more correct we can adopt a test similar to an analysis of variance. For each occupational group in each plant shown in Table 5.4 above, the percentage change in standard weekly earnings is subtracted from the percentage change in standard weekly earnings for all workers in that plant. The resultant differences are squared and summed to give V_1, thus:

$$V_1 = \sum_{j=1}^{n} [(x_{fj}-x_{wj})^2+(x_{tj}-x_{wj})^2+(x_{lj}-x_{wj})^2]$$

where x_{fj}, x_{tj}, x_{lj}, and x_{wj} represent the change in standard weekly earnings for fitters, turners, labourers and all workers in the jth plant. V_1 is, therefore, the sum of squared deviations of changes in standard weekly earnings by occupation from the change in standard weekly earnings for all workers employed by the plant, and measures the extent of changes in *intra*-plant differentials.

The extent of changes in *inter*-plant differentials by occupational groups is derived in a similar manner. In this case, for each occupational group in each plant in turn, the change in standard weekly earnings is subtracted from the unweighted average change in standard weekly earnings for that occupational group in all plants. The resultant differences are squared and summed to give V_2, thus:

$$V_2 = \sum_{j=1}^{n} [(x_{fj}-\bar{x}_f)^2+(x_{tj}-\bar{x}_t)^2+(x_{lj}-\bar{x}_l)^2]$$

where $\bar{x}_f$, $\bar{x}_t$ and $\bar{x}_l$ represent the unweighted average of changes in standard weekly earnings in all plants for fitters, turners and labourers respectively. V_2 gives the sum of squared deviations of changes in standard weekly earnings by occupation from the average change in occupational earnings for all plants.

Since V_1 measures changes in *intra*-plant differentials and V_2 measures changes in *inter*-plant differentials, the smaller is V_1 relative to V_2 the *more* will changes in earnings depend on the plant rather than on the skill of the worker under consideration. In this instance, the economic circumstances of the employing unit will be

[1] It is, of course, possible that secondary wage drift takes the form of similar absolute increases rather than similar percentage increases in earnings. We shall, however, ignore this complication.

the crucial determinant of an individual's change in earnings—it will be more important to know whether an individual works for plant A or plant B rather than whether he is a fitter or a turner. Conversely, if V_2 is small relative to V_1, the individual's occupation rather than his employer will be the more important determinant of the increase in earnings he obtains, reflecting, as competitive theory suggests, that relative wage increases are a function of changes in the demand and supply conditions facing different occupational groups rather than of the conditions facing different employers. V_1 and V_2 were calculated in the fashion described for Glasgow, and the same process, *mutatis mutandis*, was repeated for Birmingham and for both areas using gross weekly earnings. The results are as follows:

	Standard weekly earnings		*Gross weekly earnings*	
Glasgow	V_1 = 5,660	V_2 = 11,446	V_1 = 13,705	V_2 = 33,239
Birmingham	V_1 = 6,543	V_2 = 7,967	V_1 = 7,339	V_2 = 9,809

In each market, V_1 is smaller than V_2 for calculations based on standard or gross weekly earnings. Hence there is some evidence, particularly in Glasgow, to support the view that changes in earnings are dependent upon economic conditions in the plant rather than on the occupation of the worker under consideration. Moreover, it is possible that this tendency is stronger than the data in Tables 5.4 and 5.5, and the calculations above, suggest. Thus, some of the changes in the intra-plant wage structure observed may represent not the emergence of a new pattern of occupational differentials but a return to 'established' or 'recognized' relativities disturbed by prior changes in the internal wage structure. In short, when examining the behaviour of the internal wage structure the choice of period will be crucial, the more so the shorter the period examined. The longer the period the more likely it is that V_1 will be small relative to V_2. This explains the contrast between the Glasgow and Birmingham results, but even in Glasgow it is likely that some of the changes in internal wage structures over 1959–66 represent an attempt to reverse shifts in the opposite direction in a previous period.

The view that coercive comparisons within the plant result in a situation where changes in *intra*-plant differentials are smaller than changes in *inter*-plant differentials is supported by the behaviour of fitters' and turners' earnings in Glasgow plants. Turners have not improved their wage position relative to fitters as the competitive hypothesis would predict, but we can find evidence which suggests that strong pressures are exerted to maintain the accepted differential between these groups. In 1959, turners' standard weekly earnings were higher than fitters' standard weekly earnings for six of the 14 plants in Table 5.4—plants G3, G9, G10, G11, G12 and G18. In each of these plants, save G12, the earnings of the lower-paid group, fitters, increased more rapidly over 1959–66. Now consider the reverse case of the seven plants[1]—G4, G6, G7, G15, G16, G19 and G20—where fitters' earnings were higher than turners' earnings in 1959. Here again the position of the lower wage group, in this instance turners, improves in all plants save one. If gross weekly earnings are substituted for standard weekly earnings, the same result emerges: there was a distinct tendency for the lower-paid employees in 1959, be they fitters or turners, to improve their relative position over 1959–66.

One of the difficulties that arises when we refer to 'conventional' differentials is that we have little idea what the conventions actually are. The fitter–turner relationship outlined above was convenient as it appeared to fluctuate around parity—a not too unlikely benchmark! It also seems reasonable to assume that, whatever the precise conventional relativities within a plant, there is a notion of 'equity' or 'fairness', which is simply that a plant should not pay one group of employees a high wage relative to other units, while other groups of employees have low relative wages. We have already uncovered some evidence consistent with this hypothesis with our finding that, where the earnings of one group of workers in a plant are relatively high, then all other groups employed by the plant also tend to have high earnings relative to other units.[2] This is not conclusive, for high earnings across a plant could simply indicate the application of strict hiring standards to all types of employees, while low-wage plants may find it possible to secure only poor-quality labour at all skill levels. There are, however, two further aspects of the behaviour of internal wage structures over time which may indicate the force of coercive comparisons within the plant. First, changes in plant earnings tend to take place 'across the board', so that where one group of employees gains a large increase in earnings relative to that in other establishments, then all other workers in that plant also obtain a relatively large increase and vice versa. Second, where this rule does not apply it is often because one group was, in relation to other occupations in the plant, doing well or badly in the base period. Where this was so, the 'atypical' group was generally brought back more into line with the other groups in the plant.

We can illustrate this latter point by inspecting Tables 4.6 and 4.7,

[1] We exclude G14 where fitters and turners had the same standard earnings in 1959. It is interesting to observe that for this plant both groups had very similar increases in earnings over 1959–66, as we would expect on the basis of the above argument.

[2] See p. 84 above.

pp. 81–2 above. For example, in Glasgow fitters improved their ranking relative to turners in G10; in plants G12 and G16 the change in rankings was most favourable for turners while labourers lost ground relative to fitters in G16; and fitters did badly relative to labourers in G15 and G19. In each case the tendency was for rankings to come more into line across the plant, and the same pattern of behaviour is observable in Birmingham. Thus there were three plants—B4, B15 and B16—in which labourers held a low position in the inter-plant wage structure relative to toolroom workers in 1963. In each case the ranking for labourers came more into line over the subsequent period 1963–6. On the other hand, in two of the three cases, where the relative ranking was favourable to labourers in 1963 (B6 and B9), there was a sharp deterioration in their position subsequently, and the change in the remaining plant (B7), while in the opposite direction, was marginal.

Equitable comparisons within the plant are, therefore, important, so that the employer who attempts to meet labour shortages by raising wages for particular groups of employees has to bear in mind the probability that this will give rise to wage demands from other groups. Yet while most plant managements stress the importance of 'established' or 'recognized' relativities and while coercive comparisons within the plant tend to give rise to an interlinked plant wage structure which moves bodily upwards through time, it is still apparent that the internal wage structure is far from inflexible. A glance at Tables 5.4 and 5.5 will indicate that shifts in the intra-plant earnings structure can and do occur over time, but it is not possible to explain such shifts purely in terms of the difficulties experienced by plants in recruiting different types of labour. Primary drift is, of course, difficult to distinguish from secondary drift, and the employment conditions ruling in the external market are undoubtedly a factor causing primary drift, particularly in Birmingham. It is, however, difficult to square the supposed primacy of employment conditions with all the movements in internal wage structures observed, particularly those in favour of labourers where the recruitment position was always easiest.

Wage theory has, therefore, to be reconciled with the existence of very complex intra-plant earnings structures. Such a reconciliation may be possible, since there is no reason why earnings relativities should be the same in all establishments, but it is not at all clear that employment conditions have a major impact on the evolution of intra-plant earnings structures. The competitive model suggests that the rate of change of occupational earnings will be a function of the relative ease or difficulty with which different types of labour can be recruited. This provides a good explanation of changes in occupational differentials between skilled and unskilled labour in Glasgow, but no relationship can be established between changes in occupational earnings and external labour market conditions when one considers skilled fitters and turners in Glasgow or toolroom, semi-skilled and unskilled employees in Birmingham; and this despite marked differences in the demand/supply position for these types of labour.

The failure of the internal wage structure to respond to external labour market conditions in part reflects the strength of equitable comparisons within the plant. Such comparisons are not, however, strong enough to prevent any change in earnings relativities, and these changes can on occasion be very substantial. It is not clear why these shifts in the internal wage structure occur, particularly in those cases where they do not seem to be related to labour demand and supply, but one explanation lies in the systems of wage payment adopted by many engineering plants, which make it difficult to establish effective control over increases in earnings or over the internal earnings structure which emerges at any point of time. Simplicity is not necessarily a virtue, and any method of wage payment adopted must attempt to meet a number of different objectives which are often difficult to reconcile. Nonetheless, there is little evidence that the complex patterns of intra-plant earnings really reflect a series of rational responses to differing needs and pressures. In many plants, particularly those under British management, little attention has been paid to fundamental principles or to long-term considerations. In such circumstances, it is hardly surprising that managements have found it difficult to resist inflationary pressures on wage costs; nor is it to be expected that the internal wage structure will adapt readily to reflect the demand and supply conditions for different types of labour.

6. CONCLUSIONS

(i) Although, due to technological factors, pieceworking was more extensively used for males in Glasgow than in Birmingham, the majority of plants in each area employed males on both piecework and timework. It is the existence of both methods of wage payment within the plant which appears crucial, for the difficulty of controlling pieceworkers' earnings and maintaining relativities between pieceworkers and timeworkers was given equal stress by managers in both areas.

(ii) In Birmingham, the traditional method of piecework payment of a price 'per piece' was still commonly employed, although in Glasgow piecework was generally based on a system of standard times. Two broad systems of payment for timeworkers were in general use in Birmingham and Glasgow. Timeworkers could be paid a consolidated time rate (C.T.R.) equal to, or above, the district minimum. In the first case, the C.T.R. was always supplemented through lieu bonuses, merit awards or other payments. The simplest form of payment was, then, a C.T.R. above the district minimum, but even here further supplementation was common. A wide variety of other devices were employed to maintain relativities between pieceworkers and timeworkers, including longer overtime working by the latter group.

(iii) The wage structures of many engineering plants in Birmingham and Glasgow were often very complex, having evolved, over time, through a series of *ad hoc* decisions. In large measure this springs from the fact that the locus of power with regard to wage decisions rests with those who have day-to-day responsibilities for production and, consequently, are inclined to buy industrial peace rather than insist on adherence to a consistent and agreed method of wage payment. The upshot of this is that the wage structure becomes more complex and more difficult to control with the passage of time.

(iv) In contrast to the plants in Birmingham and Glasgow, a majority of the engineering establishments in the Other Scottish Areas had evolved internal wage structures based on simple and consistent principles. The contrast arises because the six plants in the Other Scottish Areas under American control employed timeworkers exclusively, and amongst these units there was a tendency to pay timeworkers on the basis of a flat hourly rate without supplementation through lieu bonuses, etc. Establishments under British management in the Other Scottish Areas were, however, likely to have wage structures as complex as those found in the two conurbations, with a substantial element of pieceworking and an hourly rate for timeworkers built up by a number of separate elements.

(v) In Birmingham and Glasgow the intra-plant wage structure showed considerable variation from unit to unit. The only generalizations which can be safely drawn are that pieceworkers were more highly paid than timeworkers of the *same* skill, and that, treating timeworkers and pieceworkers as separate groups, earnings tended to increase with skill. However, even the latter conclusion has to be modified when gross weekly earnings are taken as the unit of measurement, and the situation is further confused where the method of wage payment swamps the effect of skill. It does not appear unusual for semi-skilled, or even unskilled, pieceworkers to earn more than skilled timeworkers, especially where wages are measured by standard weekly earnings. The upshot is that it is difficult to discern any consistent pattern of occupational differentials, so that each internal wage structure is unique to the particular plant under investigation.

(vi) Changes over time in intra-plant differentials do not appear to be responsive to market forces as measured by the demand/supply conditions for different types of labour. There is one exception to this rule in Glasgow where plant skill differentials usually widened over 1959–66, but no consistent pattern of behaviour emerged for other groups in Glasgow and Birmingham despite substantial differences in the employment conditions facing different types of labour. This implies that, where competition for scarce labour gives rise to primary drift, then secondary drift will often result, leaving intra-plant differentials unchanged.

(vii) Support for conclusion (vi) above can be found in the observation that, where one group of workers employed by a plant obtained an above-average increase in earnings, then all other groups employed by that plant also obtained above-average increases, and vice versa. A more systematic test confirmed the view that the most important factor determining an individual's increase in earnings was the plant rather than the occupational group in which he was employed. In other words, while the internal wage structure was not rigid, the earnings of different groups were interrelated, so that it was difficult to modify internal wage structures in the face of coercive comparisons. The influence of such comparisons was illustrated by the tendency of all employees in the plant to occupy approximately the same relative position in the interplant wage structure. Again, if one group of employees fell behind other groups in the plant in relative terms, then pressures were set up to bring the 'errant' group into line with other employees.

PART III: LABOUR TURNOVER

CHAPTER 6

WAGES AND LABOUR TURNOVER

1. INTRODUCTION

Labour turnover occurs when workers join or leave a plant. Turnover can be expressed as a number, or as a rate or proportion of the stock of employees, and in one of its dimensions, those leaving, can be divided into a number of categories according to the reason for leaving. Such 'wastage', which we will use as an omnibus term to cover various categories of leavers, imposes costs on both parties involved. The employer loses assets embodied in the acquired experience and training of the employee and may have to bear further expenditure in recruiting and training a new employee. The employee, if he moves to new employment, must bear the costs of the job-search and any costs of adjusting to a new environment such as a loss of earnings or psychic income.

Viewed in this light, it is in the employer's interest to reduce labour wastage to a minimum. However, wastage may be a boon as well as a curse. If the employer wishes to reduce his labour force, then the normal process of attrition through wastage may be a comparatively painless method of obtaining this objective. Alternatively, wastage in other establishments provides a potential source of additional recruits, and the whole process of labour turnover is essential if the economy is to adapt to changes in labour requirements necessitated by economic growth. Turnover has, then, a dual aspect. It imposes costs on the employer and the employee, but it is a necessary feature of any dynamic economy. This being so, it is important to understand the factors which determine the volume and shape the behaviour of labour turnover.

These may be considered under two broad headings—personal and environmental. Personal characteristics, such as age, length of service and sex, are discussed in Chapter 8 while this and the subsequent chapter look at environmental influences 'external' to the individual.

In this case, they are defined narrowly to include only 'economic' variables, such as employment conditions and wage differentials, while excluding institutional and other factors. A distinction must also be drawn between cross-sectional and time series analysis. In the cross-sectional analysis of this chapter we attempt to explain why turnover rates differ between plants in the same time period and the same market area. We are concerned not with the average or 'market' turnover rate for all plants but with the dispersion of plant turnover rates about this mean. In contrast, the time series analysis of Chapter 7 investigates why the average or market turnover rate differs between areas and over time. In both this chapter and Chapter 7 we distinguish between the turnover rates of different male occupational groups. The influence and interaction of occupation and other personal characteristics on turnover is also discussed in Chapter 8.

Section 2 of this chapter, which examines the turnover data available, is also relevant to Chapters 7 and 8. Section 3 describes certain characteristics of plant turnover rates—in particular, the extent to which the voluntary quit rate differs between establishments; whether these differences are stable over time and whether they respond to changes in employment conditions; and whether all occupational groups in an establishment tend to have relatively high or relatively low quit rates. The relationships which might be expected to emerge between plant earnings and labour turnover are discussed and tested in Sections 4 and 5, and the analysis is extended in Section 6 where plant size and the rate of recruitment are introduced as additional variables which might assist in explaining differences in plant quit rates. The main conclusions reached are summarized in Section 7.

2. TURNOVER DATA

The analysis of labour turnover is based on the data sheets completed for those individuals employed as manual workers by the case-study plants at some period over 1959–66. From the data sheet the date of engagement could be obtained along with information relating to job on joining the establishment, sex, age, marital status, etc. Similar information, with the addition of 'reason for leaving', was available when an individual left a case-study plant. It was, therefore, possible to obtain a count of the number of persons who joined and left any unit in any given time period, and these flows of labour could be further subdivided to distinguish categories by occupational groups[1] and by reason for leaving.

[1] Recruits are distinguished by occupational group according to their job on

Each year was broken into four quarters, January–March, April–June, July–September and October–December. Thus, there are 32 quarters in all and these are numbered 1–32 in the following text.[1] Ten occupational groups were distinguished for both males and females. However, as female employees were almost exclusively semi-skilled, we shall confine our attention to the turnover rate for *all* female manual workers. The male labour force was much less homogeneous in terms of skills, and at the relevant points in the following analysis we therefore distinguish between the three main groups of male manual workers—skilled, semi-skilled and unskilled.

A count was obtained of the number of males and females employed in each plant and in each occupational group on the first day of each quarterly period. This yielded the 'stock' of employees to which the number of new recruits and the number of leavers, distinguished by quarter, occupational group, sex and reason for leaving, could be compared. It was then possible to obtain each plant's rate of recruitment[2] for all males and females and for males by occupational groups and by quarters. In like manner, the total separation rate, the voluntary quit rate and the redundancy rate could be calculated for each plant and each occupational group on a quarterly basis.

There are good *a priori* reasons for supposing that the voluntary quit rate will have characteristics different to those shown by the total separation rate for all leavers, and more especially to those characteristics displayed by the redundancy rate, the dismissal rate and so forth. A difficulty therefore arises when the reason for leaving was not available from plant records. In our case, incomplete data on the reason for leaving only became important in the case of Birmingham plants. In Glasgow and North Lanarkshire the reason for leaving was not known in the case of only some 3 per cent of all separations, and there was relatively little variation in this proportion by establishments.[3] This being so, no adjustment was made in these areas for those whose reason for leaving was not known.

In Birmingham, 'not-known' leavers accounted for 9·1 per cent and 12·1 per cent of male and female separations respectively, and this proportion showed greater variation between plants than was the case in Glasgow and North Lanarkshire. If no allowance was made

joining the plant. Leavers are allocated occupationally according to the job held at the time of leaving.

[1] On occasion we will refer to these periods as quarters I, II, III and IV in *each* year.

[2] The number recruited each quarter as a percentage of the stock of employees at the beginning of each quarter.

[3] One plant in Glasgow with a high proportion of not-known leavers was excluded from the detailed statistical analysis of Sections 5 and 6.

for not-known leavers this might invalidate inter-plant comparisons within Birmingham and inter-market comparisons between Birmingham and the other labour market areas. The nature of any adjustment depends on whether not-known leavers have characteristics similar to or different from leavers whose reason for leaving is known. In the event, the two groups have like characteristics. This is best illustrated by comparing, for Birmingham males, the percentage distribution of the two groups by occupation, which is important because occupational attachment has a substantial impact on labour turnover in Birmingham.

	Percentage			
	Skilled	Semi-skilled	Unskilled	All Others
Reason for leaving known	10·0	54·1	17·3	18·6
Reason for leaving not known	8·6	46·1	23·0	22·3

It can be seen that the occupational characteristics of both groups were fairly similar. Because of this it appears reasonable to assume that not-known leavers had the same proportion of voluntary quits as those whose reason for leaving was known. Therefore, an adjusted voluntary quit rate was calculated for each Birmingham plant,[1] both for males and females and for each male occupational group, on the assumption that not-known leavers contained the same proportion of voluntary quits as those whose reason for leaving was known.[2] Hereafter, all references to quit rates in Birmingham are to this rate so adjusted.

3. PLANT QUIT RATES

This section is descriptive rather than analytic. The objective is to record certain characteristics of plant quit rates rather than to attempt to explain why these characteristics arise. Having established those factors which appear to be most important, the following sections take up the question whether the actions of employees in

[1] Two plants in Birmingham with a high proportion of not-known leavers were excluded from the analysis of Sections 5 and 6.

[2] Let V_{tj} represent the 'true' or adjusted quit rate for the occupational group and quarter considered in the jth plant. Then V_{tj} is given by the formula $V_{tj} = V_{kj} + V_{kj}/(S_j - L_j).\ L_j$ where V_{kj}, S_j and L_j are the known voluntary quit rate, the separation rate and the not-known leaving rate for the occupational group and quarter considered in the jth plant.

joining or leaving plants can be explained by variables, such as earnings, which vary from one establishment to the next.

The analysis in this section is exclusively concerned with voluntary quits. The reasons for this are obvious enough. First, voluntary quits account for the bulk of all separations. Second, voluntary leavers have *chosen* to leave a plant.[1] If decisions by employees are influenced by factors such as earnings, this is more likely to show up in the behaviour of voluntary quits than in the behaviour of other groups who leave through management action—redundancies, unsuitability, misconduct—or through factors over which neither the employee nor the employer has much control—sickness, retirement and death.

The characteristics of plant quit rates are illustrated by reference to data for male and female manual workers in Birmingham and for males in Glasgow. In this and the following chapter, quit rates for other areas are not discussed in the same detail because the number of case-study plants is not sufficient to merit similar treatment. Both North Lanarkshire and New Town contain only five case-study plants, and there are only two case-study plants in Small Town. Hence detailed cross-sectional analysis is inappropriate. Female employees in Glasgow are excluded for a similar reason, as the great bulk of all female employment was concentrated in five plants. In contrast, there are 27 case-study establishments in Glasgow for which detailed male turnover data are available and 25 in Birmingham. Of the latter number, all but two employ a substantial proportion of female employees.

The extent to which quit rates differed between plants is shown by Table 6.1. For each group considered, the quit rate was calculated for each plant on a quarterly basis. These quarterly rates were then summed and divided by the number of quarters (32) to give the unweighted average of plant quarterly quit rates for the period 1959–66. For convenience, we shall call this the average plant quit rate. From this, three measures of dispersion were obtained: first, the range between the plants with the lowest and highest average quit rates; second, the standard deviation of the average plant quit rates was calculated and, third, the coefficient of variation of these rates.[2]

The range between the plant with the lowest and that with the highest average quit rate was greater for both Birmingham males and females than for Glasgow males. The dispersion of average plant quit rates, measured by the standard deviation, was also greater in

[1] Or at least the majority are likely to have done so. Some may have left because they knew or feared they would be declared redundant or dismissed.

[2] The coefficient of variation is given by dividing the standard deviation by the mean and multiplying by 100.

Birmingham than in Glasgow. If we are interested in the degree of dispersion relative to the mean, as shown by the coefficient of variation, then once again average plant quit rates showed greater variability for Birmingham males but not, in this case, for Birmingham females.

With males, we find establishments in Birmingham whose average quit rate lay below and above the minimum and maximum average rates in Glasgow. The difference was more marked at the top of the scale, and this also applies when we compare Birmingham females to

TABLE 6.1

MEASURES OF DISPERSION: AVERAGE QUARTERLY PLANT QUIT RATES: GLASGOW MALES, BIRMINGHAM MALES AND FEMALES

Average plant quit rates	Glasgow males	Birmingham males	Birmingham females
Range (lowest to highest)	1·7 to 10·1	1·1 to 15·0	2·4 to 23·5
Standard deviation	2·4	3·6	5·0
Coefficient of variation	58·2	65·5	45·3

Glasgow males. In the 'tight' Birmingham labour market, therefore, quit rates varied more from one unit to the next than was the case in the less fully employed Glasgow market. The form this took is interesting. While the average or market quit rate for all plants was higher in Birmingham than in Glasgow,[1] there was little difference between those units which had the lowest quit rate in each market. The market quit rate for all plants was higher in Birmingham because certain plants had extremely high quit rates, much above the maximum rate in Glasgow. This suggests the existence of certain 'marginal' units in each labour market which suffer a relatively rapid increase in quits as the labour market becomes more fully employed.

We can pursue this argument further by considering whether the degree of dispersion of plant quit rates varies with changes in underlying employment conditions. The previous analysis would lead us to expect that, just as the degree of dispersion varies *between* market areas according to whether the market is tight or slack, it will vary *within* a market as unemployment[2] rises and falls in that market; i.e. differences between plant quit rates will widen as unemployment falls

[1] See Table 7.1, p. 169 below.

[2] Employment conditions are not accurately described by unemployment statistics alone (see Chapter 7 below), but such data will serve to illustrate the simple point discussed above.

and narrow as unemployment rises. The hypothesis is that in plants where quit rates are already low there will be little room for a further reduction in quits as unemployment rises.[1] On the other hand, a plant with a high quit rate is likely to experience a marked fall in that rate with increasing unemployment. Those who would have left will now be reluctant to risk unemployment and will remain in jobs which they would have discarded in more favourable circumstances. It follows that in such plants quit rates are likely to rise sharply when employment conditions improve. Those dissatisfied with their work will now vent that dissatisfaction by leaving to find alternative work elsewhere.[2] To test this, the standard deviation of plant quit rates was calculated for each quarter and for the groups included in Table 6.1. This was then correlated with the appropriate quarterly unemployment rate for males and females.[3] The resulting coefficients (r) are $-0{\cdot}57$, $-0{\cdot}54$ and $-0{\cdot}34$ for Glasgow males, Birmingham males and Birmingham females respectively.[4] As expected, the degree of dispersion of plant quit rates within a market is lowest when unemployment is high. When unemployment falls, differences between plant quit rates widen.

While one can identify units which tended to have high (or low) quit rates in all, or in most, quarters, changes in ranking did occur. In the short run, these were sometimes fairly substantial, so that it is hazardous to predict whether a plant will have a low or high quit rate simply from knowledge of its 'average' position over a long period of time. This can be demonstrated by reference to Table 6.2, which was compiled as follows. For the three groups considered, plants were ranked in ascending order from low to high quit rates according to their average quit rate for the period 1959–66. Two periods of four years were then chosen running from 1959 to 1962 and from 1963 to 1966 inclusive. In each of these four-yearly periods the quarter of minimum and maximum unemployment was chosen as was the quarter where unemployment was nearest to the quarterly average

[1] In plants with low quit rates a higher proportion of voluntary leavers are likely to leave the establishment for essentially non-work reasons, e.g. age, illness, personal reasons, etc. The actions of these individuals are less likely to be affected by changes in employment conditions.

[2] Thus a plant with a high average quit rate also experienced large fluctuations in quit rates from quarter to quarter. If the average plant quit rate is correlated with the standard deviation of the plant's quarterly quit rates, then the resultant coefficients are $+0{\cdot}69$, $+0{\cdot}61$ and $+0{\cdot}85$ for Birmingham males and females and Glasgow males respectively. All coefficients are significant at the 1 per cent level.

[3] Here unemployment is defined to include only those 'wholly unemployed', i.e. those 'temporarily stopped' are excluded. The reasons for this are explained on p. 178 below.

[4] The first two coefficients are significant at the 1 per cent level.

for the whole period 1959–66. These periods are denoted as Min. 1 and 2, Max. 1 and 2 and Nor. 1 and 2. In each of these quarters, plants were ranked in ascending order by quit rates. Spearman rank correlation coefficients were then calculated by comparing each of these rankings in turn with the ranking obtained for average plant quit rates over 1959–66. The results are shown in Table 6.2

TABLE 6.2

RANK CORRELATION COEFFICIENTS: AVERAGE QUARTERLY PLANT QUIT RATES AGAINST PLANT QUIT RATES IN SELECTED QUARTERS: GLASGOW MALES, BIRMINGHAM MALES AND FEMALES

	Rank correlation coefficients		
Quarter	Glasgow males	Birmingham males	Birmingham females
Min. 1	*+0·52*	+0·48*	+0·17
Min. 2	*+0·84*	*+0·51*	*+0·74*
Max. 1	*+0·52*	*+0·76*	*+0·89*
Max. 2	*+0·50*	*+0·84*	*+0·81*
Nor. 1	*+0·54*	+0·47*	*+0·51*
Nor. 2	*+0·70*	*+0·81*	+0·37

Note: 1. Italic indicates significant at the 1 per cent level.
* Indicates significant at the 5 per cent level.

The ranking of establishments by quit rates displayed a fairly high degree of stability over time. All the Spearman rank correlation coefficients are positive and, with two exceptions, are significant at the 1 per cent or 5 per cent level. This being so, there were units which had low (or high) quit rates in most periods. Yet, as Table 6.2 shows, the rankings were never entirely stable over time, and in certain periods fairly substantial changes occurred, particularly in the case of plant quit rates for females in Birmingham. Sometimes these changes in rankings were quickly reversed; on other occasions a more fundamental shift occurred where the change in a plant's ranking was of a more permanent, long-run nature.

It is possible then to find some support for the hypothesis that 'labour turnover is the resultant of a quasi-stationary process'[1]—that there is some 'normal' level of quits and of turnover associated with

[1] J. M. M. Hill, 'A Consideration of Labour Turnover as the Resultant of a Quasi-Stationary Process', *Human Relations*, Vol. 4, 1951, p. 264. See also A. K.

each plant, and that departures from this norm in response to disturbing factors will be of a temporary nature. Nonetheless, the concept of a normal, equilibrium level of turnover specific to each plant leaves much unexplained; in particular, it does not explain why certain units undergo substantial shifts in ranking which are not quickly reversed.

One further aspect of plant quit rates merits investigation. The occupational characteristics of the work group considered may influence quit rates, so that a plant with a relatively low (or high) quit rate for one occupational group, say skilled males, may display very different characteristics for other groups such as semi-skilled or unskilled males. This was tested by calculating for each plant the average quarterly quit rate for skilled, semi-skilled and unskilled males. The plants were ranked in ascending order of average quit rates for skilled males, and this ranking was correlated with the rankings obtained for semi-skilled and unskilled males. For Glasgow plants the rank correlation coefficients obtained are +0·84 and +0·57 respectively. The corresponding coefficients for Birmingham are +0·78 and +0·47. The last coefficient is significant at the 5 per cent level; the remaining three are significant at the 1 per cent level. These results indicate that an establishment with a low (or high) quit rate for a particular skill group tended to have correspondingly low (or high) quit rates for other occupational groups, this proposition applying with especial force when comparing quit rates for skilled and semi-skilled males.

Any analysis which sets out to explain variations in plant quit rates has, then, to accommodate a number of factors. The relationship between plant quit rates displays some stability over time inasmuch as there are units which tend to have relatively high or relatively low quit rates in most quarters. Yet the pattern is not fixed and plants do change rankings in the short run; less frequently more permanent shifts appear to occur. Similarly, a unit with a high (or low) quit rate for one occupational group tends to occupy a similar position for all other groups, although this is less true when one of the groups considered is the unskilled. Lastly, the spread between plant quit rates narrows when unemployment rises and widens when unemployment falls. This occurs because, as a market approaches full employment, the rise in quit rates is sharpest for establishments whose quit rates were relatively high in the first instance. These 'marginal' units therefore bear particularly heavy costs of recruitment, selection and training in a 'tight' labour market.

Rice, J. M. M. Hill, and E. L. Trist, 'The Representation of Labour Turnover as a Social Process', *Human Relations*, Vol. 3, 1950.

4. RELATIONSHIPS BETWEEN EARNINGS AND LABOUR TURNOVER

Labour market theory assumes that labour mobility and labour turnover will be responsive to differences in net advantages, so that both pecuniary and non-pecuniary factors influence job choice, but, as Chapter 1 pointed out, it is extremely difficult to specify the non-wage conditions which might influence job choice and more difficult still to measure and weight these conditions so as to obtain some meaningful measure of net advantages. The statistical tests employed therefore set out to establish whether any relationship exists between plant *earnings* differentials and labour turnover. In Section 5, an attempt is made to broaden the basis of this analysis but it cannot be pretended that our explanatory variables—earnings, plant size and rate of recruitment—measure all the factors which impinge on job-choice.

Empirical investigations of the relationship between wages and movements of labour have usually been based on macroeconomic data and have often adopted changes in relative wages as the relevant independent variable.[1] These choices reflect the availability of data on wage earnings and employment. Published information is generally based on an industrial classification.[2] Only rarely is any distinction drawn between occupational groups, and even when occupational data are available they are obtained by aggregating earnings for establishments in very different labour market situations. Again, because earnings data are often available only in index form the *percentage* change in earnings is usually adopted as the appropriate explanatory variable, whereas labour mobility and labour turnover may be more responsive to *absolute* changes in earnings or to differences in the level of earnings.[3] Lastly, labour mobility has usually been measured by *net* changes in employment because of the dearth of information on gross flows of labour.

Some of these difficulties can be circumvented in the present study.

[1] See, for example, J. Long and I. M. Bowyer, 'The Influence of Earnings on the Mobility of Labour'; and R. Wilkinson, 'Differences in Earnings and Changes in the Distribution of Manpower in the U.K., 1948–57', *Yorkshire Bulletin of Economic and Social Research*, Vol. 5, 1953 and Vol. 14, 1962; and W. B. Reddaway, *op. cit*.

[2] The chief source of earnings information in Britain is *Statistics on Incomes, Prices, Employment and Production* (H.M.S.O.) which provides detailed occupational data only for the engineering industry.

[3] See, G. S. Becker, *Human Capital* (1964) pp. 52–5 and A. M. Ross and W. Goldner, 'Forces Affecting the Interindustry Wage Structure', *Quarterly Journal of Economics*, Vol. 64, 1950.

Earnings data are available at the level of the individual establishment, so that we can take account of the intra-industry spread of earnings and the existence of widely differing employment conditions between different labour market areas. Equally important is the ability to distinguish between occupational groups and between *gross* flows of labour into and out of establishments. Certain weaknesses remain, of which the most important is our enforced reliance on *average* earnings by occupational groups and by plant. We cannot measure the spread of intra-plant earnings about the mean, although this may influence turnover; nor can we compare an individual's earnings with one of the case-study plants to his earnings before joining or after leaving the establishment. Nonetheless, we are able to move away from the macroeconomic analysis which is inappropriate for most wage studies, to an examination of earnings and turnover at the microeconomic level.

We have still to consider whether labour turnover is likely to be related to the level of earnings or to the percentage or absolute change in earnings over some previous period.[1] Most empirical work has been based on the percentage change in earnings, but this many not be appropriate in all circumstances. Quite apart from the fact that one has to decide whether to use percentage or absolute changes in earnings, which may give different results,[2] it is possible that labour turnover may respond to this difference in earnings *levels* rather than to any change in earnings over some previous period. The reason for distinguishing between earnings levels and changes in earnings may be explained as follows. Suppose the wage differentials we have observed in Glasgow and Birmingham were offset by other non-wage factors so that net advantages were equalized in all plants. Equilibrium would then have been established. There would be no tendency for labour to move from low to high earnings units and no relationship between earnings levels and labour turnover. If the equilibrium was then disturbed, say, by unequal changes in plant earnings levels, then labour mobility would occur. In this instance, labour turnover would be influenced not by differences in earnings levels but by the *changes* in earnings levels, this being the factor which caused the labour market to depart from equilibrium. Hence changes in earnings are the appropriate consideration in explaining

[1] The importance of these distinctions is discussed at greater length by T. P. Hill, 'Wages and Labour Turnover', *Bulletin of the Oxford University Institute of Statistics*, Vol. 24, 1962; and *Wages and Labour Mobility*, pp. 85–93.

[2] Consider two plants, A and B, whose average earnings were £10 and £20 respectively. If average earnings increase by £4 in plant A and by £5 in plant B, then the absolute increase in earnings is greater for plant B while the percentage increase is greater for plant A.

labour turnover where the position from which the changes are measured is, or is close to, an equilibrium one. Changes in earnings can then be regarded as departures from equilibrium causing differences in net advantages to emerge to which labour responds.

The concept of equilibrium is, however, best regarded as a pedagogical device useful in describing the forces which may operate in a labour market rather than as a description of a position which is never realized or even approached in practice. There is abundant evidence that various imperfections exist which prevent employers and employees from adjusting instantaneously to any change in labour market conditions. The wage differentials found to exist in Glasgow and Birmingham may then reflect real differences in net advantages. The more this is so—in other words the further labour markets depart from equilibrium—the more might labour turnover be expected to respond to differences in earnings levels. The choice of the independent variable therefore depends on the view taken of labour market imperfections. If these are substantial, then the relevant explanatory variable will be the level of earnings in the period in which labour turnover is studied. On the other hand, where these imperfections are of little influence, labour turnover is likely to respond to changes in earnings over some previous period. For the moment, no final decision need be taken. Tests are carried through which distinguish between the level of, and the absolute and percentage changes in, plant earnings.

Similarly, a number of different formulations can be adopted for the dependent variable, labour turnover. Because it is difficult to obtain information on gross flows of labour between plants or industries, previous studies have usually taken the net change in employment as the dependent variable.[1] This is not very satisfactory. The change in a plant's employment depends on the difference between the rate of recruitment and the separation rate. The former is a function of job openings as well as of differences in earnings, and there is little reason to suppose that job openings are necessarily more plentiful in plants with high earnings. Hence if there is no positive relationship between earnings and recruitment, this does not necessarily imply that employees do not respond to earnings differentials. In like manner if no relationship is found between earnings and net changes in employment, this may merely reflect the fact that plants

[1] Two exceptions are V. Stoikov and R. L. Raimon, 'Determinants of Differences in the Quit Rate among Industries', *American Economic Review*, Vol. 58, 1968; and T. J. Wales, 'Quit Rates in Manufacturing Industries in the United States', *Canadian Journal of Economics*, Vol. 3, 1970.

with high earnings are not seeking to expand their labour forces relative to those of other units.

The same considerations do not apply when dealing with voluntary quits. An individual can always exercise his preference to leave a given establishment, while he can only find work at those plants which offer him job openings. Again, those leaving a plant voluntarily have exercised a choice at least in the limited sense that they have chosen to leave a given plant, and this choice is likely to have been affected by wage considerations. The behaviour of total separations is more problematical, for this includes, in addition to voluntary quits, persons who may well have wished to remain with the plant but who leave through management action or through other circumstances over which they have little control. In the event, this contrast between quits and separations may not be crucial, because the behaviour of the former group may largely determine the course followed by separation rates. Nonetheless, the distinction has to be borne in mind for analytical purposes.

We can conclude, then, that the crucial test for labour market theory is the behaviour of voluntary quits. As far as wages are concerned, the specification of the independent variable will depend on the view taken of labour market imperfections. If these are important, and all the evidence is that they are, then the relevant factor shaping behaviour is likely to be the *level* of plant wages. We should also note that where knowledge is imperfect this is likely to affect recruitment more than quits. While an individual seeking a new job may have some difficulty in distinguishing plants with high or low earnings, he is still likely to be able to recognize a 'good' or a 'bad' job when he stumbles across one. Hence, so long as some degree of information exists which stops short of total ignorance or perfect knowledge, there is likely to be a weaker relationship between recruitment and earnings than between quits and earnings.

5. EARNINGS AND LABOUR TURNOVER: EMPIRICAL RESULTS

Before proceeding to the statistical analysis it would be useful to indicate the expected nature of the relationships between our dependent and independent variables. Let us assume for the moment that labour turnover is responsive to differences in earnings levels and/or to changes in earnings levels. A simple view of labour market theory would suggest that the expected relationship between recruitment and net changes in employment on the one hand, and any of the wage variables on the other, will be *positive*. Hence if earnings signals do redistribute labour in the manner suggested by labour market theory, then high levels and/or large increases in earnings will attract new recruits, while low levels and/or small increases in earnings will repel potential employees.[1] The reverse holds for quits and separations. The relationship between these flows of labour out of plants and earnings levels or changes in earnings levels can be expected to be negative. Thus, by and large, plants which have high levels of earnings or large increases in earnings will be expected to have low quit rates and low separation rates, and vice versa.

The nature of the wage data utilized in this study has already been explained in some detail. Here we merely recapitulate the previous discussion. The first source of information is the W.E. returns collected by the Department of Employment and Productivity which provide, by plant, average male and female gross weekly earnings (including overtime) in a given working week. These returns fall in every second quarter as these are defined in this study, so that earnings data are available for 16 of the 32 quarters over 1959–66.[2] Overtime earnings are included but are not distinguished separately, and no detail is provided by occupational groups. Both these disadvantages can be overcome by using the second source of wage information, the occupational earnings returns collected by the Engineering Employers' Federation and by the D.E.P. As these latter returns are available less frequently, we begin by considering average plant gross weekly earnings (including overtime) for males and females derived from the W.E. returns.

Average gross weekly earnings were calculated for those plants[3] and for those periods for which returns were available. From this the percentage and absolute increase in gross earnings was obtained, and each of these variables was correlated with the voluntary quit rate, the separation rate, the net change of employment and the rate of recruitment by plants. Hence we relate these measures of labour turnover to the level of gross earnings in the same period and to the increase (percentage or absolute) in gross earnings over the previous period. For example, in Table 6.3, labour turnover by plant in quarter 4 is correlated to the percentage and absolute increase in plant

[1] We are assuming, in effect, that wage changes (or differences) are demand- rather than supply-induced. This hardly solves the 'identification problem', but it seems realistic enough given that we are dealing with short periods and with individual plants located in the same labour market.

[2] We assume, therefore, that earnings in the given week are representative of earnings in the quarter in which the week falls.

[3] There were 15–22 observations per quarter for Glasgow males, 12–14 for Birmingham males and 11–12 for Birmingham females.

earnings in quarter 4 as compared to quarter 2 and to the level of earnings in quarter 4.

The results obtained for Glasgow males are shown in Table 6.3. In the interests of brevity, we have excluded the correlation coefficients obtained between absolute increases in plant earnings and the various measures of plant turnover. This is done because the results do not differ significantly from those obtained by correlating percentage increases in plant earnings with plant turnover. The relationship between plant turnover and absolute changes in plant earnings is discussed below.

It is immediately apparent that there is little evidence of a systematic relationship between any of the earnings variables and plant recruitment rates. When the rate of recruitment is correlated with the percentage change in earnings, no less than 7 of the 15 coefficients have a negative sign, whereas a positive sign was expected. The correlation coefficients between absolute changes in earnings and the rates of recruitment have negative signs in 6 of 15 quarters while no less than 13 of 16[1] observations are negative when the level of plant earnings is substituted as the explanatory variable.[2] Given our previous discussion, these results are not very surprising. The lack of any systematic relationship between plant earnings and plant recruitment rates does not necessarily mean that potential recruits do not respond to wage signals. More probably, it simply indicates that plants with high earnings or with large increases in earnings were not necessarily seeking relatively large numbers of new recruits.

The relationship between earnings and net changes in employment appears to be more in line with the predictions of labour market theory. For each earnings variable, including absolute changes in earnings, there is evidence of a positive relationship, the sign of the correlation coefficient being negative on only two or three occasions in each case. There is in each case a scattering of coefficients significant at the 5 per cent level, so that on these grounds there is little to choose between the explanatory power of any of the earnings variables. It must, however, be observed that, while the coefficients are usually positive as expected, they are also small, so that the relationship between earnings and changes in employment is weak.

[1] As no earnings data are available prior to 1959, the results for absolute and percentage changes in earnings are obtained for only 15 quarters compared to 16 quarters when the level of earnings is considered.

[2] The predominantly negative relationship between recruitment and wage levels results from the negative relationship between wage levels and quits. The low rate of quits gives a low rate of recruitment provided, as argued above, that high-wage plants are not seeking to increase their labour force.

TABLE 6.3

CORRELATION COEFFICIENTS: AVERAGE PLANT GROSS WEEKLY EARNINGS, PERCENTAGE CHANGE IN PLANT EARNINGS[1] AND PLANT TURNOVER: GLASGOW MALES

	Average earnings				Percentage change in earnings			
Quarter	Quit rates	Separation rates	Recruitment rates	Changes in employment	Quit rates	Separation rates	Recruitment rates	Changes in employment
2	−0·19	−0·32	−0·15	0·11	—	—	—	—
4	−0·54*	−0·54*	−0·21	0·37	−0·48*	−0·35	−0·25	0·15
6	−0·35	−0·29	0·14	0·36	0·13	0·06	0·32	0·18
8	*−0·59*	−0·55*	−0·18	0·04	−0·24	−0·02	−0·30	−0·41
10	−0·35	−0·30	−0·04	0·27	0·35	−0·07	*0·57*	*0·83*
12	−0·48*	−0·36	0·28	0·43	−0·17	−0·20	0·17	0·30
14	−0·15	0·03	−0·32	−0·29	0·24	0·05	0·38	0·23
16	−0·18	−0·18	−0·08	0·07	−0·09	−0·41	0·14	0·43
18	−0·54*	*−0·65*	−0·10	0·53*	−0·23	−0·25	−0·26	0·06
20	−0·36	−0·46*	−0·09	0·21	−0·53*	−0·43*	−0·04	0·19
22	0·00	−0·02	−0·04	−0·07	0·43*	0·42	*0·67*	*0·54*
24	−0·57*	−0·49*	−0·18	0·29	−0·43*	−0·25	−0·19	0·01
26	−0·41	−0·40	−0·16	0·18	−0·19	−0·25	−0·53*	−0·32
28	−0·32	−0·30	−0·13	0·27	−0·04	−0·06	−0·07	0·01
30	−0·54*	−0·46*	0·13	0·49*	−0·38	−0·24	0·22	0·40
32	−0·39	−0·52*	−0·42	0·48*	0·06	−0·02	0·15	−0·06

Notes: 1. Earnings include overtime payments.
2. Italics—significant at the 1 per cent level.
* Significant at the 5 per cent level.

For reasons already stated, it appears probable that if employees do respond to wage signals this is most likely to be reflected in the behaviour of voluntary leavers and, through this group, in the behaviour of separation rates. In both cases, a negative relationship is expected—quit rates and separation rates will, in general, be lower in establishments which have high earnings or which have experienced large increases in earnings over the previous period. The latter expectation is not fulfilled. When percentage changes in earnings are correlated with quit rates the sign is positive in 5 of 15 cases, and in 3 of 15 cases when separation rates are the dependent variable. For absolute increases, positive correlation coefficients occur in four and three quarters respectively. Moreover, few of the negative coefficients are significant at the 5 per cent level.

A sharp contrast emerges when plant quit rates and separation rates are correlated with plant earnings levels. In both cases a positive coefficient is obtained in only one of 16 quarters and a high proportion of the coefficients are significant at the 5 per cent level. It is therefore clear that, if we wish to 'explain' differences in plant wastage rates, we must search for that explanation by considering the levels of plant earnings rather than changes in these levels. Much the same conclusions can be derived from a study of the relationship between earnings and turnover for Birmingham males and females. For both groups there is no sign of a systematic and positive relationship between plant earnings and recruitment, and negative correlation coefficients between any of the wage variables and net changes in employment are more frequent for both groups than for Glasgow males. Once again, a clear distinction between the influence of the earnings variables only emerges when we analyse quit rates and separation rates. As in Glasgow, a negative relationship emerges between the level of earnings and wastage but not between wastage and changes in earnings. In consequence we show in Table 6.4 only the coefficients obtained from correlating average plant earnings with plant quit rates and separation rates for Birmingham males and females.

The correlation coefficients obtained between average plant earnings and plant quit rates and separation rates are negative, as expected, in the great majority of quarters. For each group considered therefore—Glasgow males and Birmingham males and females—quits and separations were a decreasing function of average plant earnings. No such systematic relationship emerges when changes in earnings are substituted for the level of earnings. This implies that the earnings differentials observed in our labour markets were in some sense 'real'—they did *not* merely reflect the existence of offsetting non-monetary advantages in low wage units so that net advantages were equalized in all establishments. It would seem therefore that at any point in time a labour market is in a greater or lesser degree of disequilibrium. Earnings and net advantages are not equalized in all plants, so that the relevant variable influencing turnover is the level of earnings rather than changes in that level.

TABLE 6.4

CORRELATION COEFFICIENTS: AVERAGE PLANT GROSS WEEKLY EARNINGS[1] AND PLANT WASTAGE: BIRMINGHAM MALES AND FEMALES

Quarter	Males		Females	
	Quit rates	Separation rates	Quit rates	Separation rates
2	0·15	0·11	0·21	0·07
4	−0·05	0·01	−0·02	−0·15
6	−0·09	−0·12	−0·35	−0·27
8	0·09	0·13	−0·35	−0·38
10	−0·07	−0·07	−0·39	−0·47
12	−0·26	−0·19	−0·31	0·17
14	−0·33	−0·44	−0·67*	−0·61*
16	−0·23	−0·15	−0·37	−0·43
18	−0·17	−0·19	−0·40	−0·52
20	−0·41	−0·38	−0·41	−0·51
22	−0·28	−0·33	*−0·75*	*−0·79*
24	−0·22	−0·18	−0·70*	−0·65*
26	−0·37	−0·40	−0·49	−0·61*
28	−0·24	−0·26	−0·68*	−0·67*
30	−0·26	−0·38	−0·38	−0·28
32	−0·48	−0·43	−0·50	−0·40

Notes: 1. Including overtime payments.
2. Italics—significant at the 1 per cent level.
* Significant at the 5 per cent level.

This result appears somewhat incongruous at first sight. After all, if a plant experiences large absolute and percentage increases in earnings over an extended period of time it will eventually become a high-earnings unit, and if quits and separations are inversely related to earnings levels we might expect the same relationship to hold for changes in earnings. Over a long period, such a relationship might indeed emerge, for the longer the period considered the greater will be the increase in earnings relative to the original earnings level. Increases in earnings would then swamp the effect of plant differentials which existed in the first instance, so that the latter could be safely ignored. In our case, we have measured changes in earnings over six-monthly periods. Hence the change in earnings is usually small relative to the earnings level. If there are, as is the case in Glasgow and Birmingham, substantial differentials in plant earnings which reflect, at least to some extent, differences in net advantages, then any inverse relationship between changes in earnings and labour wastage will be extremely weak.[1]

We have seen that no positive relationship emerges between any of the earnings variables and recruitment rates. Even where the level of earnings is considered, negative coefficients outnumber positive coefficients. What is implied by this absence of a systematic relationship between plant earnings and recruitment? There are only three possible explanations. Either individual employees are not attracted by higher earnings; or they do not know they exist; or high-earnings plants do not wish to increase their employment relative to other units. The first possibility does not appear tenable given the negative relationship between earnings and quits. The second is at best only a partial explanation. Most labour market studies have found ignorance and imperfect knowledge to be important factors, and the engineering industry in Britain with its proliferation of wages systems is hardly likely to prove an exception. However, our previous analysis[2] suggests that imperfect knowledge is unlikely to be a sufficient explanation for the earnings differentials which arise in labour markets. It is, then, difficult to avoid the conclusion that plants with high earnings do not necessarily wish to obtain a relatively large share of the supply of potential recruits. Nor is it true that high earnings arise primarily because plants are attempting to expand their labour forces more rapidly through holding down wastage rates relative to the recruitment rate, since the relationship between earnings levels and net changes in employment is weak. Certainly if an establishment wishes to hire additional labour it may be necessary for it to raise its wage level. Nonetheless, the lack of a significant positive relationship between earnings levels and recruitment rates over an eight-year period must suggest that high wage units did not have this purpose *primarily* in mind.

High earnings may, of course, confer other benefits on the plant. The plant may be able to impose higher hiring standards so that

[1] For this reason a negative relationship between changes in earnings and wastage rates is only likely to emerge in the short run where there is a positive correlation between earnings levels and changes in these levels over a previous period. It is, of course, to be understood that the short run may be a period lasting years rather than months if wide earnings differentials exist.

[2] See pp. 85–7 above.

efficiency wages are equal to or even lower than those in low earnings plants. Such a plant is likely to have relatively low quit rates (and hence relatively low separation rates) so that it will enjoy lower costs of recruitment, selection and training. A high-earnings policy may then be perfectly sensible even where the unit is not attempting to expand employment.

Although high-earnings plants will tend to have lower quit rates the relationship between earnings and quits, while usually negative, is also, usually, rather weak. We can judge how far differences in plant quit rates are 'due' to differences in earnings levels by squaring the relevant correlation coefficients in Tables 6.3 and 6.4. At best, the coefficient of determination (r^2) is only 0·56 (Birmingham females, quarter 22) indicating that we have 'explained' just over half of the observed differences in plant quit rates. Generally the association is much weaker than this, particularly in the case of Birmingham males where the highest r^2 obtained is 0·23. While it is difficult to generalize on the basis of the above tables, it would seem that the variation in plant quit rates accounted for by differences in plant earnings is some 10–33 per cent for Glasgow males, 10–50 per cent for Birmingham females, and is seldom more than 10 per cent in the case of Birmingham males. We shall see shortly that the size of r^2 is not necessarily vital to our argument, but for the moment we should consider whether a stronger relationship would emerge if more detailed earnings and turnover data were substituted for those employed in Tables 6.3 and 6.4.

The occupational earnings returns of the E.E.F. and the D.E.P. provide earnings by male occupational groups and allow us to distinguish between average gross weekly earnings (including overtime payments) and average standard weekly earnings for a standard working week. For reasons already explained, the E.E.F. returns are used for Glasgow males and the D.E.P. returns for Birmingham males. Average gross earnings and average standard earnings were calculated by occupational groups for those plants and for those quarters for which returns were available.[1] From this the percentage and absolute change in earnings including and excluding overtime were obtained. Each of these earnings variables was then correlated with quits, separations, recruitment and net changes in employment distinguished by plant and by occupational group.

Some of the results can be dismissed quickly as they merely con-

[1] Data for seven quarters were available for Glasgow males falling in the fourth quarter of each year except 1963. There were 10–20 observations in each quarter. For Birmingham males, two returns were available for the second and fourth quarter of each year over 1963–6. There were 10–13 observations in each quarter.

firm the findings based on Tables 6.3 and 6.4. In no case is there evidence of a systematic relationship between absolute and relative changes in earnings (including or excluding overtime) and any of the measures of labour turnover, by occupational groups. Our previous conclusion based on Table 6.3 and 6.4, that in the short run absolute and percentage changes in earnings have little effect on labour turnover, is not, then, due to aggregation. It holds even when we consider earnings and labour turnover by occupational groups. Similarly, we can reject the hypothesis that rates of recruitment and changes in employment are positively related to plant earnings levels. The correlation coefficients between these variables distinguished by occupational groups are as often negative as positive, whether earnings are defined to include or exclude overtime payments.

TABLE 6.5

CORRELATION COEFFICIENTS: AVERAGE PLANT GROSS WEEKLY EARNINGS[1] AND PLANT WASTAGE: GLASGOW SKILLED AND UNSKILLED MALES

Quarters	Skilled		Unskilled	
	Quit rates	Separation rates	Quit rates	Separation rates
4	*−0·67*	−0·57*	−0·39	−0·23
8	−0·61*	−0·58*	−0·42	−0·49
12	−0·09	−0·05	−0·47	0·28
16	−0·37	−0·49	0·22	−0·02
24	−0·36	−0·54	−0·10	0·00
28	−0·24	−0·19	−0·33	−0·32
32	−0·16	−0·38	−0·08	−0·18

Notes: 1. Including overtime payments.
2. Italics—significant at the 1 per cent level.
* Significant at the 5 per cent level.

It only remains to consider whether disaggregation by occupational groups strengthens the relationship between earnings levels and wastage rates. In Glasgow, occupational earnings were calculated for skilled and unskilled males while the different source data for Birmingham males allow us to isolate three groups—skilled, semi-skilled and unskilled. Average earnings by plants for these groups were then correlated with quit rates and separation rates by plant and by occupation. The results obtained for Glasgow males are shown in Table 6.5 and those for Birmingham males in Table 6.6, with earnings defined to *include* overtime earnings (i.e. gross earnings). This choice was made simply to avoid unnecessary duplication. If occupational earnings are defined to *exclude* overtime earnings, then the results obtained do not differ in any important respect from those shown in Tables 6.5 and 6.6.

For male occupational groups in both Glasgow and Birmingham, quit rates and separation rates are usually inversely related to plant earnings levels, but the strength of this relationship is no greater than when no differentiation is made by occupational groups. This can be seen by contrasting the results for the quarters shown in Tables 6.5 and 6.6 with those for the same quarters in Tables 6.3 and 6.4. One other interesting result emerges from Tables 6.3 to 6.6. This is the

TABLE 6.6

CORRELATION COEFFICIENTS: AVERAGE PLANT GROSS WEEKLY EARNINGS[1] AND PLANT WASTAGE: BIRMINGHAM SKILLED, SEMI-SKILLED AND UNSKILLED MALES

Quarters	Skilled		Semi-skilled		Unskilled	
	Quit rates	Separation rates	Quit rates	Separation rates	Quit rates	Separation rates
20	−0·32	−0·23	−0·32	−0·25	−0·11	−0·13
22	−0·04	−0·02	−0·46	−0·43	−0·34	−0·42
24	0·19	0·14	−0·20	−0·22	−0·47	−0·31
26	−0·02	0·00	−0·37	−0·42	−0·50	−0·71
28	−0·32	−0·54	−0·38	−0·48	−0·29	−0·39
30	0·06	0·10	−0·39	−0·53	0·18	0·18
32	0·05	0·06	−0·52	−0·48	−0·30	−0·45

Note: 1. Including overtime payments.

lack of any evidence to support the hypothesis that labour mobility and turnover is likely to be more responsive to differences in earnings in 'tight' labour markets. Indeed, whichever group of males is considered, a more marked inverse relationship between earnings and labour wastage emerges in Glasgow, despite a higher level of unemployment in Glasgow in all the quarters considered.[1] In the next section we shall look at this question in greater detail; we shall suggest that in some respects we might *expect* a weak relationship between wastage and earnings in 'tight' labour markets.

[1] It is worth while noting that T. P. Hill's study of earnings and labour turnover by collieries found the association weakest in those regions (the West and East Midlands) where labour market conditions were tightest, *op. cit.*, pp. 226–7. We should also note that the relationship between plant quits and earnings does not strengthen as unemployment in the market falls. Correlating the coefficients from Table 6.3 with the quarterly rate of male unemployment in Glasgow yields a new coefficient of +0·19. The same procedure gives coefficients of −0·33 and +0·18 for Birmingham males and females. None are significant at the 5 per cent level and different definitions of unemployment have no effect on the results.

The results of the above analysis can be summarized as follows. In the short run, there was no systematic relationship between changes in plant earnings and plant turnover rates. Again, differences in plant earnings appear to have had little effect on the rate of recruitment or on the plant's change in employment. This may be ascribed partly to imperfect knowledge, but the fact that high-wage plants were not necessarily seeking to increase their labour forces is also likely to be important. When quit rates and separation rates are correlated with plant earnings, a negative relationship clearly emerges. High-wage plants may, then, have enjoyed certain advantages through lower wastage rates and may also have been able to enforce higher hiring standards. However, while employees do appear to respond to economic incentives, variations in earnings do not explain the major part of differences in plant wastage rates. The relationship between earnings and labour wastage is, then, a weak one, but is it weak enough to be dismissed as insignificant?

Weak coefficients might have been expected for three reasons. First, almost all microeconomic analysis is based on some sample of the population. This study is no exception and, this being so, the correlation between earnings and turnover is unlikely to be perfect even where all turnover takes place in response to differences in plant earnings.[1] Second, because we are measuring turnover over fairly short periods of three months fluctuations in wastage due to random factors may be expected to reduce the coefficients obtained. Third, the theory of net advantages, and indeed everyday observation, suggests that labour turnover is not simply a function of strictly pecuniary considerations. Many factors other than earnings may influence the decision to stay with or leave an employer. Unless these are dismissed as unimportant, we would not expect the relationship between earnings and turnover to be strong. For these reasons, it has been suggested that 'the strengths of the correlations observed in a microeconomic enquiry of this kind are almost irrelevant in assessing the economic significance of the relations examined, and weak correlations are not regarded as particularly disturbing provided they are not so small as to raise serious doubts that they may have arisen purely by chance'.[2]

It would be wrong to judge the relationship between earnings and labour wastage by treating the individual coefficients in Tables 6.3 to 6.6 in isolation from each other. In most cases, when inspecting the coefficients singly we would have to accept the null hypothesis that there was no systematic relationship between the level of earnings on the one hand and plant quit rates and separation rates on the other.

[1] T. P. Hill, *op. cit.*, pp. 228–9.

[2] *Ibid.*, p. 199.

However, inspection of all the coefficients obtained by correlating wastage and earnings levels will indicate that the sign is, as expected, negative in almost all cases. The coefficients obtained do not, therefore, differ from zero through chance. We can, then, *reject* the null hypothesis that these variables are unrelated. In plain language, we find that high-wage plants do, indeed, tend to have relatively low quit rates and hence low separation rates, and vice versa. Nonetheless, it has to be admitted that differences in plant earnings levels do not explain all, or even the greater part of, observed variations in plant wastage rates. It is necessary to consider, therefore, whether there are any other factors which may cause labour wastage to differ between establishments.

6. QUIT RATES, EARNINGS, PLANT SIZE AND RECRUITMENT RATES

This section examines in greater detail the behaviour of plant quit rates. In the previous section we found an inverse relationship between average plant earnings (including overtime) and quit rates distinguished by sex. The strength of this relationship did not change when occupational groups were considered and when occupational earnings were analysed with and without overtime payments. This is convenient, for it allows us to utilize the W.E. returns, on which Tables 6.3 and 6.4 are based, to explore further the behaviour of quit rates.

In addition to earnings, what other factors might be taken into account? First, quit rates may be a function of plant size, which we take as the number of male *or* female employees as the case may be. The evidence pertaining to this relationship is extremely sketchy and occasionally conflicting.[1] It is seldom possible to distinguish the effect of plant size from other factors which may be associated with size and with quits. Hence, although it has been suggested that quit rates are inversely related to size so that large units tend to have relatively low quit rates, the apparent association may be spurious. It may arise because large plants tend to be concentrated in industries with low turnover, so that the 'true' relationship is between industry and quit rates and not between plant size and quit rates; or plant size

[1] See *Wages and Labour Mobility*, pp. 58–9; J. R. Greystoke, G. W. Birks and T. Murphy, 'Surveying Labour Turnover in the Sheffield Region', *Yorkshire Bulletin of Economic and Social Research*, Vol. 3, p. 86; J. R. Long provides evidence which suggests that plant size and wastage may be negatively or positively related, see *Labour Turnover Under Full Employment* (University of Birmingham, Studies in Economics and Society, 1951), pp. 75–80.

and earnings may be positively associated,[1] so that large plants have low quit rates not because they are large but because they are high-earnings units.

The reasoning behind the supposition that plant size and quit rates are inversely related has seldom been made explicit, but two factors appear to be given particular weight. First, the large plant may offer better promotion and wider employment opportunities. Job changes can then take place within a large unit, whereas the same job change may require a change of employer if the plant is a small unit. In consequence, large units may have lower quit rates accompanied by greater mobility within the plant. Second, the large plant may offer greater security of employment. Employment may be less sensitive to short-run changes in economic conditions and, secularly, employment prospects may also be more favourable. It is, indeed, possible to discern over the long run a greater growth of employment in larger units,[2] but the pattern of employment changes very slowly, so slowly that it is difficult to envisage it having much impact on quit rates. The cyclical relationship between plant size and employment stability is one of the neglected areas of economics, a subject area where there has been little or no discussion of a theoretical or empirical nature. However, we shall see subsequently that units with more than 1,000 male employees in Glasgow have a very low rate of male redundancies.[3] Although the evidence is far from conclusive, it does at least suggest that plant size may have some influence on the security of employment and, through this, on quit rates.

The unsatisfactory nature of the available evidence provides ample reason for exploring the relationship between plant size and quit rates.[4] In this case, we can, as it were, hold industry and earnings 'constant' so as to examine whether size has an independent influence on quits. Where the relationship between quit rates, earnings and plant size is concerned, this is achieved through multiple regression analysis. However, no variable is introduced to represent the

[1] British evidence indicates that there is 'a tendency for both average weekly earnings and average hourly earnings to rise according to the size of the establishment', *Ministry of Labour Gazette*, Vol. 67, 1959, p. 126 and pp. 245–7. See also R. A. Lester, 'Pay Differentials by Size of Establishment', *Industrial Relations*, Vol. 7, 1967.

[2] Over time, the average size of establishment has increased (see R. Evely and I. M. D. Little, *Concentration in British Industry* (1960), p. 171) and since 1939 there has been a marked rise in the proportion of the manufacturing labour force employed in plants with more than 1,000 employees.

[3] See Table 14.1, p. 372 below.

[4] Unlike economists, industrial sociologists have argued that size and quit rates are *directly* related. The proposition is best analysed by G. K. Ingham, *Size of Industrial Organisation and Worker Behaviour* (1970).

industry in which the plant is engaged. We therefore assume that differences in product between plants, all of which are in the engineering industry, are not sufficient to cause, by themselves, significant fluctuations in quit rates. Output is not, of course, homogeneous, but techniques of production and work organization are sufficiently similar to suggest that product differences should be of only minor importance in affecting quit rates.

Plant quit rates may also be influenced by the distribution of the work force by length of service. It has been suggested that, of the 'personal' characteristics[1] which influence turnover, 'One criterion . . . outweighs all others as a predictor of staff loss in every one of the classes of labour we have studied: it is tenure.'[2] For any group of new recruits the proportion of quits will be extremely high in the first few months or even weeks. The 'induction crisis', as Rice has aptly labelled this period,[3] is unlikely to exceed three months. After the induction crisis the proportion of recruits leaving the plant will continue to fall, but at a diminishing rate as length of service increases. The survival pattern of any group of recruits can be represented by a frequency distribution of length of service at termination of employment. This can be described by a hyperbolic function of the form $y = ax^{-b}$, where a is a constant measuring the percentage of recruits leaving in the first period and b is the slope of the hyperbolic function, i.e. a measure of the deceleration in the percentage leaving over time.

Since the relationship between quits and tenure is well established,[4] it is necessary to consider how we might allow for it in our model. Ideally, an attempt to predict the quit rate from a given stock of employees should take into account the time pattern of recruitment of the stock over all past periods. We write,

$$Q_{j(t)} = f[A_{j(t)}, A_{j(t-1)} \ldots A_{j(t-n)}]$$

where Q_j represents the quit rate and A_j the recruitment (accession) rate in the jth plant and (t), $(t-1)$ . . . $(t-n)$ represent equal time periods.[5]

[1] By which we mean age, skill, sex, length of service and so forth.

[2] A. Young *op. cit.*, p. 30.

[3] Rice, Hill and Trist, *op. cit.*, p. 359. See also J. M. M. Hill, *op. cit.*

[4] In fact, most studies, including those of the Tavistock Institute, have been based on separation rates. However, as voluntary leavers account for the bulk of all separations, it is unlikely that their behaviour will be significantly different. We shall see later (pp. 220–8 below) that a high proportion of quits are accounted for by short-service employees.

[5] The reader should note that as long as we are considering quits over some finite time period we have conceptually to allow for quits amongst those re-

This formulation takes cognizance of the possibility that, as Silcock puts it, 'the value of the [wastage] rate in any year is likely to depend more upon past history than on the present state of industrial relations'.[1] In practice, it is difficult to take all of 'past history' into account, especially as survival rates vary between plants and in the same plant over time as employment conditions change. This does not mean that the problem must be abandoned; merely that we must consider a simpler formulation of the proposition that quits are a function of tenure and hence of the time distribution of recruitment.

We know that wastage is particularly high amongst short-service employees and that after the period of induction crisis, which does not seem to exceed three months, wastage drops off sharply.[2] As turnover rates in this study have been calculated on a three-monthly, quarterly basis, we can test whether plant quit rates in a given quarter are influenced by plant recruitment rates in the previous quarter. This is, of course, far from perfect. As we have suggested, quit rates will be affected by the time distribution over which the entire stock of employees has been recruited. Again, some of those recruited in the previous quarter will have left the plant before the current quarter commences. More important, we ignore the probability that some of those recruited in the current quarter will also leave in the current quarter, so that there is a relationship between the quit rate in a given quarter (Q_t) and the recruitment rate in that same quarter (A_t). The reason for excluding A_t from our analysis is that a positive relationship between Q_t and A_t would not demonstrate that high (or low) quit rates were *caused* by high (or low) recruitment rates. The reverse may equally well have been true, i.e. A_t would be high (or low) according to whether Q_t, and with this the need to recruit replacements, was high (or low).

Because of this, there is little point in introducing A_t as an additional explanatory variable. Its introduction would certainly bring an impressive statistical improvement in the 'fit' obtained, but this would hardly assist us in explaining the forces at work. We have then the regression model:

$$Q_{js(t)} = a+b_1\,W_{js(t)}+b_2\,S_{js(t)}+b_3\,A_{js(t-1)}+\varepsilon \qquad (1)$$

where Q_j, W_j, S_j and A_j are respectively the quit rate, average earn-

cruited *within the 'present' time period.* We shall see shortly that we cannot test this relationship between Q_t and A_t without making a heroic, and unjustified, assumption about the direction of causation.

[1] H. Silcock, 'The Phenomenon of Labour Turnover', *Journal of the Royal Statistical Society*, Series A (General), Vol. 117, 1954, p. 432.

[2] Rice, Hill and Trist, *op. cit.*, p. 363; J. M. M. Hill, *op. cit.*, p. 258; and Greystoke, Birks and Murphy, *op. cit.*, pp. 89–92.

ings including overtime payments, size as measured by the number of employees at the beginning of the quarter, and the rate of recruitment in the jth plant. The error term is denoted by ε; subscript s denotes sex and (t) and $(t-1)$ the 'present' and the 'previous' quarters.[1]

The regression equation was calculated for Glasgow males and for Birmingham males and females for each of the quarters shown in Tables 6.3 and 6.4. Inspection of the results indicated no systematic relationship between plant size (S) and the quit rate (Q). In the case of Birmingham females, the partial regression coefficients for plant size in the above equation were positive in 11 of the 16 quarters for which results were available. For Glasgow males, positive signs also predominate, the regression coefficient for S being negative in only 2 of 16 quarters. This would indicate that, when allowance is made for earnings (W) and the recruitment rate in the previous quarter (A), there is a positive relationship between plant size and quit rates; i.e. if other things are held constant the effect of increasing plant size is to increase rather than to reduce the quit rate. The position is, however, confused, for the pattern for Birmingham males is the reverse of that found for Glasgow males. Thus the regression coefficients for S are negative in 14 of 16 quarters when the regression equation is calculated for Birmingham males.

It would be wrong to conclude from the above that, although the effect may run in either direction, plant size is an important determinant of the plant quit rate. On the contrary, the partial regression coefficients for S are always extremely small for all groups and periods considered, and only rarely do they exceed their standard error. In other words, plant size is simply unimportant in explaining variations in quit rates between establishments. There is no convincing evidence that large units have relatively low quit rates *because* they are large.

The evidence available suggests that attempts to quantify the relationship between quit rates and plant size may be positively misleading if other variables which also influence quits are not taken into account. Thus, if the simple correlation coefficient between Q and S is calculated for Birmingham males and Glasgow males, the sign is negative in all cases for Birmingham males and in 13 of 16 cases for Glasgow males. This inverse relationship is produced because plant size and earnings are positively related. When account is taken of earnings, the negative relationship between Q and S disappears alto-

[1] The equation was calculated for every second quarter over 1959–66; it therefore refers to quarter II or IV in each year and $t-1$ to quarter I or III as the case may be.

gether or becomes insignificant. The inverse relationship which appears when only Q and S are considered is, therefore, spurious.[1]

One interesting by-product of the above analysis is the positive relationship which emerges between plant size and the level of plant earnings. The simple correlation coefficient between W and S is, with one exception, positive in all periods for males in both Glasgow and Birmingham. In the latter area, the relationship is particularly strong; the coefficient is usually significant at the 5 per cent level. We shall see shortly that the relationship between earnings and quit rates is particularly weak in Birmingham. This raises the question why large units offer higher earnings if these have relatively little effect on quits. One answer would appear to be that large units *must* offer higher earnings simply to keep quit rates at the *same* level as smaller units. This is because large units draw their labour over a relatively wide area of the labour market.[2] Higher earnings are necessary to offset the cost and inconvenience of longer travel-to-work journeys. Higher earnings may be sustained and in part be a consequence of economies of scale, but any advantage large units enjoy in this fashion is, to some extent, dissipated by the need to attract labour from over a wider area.

Having established that plant size and quits are not systematically related, we are left with the following equation:

$$Q_{js(t)} = a + b_1 W_{js(t)} + b_2 A_{js(t-1)} + \varepsilon \qquad \text{(1a)}$$

The partial regression coefficients obtained for W and A are shown in Table 6.7 along with the coefficient of multiple determination (R^2) for the regression equation. The results are shown for each quarter. Within each quarter the first, second and third rows show the regression coefficients for Birmingham males, Birmingham females and Glasgow males respectively.

The R^2 obtained from equation (1a) is generally lower for Glasgow males than for Birmingham males and females. In the former case, R^2 is significant in 12 of 16 quarters at the 5 per cent level and in 6 cases at the 1 per cent level. These levels of significance are achieved in 15 and 13 cases for Birmingham males and in 14 and 9 cases for Birmingham females. Hence, while the results obtained from (1a) are generally statistically significant for all groups, the equation 'explains'

[1] It may be objected that too few small establishments are included to warrant such a conclusion. Yet the spread of plants by size was fairly considerable. In the case of Birmingham males, data for 14 plants were generally available to calculate the regression equations. Their distribution by number of male employees was, 3, less than 100; 7, 100–499; 2, 500–999; 2, 1,000 or more employees. The relevant figures for Glasgow were 3, 9, 3 and 2 respectively.

[2] See pp. 251–2 below.

a greater proportion of the variance in plant quit rates in Birmingham. This is due to the stronger relationship in Birmingham between quits (*Q*) and the recruitment rate in the previous quarter (*A*).

TABLE 6.7

RESULTS OF REGRESSION EQUATION: PLANT QUIT RATES, AVERAGE EARNINGS AND RATES OF RECRUITMENT: BIRMINGHAM MALES AND FEMALES, AND GLASGOW MALES

Quarter	*W*	*A*	R^2	Quarter	*W*	*A*	R^2
	+0·73	+0·76*	0·54*		0·00	+0·50*	0·50*
2	−0·08	+0·34*	0·33*	18	−0·18	+0·25*	0·58*
	−0·35	+0·19	0·07		−0·53*	+0·20	0·32*
	−0·38*	+0·50*	0·81*		−0·22	+0·42*	0·67*
4	−0·04	+0·42*	0·52*	20	−1·26	+0·19	0·46*
	−0·98*	+0·08	0·30*		−0·09	−0·14	0·12
	−0·21	+0·77*	0·90*		−0·66*	+0·58*	0·80*
6	−0·34	+0·47*	0·55*	22	−1·14	+0·40	0·62*
	−0·42	+0·37*	0·44*		−0·07	+0·05	0·04
	+0·28	+0·30*	0·56*		+0·33	+0·46*	0·36*
8	+0·41	+0·50*	0·58*	24	−0·66	+0·45*	0·75*
	−0·59*	+0·13*	0·57*		−0·76	+0·32*	0·41*
	+0·87	+0·83*	0·65*		−0·51	+0·02	0·14
10	+0·40	+0·32*	0·76*	26	−0·04	+0·41*	0·46*
	−1·33*	+0·29*	0·35*		−0·58	+0·23	0·30*
	−0·36	+0·17	0·29*		+0·47	+0·55*	0·53*
12	−0·40	+0·11	0·17	28	−2·72*	+0·13	0·52*
	−0·71*	+0·05	0·51*		−0·99*	+0·32	0·31*
	−0·21	+0·32*	0·61*		−0·06	+0·37*	0·50*
14	−1·89	+0·20	0·47*	30	−0·93	+0·19	0·29*
	−0·14	+0·17*	0·24*		−0·81*	+0·43*	0·86*
	−0·09	+0·17*	0·40*		−0·32	+0·17*	0·43*
16	−1·02	+0·34	0·18	32	−1·04	+0·09	0·28*
	−0·13	+0·22	0·08		−0·43*	+0·19	0·43*

Note: * Significant at the 5 per cent level.

Although the partial regression coefficients for *A* are almost invariably positive, the coefficients are larger for Birmingham males and females and are more frequently statistically significant at the 5 per cent level. Hence a given difference between plant recruitment rates will produce more substantial differences in plant quit rates in the subsequent quarter in Birmingham than in Glasgow. This arises because a higher proportion of new recruits in Birmingham leave after only a short period of service.[1] Or, to put it the other way around, a smaller proportion survive the induction crisis.

The stronger relationship between Q and A in Birmingham also arises from another related reason. Birmingham plants tend to have higher quit rates than establishments in Glasgow[2] and, therefore, have to recruit more heavily to maintain their labour forces at their existing levels. It follows that Birmingham plants have a greater proportion of employees who are likely to, and do, leave within a fairly short period. The contribution of A towards R^2 is, therefore, greater in Birmingham than in Glasgow, partly because the same rate of recruitment will produce a higher quit rate in the short run and partly because recruitment rates are higher in the former area. This has all the ingredients of the classic vicious circle. A high quit rate produces a high recruitment rate which, in turn, begets a high quit rate. The evidence of the regression equation indicates that this phenomenon is especially important in 'tight' labour markets, and this receives further confirmation in Chapter 8.

The relationship between plant quit rates and earnings is strongest for Glasgow males. For this group, the partial regression coefficients for W are always negative, are usually larger than those found for Birmingham males and females and are more often significant at the 5 per cent level. On average, this level of significance is achieved in one quarter in two for Glasgow males but only twice for Birmingham males and only once for Birmingham females. The weak relationship between plant earnings and quit rates in the 'tight' Birmingham market can be illustrated by considering the variance in Q 'explained' by W after allowance has been made for A. The introduction of W as an additional explanatory variable to A increases the coefficient of determination by more than 0·10 in only two and three quarters for Birmingham males and females respectively, compared to 11 quarters for Glasgow males.

These findings, instead of supporting the hypothesis that quits will be more sensitive to earnings differentials in 'tight' labour markets, suggest precisely the opposite conclusion. This occurs because, as labour markets tighten, quit rates are increasingly composed of employees who change employers frequently.[3] These groups appear to be little influenced by earnings differentials. In 'tight' markets like

[1] See pp. 211–2 below. [2] See Table 7.1, p. 169 below.

[3] See Parnes, *op. cit.*, pp. 65–6 and 96–7; Palmer, *op. cit.*, pp. 35–8; and Jefferys, *op. cit.*, pp. 55–7.

Birmingham, a plant recruiting labour is likely to lose a high proportion of these new recruits within a short span of time. This appears to hold almost irrespective of whether a plant offers high or low average earnings. The effect of past recruitment on current quit rates is, therefore, strong. Indeed, it must be stronger than our statistical results suggest, for these are derived by representing past recruitment by the simple but rather unsatisfactory process of measuring the rate of recruitment in the previous quarter.

The behaviour pattern which underlies our results seems to be as follows. In a 'tight' labour market, plants seeking new labour will be forced to recruit workers who are extremely mobile. For other groups also the induction crisis produces a higher quit rate because the individual who is dissatisfied with his new employment can more easily 'vote with his feet'. A high recruitment rate, therefore, leads to a high quit rate and this, in turn, to a high recruitment rate so that the cycle is repeated. A plant offering higher earnings is likely to obtain some advantage through lower quit rates and hence through lower recruitment, selection and training costs. Again, if an employer wishes to increase his labour force he may first have to improve his plant's position in the inter-plant earnings structure. But there are no grounds for supposing that such an adjustment is a sufficient condition for obtaining and retaining new recruits. Earnings may be important, but they are by no means the only factor at work. Other policies may also have to be adopted to ensure success—more intensive recruitment policies, better selection methods, improved induction procedures and so forth. Without these, the effect of higher earnings may be nullified or, at least, largely frustrated.

7. CONCLUSIONS

(i) The degree of dispersion displayed by average plant quarterly quit rates was greater for Birmingham males and females than for Glasgow males. There was little difference between the lowest plant quit rate in each market, but certain units in Birmingham had extremely high quit rates, much above the maximum rate in Glasgow. Within each labour market, differences between plant quit rates were greatest at low levels of unemployment and were lowest when unemployment was high. Plant rankings by quit rates showed a fairly high degree of stability over time, but substantial changes in rankings did occur and these were not always reversed quickly. Plants with low quit rates for skilled males tended to have low quit rates for semi-skilled males and vice versa, but the relationship between plant quit rates for skilled and unskilled males was weaker.

(ii) There was no systematic relationship between average weekly plant earnings (including overtime) for males and females and plant recruitment rates. This holds whether the independent variable, earnings, is defined as the level of earnings or as the percentage or absolute change in earnings. For Birmingham males and females, none of the wage variables was systematically related to net changes in employment. There is some evidence for Glasgow males of a positive relationship between plant earnings variables and changes in employment, but the relationship was extremely weak. Hence plants do not offer high earnings *primarily* because they wish to obtain a large proportion of the available labour supply and/or because they wish to increase employment relative to other units.

(iii) For each group considered—Glasgow males and Birmingham males and females—plant quit rates (and separation rates) were a decreasing function of the *level* of plant earnings (including overtime). Quit rates (and separation rates) were *not* systematically related to absolute or percentage *changes* in plant earnings. Thus, differences in plant earnings are not wholly offset by non-pecuniary factors which equalize net advantages. At any point in time, the market is in a greater or lesser degree of disequilibrium, and the appropriate variable which influences behaviour is the level of earnings rather than changes in that level. Nonetheless, while the negative relationship between the level of plant earnings and plant quit rates emerges quite clearly, it is apparent that differences in plant earnings do not explain all, or even the major part of, variations in plant quit rates.

(iv) The conclusions of (ii) and (iii) above, which are based on data distinguishing between male and female employees only, are reinforced by a more detailed analysis by male occupational groups. No systematic relationship emerged between occupational recruitment rates and occupational earnings, whether earnings are measured by including or excluding overtime or whether the earnings variable is defined as the level of earnings or changes (percentage or absolute) in that level. Similarly, there was no systematic relationship between changes in employment by occupational groups and any of the earnings variables, and occupational quit rates and separation rates were not responsive to absolute and percentage changes in occupational earnings. A negative relationship once again emerged between occupational quit rates (and separation rates) and the level of occupational earnings, but the strength of this relationship was no greater than that found for male and female employees in aggregate.

(v) Large plants did tend to have lower quit rates than smaller units, but this arose because larger plants tended to have higher earnings levels. When the latter factor is taken into account, it is apparent that size by itself has no effect on quit rates. The relationship between plant size and earnings may be due to economies of scale, but any advantage enjoyed by larger units in this respect must to some extent be dissipated, because such units must offer higher earnings to offset the cost and inconvenience of the longer travel-to-work journeys of their employees.

(vi) A positive relationship emerged between the plant quit rate in the current quarter (Q) and the rate of recruitment in the previous quarter (A). The relationship was strongest in Birmingham where the level of plant earnings (W) had relatively little explanatory power. In Glasgow, plant quit rates were much more sensitive to variations in plant earnings. There is, then, no evidence to support the hypothesis that quit rates are more sensitive to differences in plant earnings in 'tight' labour markets. On the contrary, it can be argued that the relationship between Q and W may be weaker in more fully employed labour markets. It also follows that the current level of quits is a function of the time distribution of past recruitment. In the model tested in this chapter, a strong relationship between these variables emerged, even although we can only represent past recruitment by the crude approximation of the recruitment rate in the previous quarter.

CHAPTER 7

EMPLOYMENT CONDITIONS AND LABOUR TURNOVER

1. INTRODUCTION

The preceding chapter described and analysed variations in turnover rates between plants in the same labour market area. The discussion was concerned with explanatory variables such as wage levels which varied from one establishment to the next and which could, therefore, be expected to give rise to differences in plant turnover rates. The average, or 'market',[1] rate of turnover for all plants in a given labour market area was taken as given, and no attempt was made to establish whether this market rate differed in the separate labour market areas or in the same labour market area over time. These latter aspects of labour turnover form the subject-matter of the present chapter. We are concerned with those factors which can be expected to influence turnover rates in all or most plants in a systematic fashion, thus producing high or low turnover rates in the market as a whole. In other words, differences by area in the market rate of turnover are likely to be due to influences 'external' to the individual establishments, particularly the employment conditions prevailing in the labour market—the ease or difficulty with which individual employees can obtain alternative employment.

We must immediately distinguish between the voluntary quit rate and the total separation rate. Employment conditions are likely to affect the market quit rate, because the decision to leave a plant will be influenced by the ease or difficulty with which alternative employment can be obtained. Thus when the market is 'tight' those dissatisfied with their present job are likely to seek and to be able to obtain work in other establishments. In consequence, the market quit rate will rise. On the other hand, when the market is 'slack' and when few plants are recruiting additional labour, individuals who leave their current job run a greater risk of unemployment. This will tend

[1] Throughout the following text, the terms 'market' quit rate or 'market' separation rate are merely a convenient shorthand for the weighted average quit or average separation rate for the case-study plants in a given area.

to discourage quits even amongst those who are dissatisfied with their current job. Our basic hypothesis is, then, that the voluntary quit rate will be high when employment conditions are favourable and low when employment conditions are unfavourable. We would, therefore, expect the market quit rate in Birmingham to have been, in general, higher than the market quit rate in the other labour market areas. Again, in each area the market quit rate is likely to have varied over time along with changes in employment conditions.

Voluntary quits do not, of course, account for all separations. There are other groups of persons who leave a plant for quite different reasons—redundancy, unsuitability, misconduct, retirement, sickness or death. None of these groups can be supposed to make a choice between staying with or leaving their present employment. They leave establishments through force of circumstance and, unlike voluntary leavers, have little control over the timing of their departure. The number of redundancies is likely to rise when employment conditions worsen, and it may be that the number dismissed for unsuitability or misconduct will also increase in such circumstances, as might the number of retirements. If such separations increase while voluntary quits fall, then the change in the total separation rate will depend on the strength of these opposing tendencies. The outcome cannot be predicted by reference to theoretical considerations,[1] and the empirical data from other studies do not reveal any clear-cut relationship between total separations and employment conditions.[2] Hence this relationship requires investigation, as does the relationship between voluntary quits and separation rates.

The data on which this chapter is based are briefly described in the following section. Section 2 also provides a preliminary analysis of the levels of, and the relationship between, voluntary quits and separations, and suggests explanations for the differences observed between the labour market areas. Sections 3, 4 and 5 deal with variations in market quit rates and separation rates over time by labour market area. The analysis is focused on Birmingham and Glasgow and particularly on the relationship between voluntary quits and employment conditions. Having established the nature of this relationship, we then consider whether the same relationship holds between total separations and employment conditions. Section 3 investigates how employment conditions might be defined and measured, while Section 4 specifies the model in greater detail by con-

[1] Voluntary quits were, however, more important than involuntary separations in our labour market areas and accounted for some 70 per cent of all male separations in each area. See Tables 7.1 and 7.2, p. 169 below.

[2] *Wages and Labour Mobility*, pp. 65–7.

sidering the manner in which changes in employment conditions might be expected to influence quit rates. Alternative hypotheses are established in Section 4, which are then tested in Section 5 to establish which variable or group of variables provides the best explanation of fluctuations in quit rates. After certain preliminary ideas have been rejected, Section 6 describes the empirical results obtained and their implications. The main conclusions are summarized in Section 7.

2. QUIT RATES AND SEPARATION RATES

The nature of the data on which the following analysis is based has already been described in some detail, but we should explain what we mean by the phrases 'market' quit rate or 'market' separation rate. In the last chapter, plant turnover rates were obtained by expressing the number of persons leaving or joining an establishment as percentages of the number of persons employed by the plant at the beginning of each quarter. In like manner, 'market' turnover rates can be calculated for all plants in a given labour market area. For example, the market separation rate is simply the total number of leavers from all plants in a quarter as a percentage of the total number of persons employed by those plants at the beginning of that quarter. As in Chapter 6, distinctions can be made between male and female employees and between various occupational groups.

Here, we are exclusively concerned with market quit rates and market separation rates which, following past practice, we shall term for convenience 'wastage' rates. Where the following discussion deals with the market separation rate for all leavers, no special problems of analysis arise. The measurement of the voluntary quit rate does, however, raise in a slightly different form a problem encountered in the previous chapter. This is the existence of a number of persons whose reason for leaving was not known. In Glasgow and North Lanarkshire, only small numbers were involved and no adjustment of the known voluntary quit rate was thought to be necessary. In Birmingham, however, 'not-known' leavers amounted to 9·1 per cent and 12·1 per cent of male and female leavers respectively. If this group was simply ignored, the voluntary quit rate calculated for Birmingham would be less than the 'true' quit rate, and it would be difficult to make valid comparisons between Birmingham and the other labour market areas. The above problem was dealt with, as in Chapter 6, by assuming that for any group under consideration—males by skill groups, total males or females—not-known leavers contained the same proportion of voluntary leavers as those whose reason for leaving was known. Quarterly quit rates for all quarters were

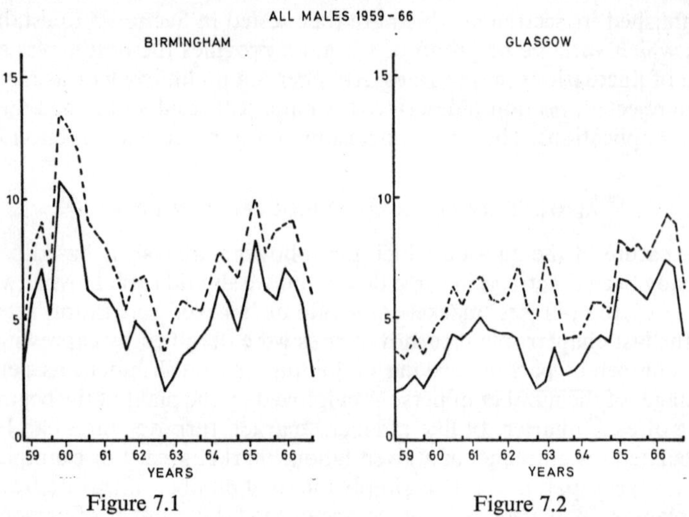

Figure 7.1

Figure 7.2

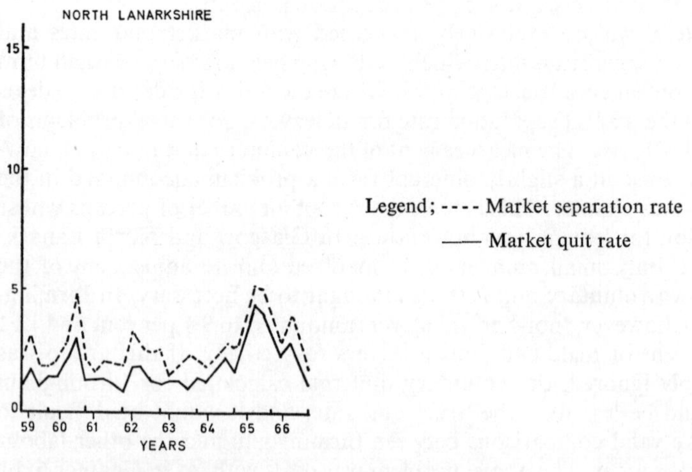

Figure 7.3

Figures 7.1–7.3. Quarterly market quit rates and separation rates: all males, 1959–66.

adjusted on this basis, and all quit rates used for Birmingham are to be understood to include an allowance for those whose reason for leaving was not known.[1]

We begin our examination of separation rates and quit rates with Figures 7.1 to 7.3. The figures show, for males in Birmingham, Glasgow and North Lanarkshire respectively, the market quit rate and market separation rate for each quarter over 1959–66.

TABLE 7.1

AVERAGE QUARTERLY MARKET QUIT RATES

Area	Males				Females
	Skilled	Semi-skilled	Unskilled	All	
Birmingham	3·1	6·6	8·0	5·5	9·0
Glasgow	3·8	5·1	4·1	4·2	5·9
North Lanarkshire	2·4	1·8	1·7	1·9	4·2
Small Town	1·6	1·5	1·3	1·4	1·6

TABLE 7.2

AVERAGE QUARTERLY MARKET SEPARATION RATES

Area	Males				Females
	Skilled	Semi-skilled	Unskilled	All	
Birmingham	4·1	8·6	12·2	7·6	12·9
Glasgow	5·6	6·9	6·7	6·0	8·4
North Lanarkshire	3·0	2·6	3·8	2·7	6·5
Small Town	2·6	2·3	2·9	2·3	4·4

The level of separation and quit rates varied substantially from area to area. This is most obvious when comparing Birmingham with North Lanarkshire. The market separation rate in the former area fell below 4 per cent in only one quarter and the voluntary quit rate was below that level in only one quarter in four. In contrast, market quit and separation rates of less than 4 per cent were the rule rather than the exception in North Lanarkshire. This is consistent with our

[1] The formula applied is $V_t = V_k + \frac{V_k}{S-L} . L$, where V_t is the 'true' voluntary quit rate; V_k the quit rate for those *known* to have left for voluntary reasons; S the separation rate and L the not-known leaving rate.

expectations. 'Tight' labour market conditions encouraged employees to change employers more readily in Birmingham, while in North Lanarkshire employment opportunities were much less favourable, so that employees showed a greater reluctance to discard a job once employment had been obtained.

It can also be seen that the market quit rate and the market separation rate were generally lower in Glasgow than in Birmingham. Again, this is in accordance with expectations, but wastage rates were substantially higher in Glasgow than in North Lanarkshire despite the fact that the level of male unemployment was similar in both areas.[1] We examine this further and break down the rather aggregative figures on which Figures 7.1 to 7.3 are based in Tables 7.1 and 7.2. Table 7.1 shows the average quarterly market quit rates over 1959–66[2] for each category enumerated and for each labour market area, save New Town.[3] Comparable information for average market separation rates is contained in Table 7.2.

In all labour market areas both the quit rate and the separation rate were highest for females. The difference between male and female quit rates was substantial in all areas save Small Town and is not to be explained away simply by differences in the occupational composition of the labour force. Females, who were almost exclusively semi-skilled, had a higher quit rate than semi-skilled males in each market area. Recruitment, selection and training costs are, therefore, likely to have been particularly heavy in the case of female workers.

The average of the market quit rate and the market separation rate for females and for all males was higher in Birmingham than in any other area. Whichever group is considered, quit rates and separation rates were always lowest in Small Town. To this extent, then, the results are consistent with our hypothesis that leaving rates are influenced by employment conditions, for male and female unemployment was lowest in Birmingham and highest in Small Town over 1959–66. Yet for all groups, quit rates and separation rates were substantially higher in Glasgow than in North Lanarkshire although employment conditions, as measured by unemployment and vacancy statistics, were very similar in these areas.

This raises the prospect that some other factor was exerting a disturbing influence. One possibility is that the nature of the labour

[1] The quarterly rate averaged 5·2 per cent in both areas over 1959–66. Engineering unemployment, which averaged 1·9 per cent, was *lower* in North Lanarkshire than in Glasgow at 2·7 per cent.

[2] An unweighted average is used, i.e. the market rate for the 32 quarters is simply summed and divided by 32.

[3] Throughout this chapter, New Town is excluded as, with one exception, the case-study plants did not employ significant numbers prior to 1964.

market, in particular its size, compactness and availability of alternative employers, may affect wastage rates independently of conventional measures of employment opportunities such as the rate of unemployment and vacancies in the market. Because of its size (in spatial terms) and less developed internal transportation networks, North Lanarkshire is not a labour market within which labour moves as freely as in the conurbations. The employee has a wide choice of alternative employment only if he is prepared to move house, whereas in Glasgow (or Birmingham) most employees can choose from a greater variety of jobs and alternative employers simply by varying their travel-to-work journey. As we shall see later, it is possible to distinguish sub-markets even within the conurbations, but as individuals change their travel-to-work habits more readily than their homes, the conurbations are more cohesive labour markets than North Lanarkshire.

An individual wishing to change his job and yet remain working within the same labour market area therefore has a more restricted choice amongst alternative employers in North Lanarkshire than in Glasgow or Birmingham. *A fortiori* the same argument applies to Small Town, especially if the employee wishes to continue working in the engineering industry in which there are only two large employers within reasonable travel-to-work distance. The exceptionally low wastage rates in Small Town are, then, only partially explicable in terms of differences in unemployment and job vacancy rates. It appears highly probable that wastage rates would remain substantially lower in Small Town even if unemployment was pushed down to the level prevailing in the Birmingham conurbation. In other words, when considering differences in wastage rates between labour markets, our analysis will be incomplete if we refer simply to the conventional measures of employment conditions. Differences in unemployment and in job vacancies will be important, but so, too, will the nature of the labour markets considered, particularly their size and spatial characteristics. At similar levels of unemployment, wastage rates will be lower in small, isolated markets and higher in large, compact markets where there is a wide choice amongst alternative employments.

From Tables 7.1 and 7.2, we can see that the pattern of male leaving rates within labour market areas shows interesting differences. In Birmingham, wastage rates rose as the skill of the group considered diminished. In the other labour market areas such a progression was not in evidence at all or, at least, was much more muted. Quit rates and separation rates in Glasgow were lower for skilled men than for semi-skilled and unskilled, but these differences were much less marked than in Birmingham. Moreover, in both North Lanarkshire and Small Town the quit rate was higher for skilled males than for either semi-skilled or unskilled. This is an important finding because previous labour market studies have suggested that the occupational pattern of wastage rates is always similar to that found in Birmingham. It is therefore necessary to establish whether the results for our other labour market areas, and especially for Glasgow, were mirrored at the microeconomic level of the individual plant.

This is done by calculating for each plant, and for skilled, semi-skilled and unskilled males, the average quarterly quit rate over 1959–66. In only 15 of the 27 establishments in Glasgow was the quit rate lowest for the skilled, and the unskilled quit rate was highest in only half of the units. By way of contrast, skilled males had the lowest quit rate in 22 of the 23 Birmingham units,[1] and in 18 cases the quit rate was highest for the unskilled. In both markets, therefore, the pattern of market quit rates merely echoed microeconomic behaviour. Amongst Birmingham plants quit rates were inversely related to skill, but this did not apply in Glasgow units.

The frequency distribution of plant quit rates by occupational groups is shown in Figure 7.4. In Glasgow, the histograms of average plant quit rates have much the same shape for skilled, semi-skilled and unskilled males. For each group, the modal value was low at 1–2 or 2–3 per cent and plant quit rates were heavily bunched within the range 1–8 per cent per quarter. A similar bunching can be observed for skilled males in Birmingham, but the degree of dispersion of plant quit rates for semi-skilled and unskilled was much greater. In every second plant in Birmingham, average quit rates for semi-skilled males exceeded 8 per cent per quarter, and no less than 16 of 23 establishments had unskilled quit rates above this level. In Glasgow, only four of 27 plants experienced semi-skilled or unskilled quit rates in excess of 8 per cent. The 'spread' in plant quit rates was much greater for semi-skilled and unskilled than for skilled males in Birmingham. For the former groups, there was no concentration of plant quit rates at the lower end of the distribution, so that market quit rates for these groups were high, as we have seen from Table 7.1.

This contrast between Birmingham and Glasgow would appear to result from the very 'tight' labour market conditions in the former area. All types of labour have at times been difficult to recruit in Birmingham. It seems plausible that in a 'tight' market quit rates and separation rates are highest for those who are less skilled. The reason

[1] Two plants with a high proportion of persons whose reason for leaving was not known have been excluded.

for this, on the supply side, is that skilled males are likely to possess attributes which are in greater or lesser degree not directly transferable. In other words, when changing employers they may have to accept in the short run a loss of earnings while they acquire further experience more directly relevant to their new work. On the other hand, semi-skilled and unskilled males undertake a narrower range of work which is duplicated in many other establishments. Even if this is not so they can quickly acquire the new skills required by an

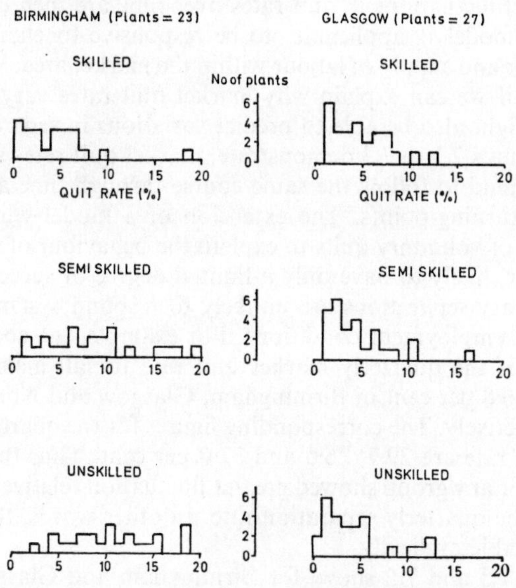

Figure 7.4. Distribution of average plant quarterly quit rates, by occupational groups: Birmingham and Glasgow males

employer. Demand side considerations point in the same direction, for employers, particularly in a 'tight' market, are likely to make special efforts to retain skilled employees who are difficult to replace. Hence a given change in employment conditions will have a greater impact on the wastage rates of less skilled groups.[1]

We began with the hypothesis that the quit rate will vary between market areas according to differences in employment conditions

[1] One can find support for the conclusion by examining within each labour market area variations in quit rates over time. The coefficient of variation of the quarterly market quit rate is in all markets lower for skilled than for either semi-skilled or unskilled males.

within those areas. The evidence put forward above is, in a general sense, consistent with that view, although it is apparent that the nature of the labour market area might exercise an influence independent of employment conditions as measured in the conventional sense by unemployment and vacancy statistics. When we consider changes in quit rates over time *within* the same labour market area, the influence of employment conditions defined in this narrow sense is likely to be more decisive. We can assume, for practical purposes, that the character or nature of a market does not change much in the short run. Fluctuations in quit rates over time are then likely, if the economic model is applicable, to be responsive to changes in the demand for and supply of labour within the market area. We can also note that, if we can explain why market quit rates vary over time, then we might also be able to predict variations in separation rates, for, as Figures 7.1 to 7.3 demonstrate, market quit rates and separation rates tend to follow the same course through time and to have the same turning-points. The extension of a model which fits the behaviour of voluntary quits to explain the behaviour of separations is, however, likely to have only a limited degree of success, because *non*-voluntary separations are unlikely to respond systematically to changes in employment conditions. For example, the coefficients of variation of the quarterly market quit rate for all males are 37·5, 37·1 and 46·8 per cent in Birmingham, Glasgow and North Lanarkshire respectively. The corresponding figures for the quarterly market separation rate are 29·7, 26·5 and 37·0 per cent. Thus the quarterly quit rate for any group showed greater fluctuation relative to its mean than did the quarterly separation rate. In other words, the quit rate was less stable cyclically.

Figures 7.5 and 7.6 show, for Birmingham and Glasgow respectively, the quarterly market quit rate for skilled, semi-skilled and unskilled males. Percentage male unemployment for the wholly unemployed in each market has been plotted on an inverse scale, so that a fall in unemployment will show as a rise in the unemployment graph as charted and vice versa. The results confirm our earlier observation that skill had a greater impact on quit rates in Birmingham than in Glasgow. In Birmingham, skilled males had a low quit rate relative to the semi-skilled and unskilled. Such quit rates differed remarkably little in Glasgow. It is also clear that cyclical fluctuations in quit rates were closely related so that, within the same market, quit rates moved in the same direction over time and shared the same turning-points. From this, it follows that the factors which determined quit rates must have been the same for all groups or, if different, must have moved in the same direction through time. It is interesting to note that, in both Birmingham and Glasgow, quit rates appear to have been inversely related to the level of male unemployment, falling when unemployment increased and vice versa. The same pattern held in the other areas and between female

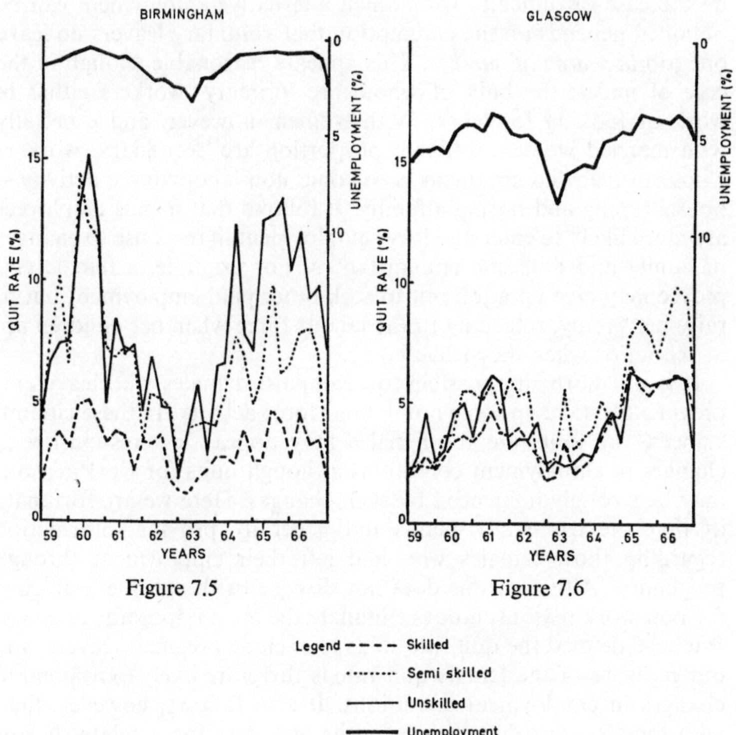

Figure 7.5

Figure 7.6

Figures 7.5 and 7.6. Quarterly quit rate, by male occupational groups; and male unemployment: Birmingham and Glasgow, 1959–66

quit rates and female unemployment. The reasons for this relationship and the possible interactions between employment conditions and quit rates are discussed in the following section.

3. THE MEASUREMENT OF EMPLOYMENT CONDITIONS

So far we have discussed the relationship between employment conditions and quit rates only in the most general fashion. To examine this relationship, if indeed one exists, we have to be more specific in defining the meaning of the term 'employment conditions' and in considering whether this concept can be related to quit rates in a manner which allows us to bring statistical evidence to bear on the problem. The hypothesis that market quit rates will be influenced by the ease or difficulty with which alternative employment can be obtained depends on the assumption that voluntary leavers do leave one job *in search of another*. This appears reasonable enough in the case of males, the bulk of whom are 'primary' workers either in work or looking for work. With women, however, and especially with married women, a higher proportion are 'secondary' workers whose primary commitment is to some non-labour force activity—housekeeping and raising a family. It follows that female employees are more likely to enter and leave employment in response to changes in family and domestic circumstances. For example, a female employee may give up a job not to seek other paid employment but to raise her family, returning to the labour force when her children are at school or when they leave home.

It is not normally possible to distinguish females who leave employment to take up some non-labour force activity. In these circumstances, the quit rate for females may appear less responsive to changes in employment conditions although quits for work reasons may be strongly influenced by such changes. Here we are fortunate in that the case-study plants did generally provide information regarding those females who had left their employment through pregnancy. Although this does not dispose of the problem of quits for non-work reasons, it does eliminate the most important category. We have defined the quit rate so as to exclude pregnant leavers, and our measure of the female quit rate is therefore likely to respond to changes in employment conditions. It also follows, however, that, with females, our definition of quits accounts for a relatively low proportion of all separations. Hence fluctuations in female *separation* rates may be less responsive than male separation rates to changes in employment conditions.

Having considered our dependent variable, the quit rate, the variable whose behaviour we set out to 'explain', we must now consider our independent variables—the factors which might be expected to determine the behaviour of the dependent variable. These factors are held to be employment conditions in the market. How are employment conditions to be defined? Our first task is to set some limits to our market. To distinguish between Birmingham and Glasgow, or between these areas and any other area, implies that job choices are influenced by some factors which are 'local' in content;

that manual workers changing jobs generally do so within a fairly restricted geographical area. This supposition is supported by numerous studies which show that most job changes by manual employees involve only short-distance mobility. There still remains the problem of defining the exact boundaries of the labour markets under investigation,[1] but provided the definitions adopted yield areas in which most employees normally seek work, then we can sensibly treat our areas as in some meaningful sense separate entities. This condition appears to be met, as we shall see later when we consider geographical mobility.

Our hypothesis is, therefore, that voluntary quits will be influenced by the ease or difficulty with which alternative jobs can be obtained within the labour market areas as defined in this study. This, in turn, depends on the extent to which employers are seeking new recruits and on the availability of supplies of labour not currently in employment, and we therefore require some measure of the demand of employers for additional labour and of the supply of labour seeking employment. Here the percentage of unfilled vacancies (V) is taken as a measure of the demand for labour and the percentage unemployed (U) as a measure of the supply of labour seeking new employment. These represent the only indicators of labour demand and supply readily available on a national and local basis, and they have been extensively used to depict the nature of prevailing employment conditions. However, neither series is ideal for the purpose in hand and we must outline the problems involved in their use.

Unfilled vacancies are those job openings notified by employers to the public employment exchanges, which remain (or appear to remain) unfilled at a specified date. Assuming that vacancies do reflect the demand for labour, as employers attempt to recruit more labour the number of unfilled vacancies will rise, and unfilled vacancies will fall as employers reduce recruitment. The initial assumption has often been questioned, on the grounds (confirmed by the evidence available to this study) that the number of unfilled vacancies is likely to understate the demand for labour in most labour market situations, and it is not at all certain that the ratio of unfilled vacancies to the 'true' demand for labour is stable over the cycle or between local labour markets. Further, there appears to be little correspondence between an individual plant's true labour requirements and the vacancies which have been notified by the plant to the employment exchanges.[2] There is some evidence that the areas of

[1] See J. F. B. Goodman, 'The Definition and Analysis of Local Labour Markets: Some Empirical Problems', *British Journal of Industrial Relations*, Vol. 8, 1970.

[2] See pp. 350–6 below.

labour shortages which one could identify from a study of unemployment and vacancy data were, indeed, those in which plants reported greatest recruitment difficulties,[1] but given the 'mis-match' between notified vacancies and actual labour requirements at the level of the individual plant, it would appear that if vacancies do provide a good indicator of employment conditions at the more aggregative local market level, they can only do so through a series of compensating errors. While it would be unwise at this stage to discard unfilled vacancies as an indicator of the unsatisfied demand for labour, it is as well to bear in mind the comment that 'there are good *prima facie* reasons for distrusting the statistics of unfilled vacancies since they neither record transactions nor register decisions. . . .'[2]

On the supply side the first problem is the definition of unemployment which should be adopted. In Birmingham very substantial fluctuations in unemployment occurred over 1959–66 when workers were 'laid-off' by motor car assembly and component plants. These individuals are counted as 'temporarily stopped'[3] in unemployment statistics, and their exclusion to yield that category known as the 'wholly unemployed' would reduce the degree of cyclical instability shown by those statistics. This can be justified on the following grounds. The temporarily stopped, at least in the short run, are persons who expect to return to their previous jobs and are unlikely to look for new ones. Other persons considering leaving their present employment may not, then, regard the temporarily stopped as potential competitors for jobs. Of course, the longer the period of lay-off the more difficult will it become to distinguish the temporarily stopped from the wholly unemployed. Yet in our period the sharp rises in male unemployment in Birmingham caused by increases in the number of temporarily stopped were always reversed in whole or in large part by the following quarter. In other words, the lay-offs were of short duration. Because of these circumstances, employment conditions facing male workers in Birmingham are likely to be better represented by trends in the number wholly unemployed. On theoretical grounds, therefore, there is good reason for preferring a definition of unemployment which excludes the temporarily stopped. This definition was, therefore, adopted for all areas,[4] and all subsequent references to unemployment must be understood in this light.[5]

[1] See pp. 61–3 above. [2] Dow and Dicks-Mireaux, *op. cit.*, p. 2.

[3] The definition for 'temporarily stopped' is given on p. 57, n. 1 above.

[4] In fact, unemployment rates in Glasgow remain the same whether they include or exclude the temporarily stopped. For Birmingham females the exclusion of the temporarily stopped does have an effect on unemployment rates, but this is much less important than in the case of Birmingham males.

[5] For Birmingham males and females the regression analysis described below

Unemployment statistics are likely to be more reliable than vacancy data because an unemployed person has to satisfy an 'activity' criterion—registration at an employment exchange as looking for work—and because non-registration would in many cases involve a financial loss (i.e. loss of unemployment benefit). Nonetheless, unemployment statistics may not, as we have seen, include all active job-seekers, particularly in the case of married women who have opted out of National Insurance schemes and who may seek work through channels other than the exchanges. Again, because of the 'discouraged worker' effect, it appears likely that the slacker the market the more will unemployment underestimate the potential supply of new recruits.[1]

Since unemployment provides safer ground than vacancy data, would it not be preferable to use unemployment alone as a measure not merely of labour supply but of employment conditions in general? After all, when we use the term employment conditions we in some sense mean the balance between labour demand and supply, and if this balance could be represented by U, then V would be superfluous. The level of unemployment is used in precisely this manner in Phillips curve analysis.[2] U is taken to represent the excess demand for labour,[3] i.e. the extent to which labour demand exceeds labour supply. However, unemployment should be taken as a proxy for excess demand or labour market balance only if no practical alternative is available.[4] Its use involves certain propositions regarding the behaviour of different types of unemployment, particularly frictional unemployment, which cannot easily be tested.[5] Possibly more important, if U is used to measure excess demand this assumes that the cause of any change in excess demand, whether it originates

was carried through using both definitions of unemployment. For males, particularly, the results were always better if the 'temporarily stopped' were excluded.

[1] Unpublished American research suggests that a wider definition of unemployment may be appropriate in Phillips curve analysis (see E. S. Phelps, 'Money-Wage Dynamics and Labor Market Equilibrium', *Journal of Political Economy*, Vol. 76, 1968, p. 684). The same may well be true in our field of interest.

[2] A. W. Phillips, 'The Relation between Unemployment and the Rate of Change of Money Wage Rates in the United Kingdom, 1861–1957', *Economica*, N.S. Vol. 25, 1958.

[3] We have continued to use the original terminology to avoid confusion. The reader will note, however, that the phrase 'excess demand for labour' is a rather peculiar one to adopt in this context. 'Excess supply of labour' or 'the balance between labour demand and supply' would have been better.

[4] R. G. Lipsey, 'The Relation Between Unemployment and the Rate of Change of Money Wage Rates in the United Kingdom, 1862–1957: A Further Analysis', *Economica*, N.S. Vol. 27, 1960, p. 14, n. 1.

[5] See B. Corry and D. Laidler, 'The Phillips Relation: A Theoretical Explanation' *Economica*, N.S., Vol. 34, 1967.

from the demand or supply side, is irrelevant.[1] There appears to be little warrant for this assumption.

In this study, we are not forced to rely on U as a measure of excess demand. Vacancy data *are* available, and it would only be sensible to use U as a measure of employment conditions if the inverse relationship between U and V, which is likely to exist, were perfect. The use of both U and V does raise problems of collinearity between our independent variables, but, as will be seen later, it is more satisfactory than a procedure relying on U or V alone. For rather similar reasons, we reject the traditional method of measuring employment conditions by means of some ratio of U and V such as V/U or $(V-U)/U$. Whereas a ratio overcomes the problem of collinearity between U and V, it is not satisfactory if, as appears likely, errors of measurement are greater in one of the terms, in this case the numerator. Moreover, a ratio does not allow us to investigate which of the variables U or V contributes most to an explanation of quit rates. To adopt a ratio to measure employment conditions assumes, as does the use of one variable (e.g. U), that a change in the ratio has the same effect irrespective of whether it is produced by a change in V or the opposite change in U. In practice, this assumption is not justified. Unfilled vacancies seem considerably to understate the true demand for labour, and the discrepancy in this case appears likely to be much larger than any discrepancy between unemployment and the actual number seeking new work. This being so, a given change in V/U or in $(V\text{-}U)/U$, caused by a change in V, is likely to denote a much more substantial shift in employment conditions than the *same* change in these ratios cause by a change in U.

4. THE MODEL

Labour mobility involving a change of employer is often analysed under three headings—geographical, industrial and occupational. We have already considered the former aspect and have concluded that, since manual workers generally seek work within a fairly restricted locality, we can represent the employment conditions which influence job choice by utilizing unemployment and vacancy data based on our local market areas. Which sets of unemployment and vacancy data are appropriate, however, depends on the industrial and occupational characteristics of job changes. If individuals were indifferent between jobs in different industries and occupations, then

[1] '. . . a given excess demand should cause a given rate of change of price *whatever the reason for the excess demand*—whether demand shift, a supply shift, or a combination of both'. Lipsey, *op. cit.*, p. 13 (italics in the original).

those leaving engineering plants would be scattered amongst alternative employments according only to their importance in the local market. In this circumstance, the factors which determined the voluntary quit rate of engineering workers would not be specific to particular industries or occupations. The crucial determinants of voluntary quits would be the *general* availability of work and the unemployment rate in the market as a whole. If we treat males and females separately, on the assumption that they do compete in different labour markets, then the relevant variables would be the unemployment and vacancies rates for all males and females respectively. We can represent these variables by U_s and V_s where the subscript s denotes sex; the subscript m and f will be substituted for s when we wish to distinguish between male and female unemployment and male and female vacancies.

It follows from the above that voluntary quits will become less sensitive to changes in general employment conditions, the greater is the tendency for individuals to remain in the same industry when changing employers. If employees who leave engineering establishments tend to look for new work with another engineering unit, the relevant factors influencing quits would be the rate of unemployment and the percentage of unfilled vacancies in the *engineering industry*. Thus we can write $Q_{es} = f(U_{es}, V_{es})$ where the subscript e represents engineering, and to examine whether U_s and V_s rather than U_{es} and V_{es}[1] determine quits, we have the regression equation:

$$Q_{es} = a+b_1U_s+b_2V_s+b_3U_{es}+b_4V_{es}+\varepsilon \qquad (1)$$

Q_{es} represents the engineering quit rate by sex; U_{es} and V_{es} the rate of unemployment and the percentage of unfilled vacancies in the engineering industry by sex and ε the error term.

Equation (1) is used to predict quit rates of engineering workers when the only distinction made is between all male and all female workers. Male engineering workers do, however, possess a wider range of skills than females, who are overwhelmingly semi-skilled, and it is therefore desirable to consider whether occupational attachment has, in the case of males, an influence on the pattern and timing of quits. If mobility between skill groups is restricted for whatever reason, and if individuals on leaving an engineering plant look for work of similar skill within or without the engineering industry, then behaviour will be influenced by the employment conditions faced by the different occupational groups.

[1] At a later point in this analysis we will find it necessary, because of collinearity, to subdivide equation (1) into two distinct equations. For ease of exposition, however, it is best to continue for the moment with equation (1).

Here a problem of measurement arises. It was possible to obtain the number of unemployed and the number of unfilled vacancies in occupational groups corresponding to our definitions of skilled, semi-skilled and unskilled engineering workers. Unfortunately, no data exist for local labour market areas of the total number of insured employees in these occupational groups. It is not possible to calculate percentage unemployment and percentage unfilled vacancies by occupational groups, and the use of absolute numbers is clearly inappropriate, but if occupational attachment is significant, it is important to find some measure of the employment conditions facing occupational groups.

This can be provided by expressing the number unemployed in any occupational group as a ratio of the number of unfilled vacancies in that occupational group. For reasons already stated, the procedure is not ideal, but it has a certain logic as a second-best solution. If the number of unemployed is greater than the number of vacancies for a particular occupational group, then individuals at that skill level will be reluctant to leave their current employment if they wish to obtain a job of similar skill. Conversely, if the number unemployed is less than the number of vacancies, then competition for jobs at that skill level will be less severe and quit rates will tend to rise.

In considering the market quit rate for skilled, semi-skilled and unskilled male workers, we can express the quit rate for any group as a function of employment conditions in the market as a whole, in the engineering industry and in the particular occupational group under consideration. We have therefore:

$$Q_{eo} = a+b_1U_m+b_2V_m+b_3U_{em}+b_4V_{em}+b_5(U/V)_o+\varepsilon \qquad (2)$$

Q_{eo} represents the engineering quit rate by male occupational group and $(U/V)_o$ represents the ratio of the number of unemployed to the number of vacancies by male occupational group.

Before proceeding to an examination of the statistical results obtained, we have to consider certain modifications which might be made to equations (1) and (2). The first is whether the quit rate will be influenced by the rate of change of unemployment and/or the rate of change of vacancies,[1] as well as, or instead of, the level of unemployment and of vacancies. We introduce, in other words, a set of 'expectational' variables on the following hypotheses. For any

[1] Strictly speaking, the following formulation measures the direction *and* the rate of change of a variable. Subsequently dummies were introduced to represent the *direction* of change, only, of the independent variables. These dummies had no explanatory power.

level of unemployment there will be two quit rates: a higher quit rate which will occur when unemployment has *fallen* to its present level and a lower quit rate when unemployment has *risen* over the preceding period to reach its present level. The position is reversed where vacancies are considered. The lower quit rate is associated with a *falling* vacancy rate and vice versa.

The reasoning underlying the above propositions is easily grasped. Assume that employees expect trends over some previous period to be continued in the future. If the level of unemployment has been rising, they will expect it to continue to increase in the future, and they will be more reluctant to leave their current employment because they will expect more intense competition for jobs and an increased risk of unemployment. When unemployment is falling, expectations are favourable, thus encouraging a greater volume of voluntary quits. By a similar process we assume favourable expectations when unfilled vacancies are increasing, and unfavourable expectations when unfilled vacancies are falling. We therefore expect a negative relationship between quit rates on the one hand and the rate of change of unemployment and the rate of change of $(U/V)_o$ on the other. A positive association is postulated between quit rates and the rate of change of unfilled vacancies. To test these propositions, the independent variables in equations (1) and (2) can be transformed to yield rates of change measured both by first differences and by half the first central differences.[1]

The second modification of equations (1) and (2) arises because of the possibility that there is a lagged relationship between quit rates and labour market conditions. Up to this point, we have assumed that quit rates will respond to employment conditions currently ruling in the market. Individuals may, however, take some time to recognize that a change in employment conditions has occurred. Again, an employee may be reluctant to leave his current job without first securing an offer of alternative employment, and the process of the job-search may introduce a lagged response to changes in employment conditions even where changes in employment conditions are immediately recognized. For these reasons, we introduce a simple one-period time-lag to test whether the quit rate in the current period (t) is a function of employment conditions in the previous period ($t-1$).

[1] Let t represent the 'current' period, $t-1$ the 'preceding' period and $t+1$ the 'next' period. Then the rate of change of male unemployment (U_m) can be measured by $U_{m(t)} - U_{m(t-1)}$ or by $\frac{U_{m(t+1)} - U_{m(t-1)}}{2U_{m(t)}} . 100$.

5. LAGS, EXPECTATIONS AND OCCUPATIONAL CHARACTERISTICS

Equation (1) assumes that the quit rates of males and females are responsive to employment conditions currently ruling in the market,[1] but we have indicated that quit rates may respond to two other sets of variables relating to the rate of change of employment conditions and employment conditions in the previous period. It would be possible, therefore, to rewrite equation (1) to yield a dozen independent variables, while equation (2) could be expanded to include 15 independent variables. However, in view of our limited number of quarterly observations—32 in all—such an approach is not practicable. It is necessary to proceed by testing independently the relationship between quit rates and each of the three sets of independent variables. We do this for the following groups only: all females in Birmingham, and males in Birmingham and Glasgow both in total and distinguished by occupational groups. The other areas are excluded because the small number of plants in each area would introduce excessive sampling errors in quit rates over time not associated with changes in employment conditions.

The procedure adopted to determine which were the significant explanatory variables was one of backward elimination.[2] For example, when investigating the relationship between employment conditions and the quit rate for all males or for females in a labour market area, four regression equations were calculated: (1) as written above; a lagged form of that equation; and (1) transformed to yield the rate of change of the independent variables measured by first differences and by half the first central difference. Inspection of the t-statistic indicated which variable in *each* equation appeared least significant. This variable was then dropped and the equation re-run. If all the resulting t-statistics then proved to be significant at the 5 per cent level, then that specification of the equation was accepted. If, however, one or more t-statistics were not significant the process of elimination was repeated. The factors affecting the quit rate of males by occupational groups were investigated by an identical process.

In each labour market and for each group examined, the above process revealed no systematic relationship between quit rates and

[1] As explained previously, quit rates are calculated for quarterly intervals. The quarters are January–March, April–June, July–September and October–December. Unemployment and vacancy data always relate to observations close to the beginning of the third month in each quarter.

[2] For a discussion of step-wise regression techniques see N. R. Draper and H. Smith, *Applied Regression Analysis* (1966), esp. Chapter 6.

those variables chosen to reflect 'expectational' factors. Hence, when the rate of change of the independent variables is substituted in equations (1) and (2), the regression coefficients for the transformed variables are always extremely small, seldom in excess of their standard error, and the sign is frequently not in the expected direction. This holds irrespective of whether expectational variables are represented by first differences or by half first central differences. We can conclude that quit rates are not influenced by the direction and rate at which employment conditions are changing, and that employees, when considering whether to leave their current employment, do *not* form expectations as to what future employment conditions are likely to be by contrasting employment conditions in the present period with those in a previous period. If expectations are important, they are not revealed by our methods of formulating expectational variables and must be shaped in some other fashion. For example, employees might assume that employment conditions currently ruling will continue in the future. In reaching such a judgment they are not influenced by the direction and rate of change of employment conditions. We can, then, reject the hypothesis that at any given level of unemployment the quit rate will be higher when the level of unemployment is falling and lower when the level of unemployment is rising. Similarly, the quit rate is not responsive to the rate of change of unfilled vacancies.

This leaves two possibilities; that quit rates react to employment conditions currently ruling, or that they are a lagged response to employment conditions in some previous period. In fact, a lagged form of equations (1) and (2) usually yields a multiple coefficient of determination (R^2) which is statistically significant. There are, however, problems of interpretation as a high degree of collinearity exists between any variable in its lagged and unlagged form. Furthermore, whatever group or market is considered, R^2 is always substantially higher when equations (1) and (2) are calculated in their original, *unlagged*, form than when a one-period time-lag is introduced. To test whether the introduction of lagged variables into (1) would improve R^2 the process of backward elimination was applied separately to equation (1) in its lagged and unlagged form, and variables were excluded from each equation until the exclusion of any further variable would have produced a statistically significant fall in R^2. At this point, the remaining independent variables were combined in a new equation with a mixture of lagged and unlagged variables. The R^2 obtained from this new equation was seldom much greater than that obtained from (1) in its unlagged form. The same procedure applied to equation (2) yields a similar conclusion: that the introduction of lagged variables does little to improve the fit of the regression equation.

The increase in R^2 from introducing lagged variables in (1) or (2) for Glasgow males is barely perceptible and never approaches a figure which would be statistically significant. In the case of each group of Birmingham males the situation is somewhat different. Lagged forms of V_m, V_{em} or $(U/V)_o$ are not systematically related to quit rates, but there is some slight evidence of a weak, negative relationship between quit rates and U_m and U_{em} lagged by one time period. This was tested at a subsequent stage in the analysis by introducing for Birmingham males a lagged form of U_m into equation (1a) and a lagged form of U_{em} into (1b).[1] Regression coefficients for U_m have the sign which accords with expectations but are never significant at the 5 per cent level. Regression coefficients for U_{em} *are* significant at the 5 per cent level for all male groups, and the improvement in R^2 passes the sequential F test in three of four cases at the 5 per cent confidence level.

The above findings raise the question whether there is sufficient evidence to warrant the conclusion that there is a lagged relationship between the level of unemployment and quit rates. There was no evidence of a lagged relationship for Glasgow males and Birmingham females, and the evidence for Birmingham males is mixed. That any relationship, if it did exist, would be a weak one seems a reasonable conclusion in the case of manual workers. Previous studies suggest that most job changes by manual workers involve a change in workplace rather than residence; that the job-search is usually carried through quickly and that little notice is required, or given, of termination of employment. It therefore seems unlikely that employment conditions in some past period can have more than a marginal effect on quit rates. This being so, we can return to equations (1) and (2) which, as originally formulated, were based on the implicit assumption that quit rates were a function of current employment conditions.

In each market area, equations (1) and (2) were tested separately. Independent variables were discarded on the basis of the sequential F test until only those which were statistically significant were included. Inspection of the variables remaining in the separate regression equations indicated that only one variable, the current level of unemployment in the market (U_s), was included in all equations. However, V_s, U_{es} and V_{es} also appeared frequently and in each market there was a strong positive correlation between U_s and U_{es} and between V_s and V_{es}. One further conclusion arising from an inspection

[1] The equations are set out on p. 187 below.

of the remaining variables was that the independent variable $(U/V)_o$ added little to an explanation of male quit rates by occupational groups over and above that contributed by measures of unemployment and vacancies for all males and for males in the engineering industry.

These findings suggest a reformulation of equation (1). The degree of collinearity between U_s and U_{es} and between V_s and V_{es} is extremely high—the lowest correlation coefficient (r) is $+0{\cdot}89$ between U_m and U_{em} for Glasgow males—indicating that movements in total unemployment and total unfilled vacancies are closely paralleled by movements in unemployment and vacancies in engineering. This is not surprising as in both Birmingham and Glasgow engineering employment accounts for a large slice of all employment, but it does mean that we cannot distinguish whether quit rates are primarily responsive to employment conditions in the market as a whole or in the engineering industry in particular. The high degree of collinearity between the independent variables in equation (1) means that we have to separate out variables relating to the engineering industry from those relating to the market as a whole. Thus we have two sets of independent variables U_s and V_s, and U_{es} and V_{es}, and we can rewrite equation (1) to give

$$Q_{es} = a+b_1U_s+b_2V_s+\varepsilon \qquad (1a)$$

and

$$Q_{es} = a+b_1U_{es}+b_2V_{es}+\varepsilon \qquad (1b)$$

It follows from our previous argument that both equations are likely to yield very similar explanations of variations in quit rate, and for the sake of simplicity we will concentrate on equation (1a), if only because the data for the independent variables are generally more readily available in this case.[1] Reference will, however, be made to equation (1b) when this is thought to be appropriate.

We have already seen that the independent variable $(U/V)_o$, representing the ratio of unemployment to vacancies for the male occupational group under consideration, appears to contribute little towards the explanation of quit rates for male occupational groups, once allowance has been made for the influence of employment conditions measured by unemployment and vacancies for all males and for males in the engineering industry. To test the influence of occupation further, equations (1a) and (1b) were calculated for the three male skilled groups in Glasgow and Birmingham and then

[1] *The Employment and Productivity Gazette* regularly publishes unemployment data by local labour market areas, but no industrial breakdown is provided.

recalculated with $(U/V)_o$ added. In all but one of the twelve equations the increase in R^2 was barely perceptible; in no case was the increase statistically significant.

To a certain extent a weak relationship between quit rates and $(U/V)_o$ has to be expected, since most indicators of labour market conditions are fairly closely related and the relationship between $(U/V)_o$ on the one hand, and U_m and V_m, or U_{em} and V_{em}, on the other, is no exception to the rule. This can be illustrated by considering the relationship between the $(U/V)_o$ and the former pair of variables. For Glasgow males, fluctuations in U_m and V_m 'explain' 72, 61 and 42 per cent of the variation in $(U/V)_o$ for skilled, semi-skilled and unskilled males. Hence, once allowance has been made for U_m and V_m, $(U/V)_o$ is not likely to be strongly related to quit rates. Yet the degree of collinearity between $(U/V)_o$ and the other independent variables is not such as to prevent $(U/V)_o$ moving independently, and is certainly not high enough to explain the extremely weak relationship which emerges between $(U/V)_o$ and quit rates. In the case of Glasgow skilled, semi-skilled or unskilled males, the addition of $(U/V)_o$ to equation (1a) improves R^2 by a maximum of only 3 per cent. No statistically significant relationship exists even in the case of unskilled males in Glasgow where U_m and V_m 'explain' less than one-half of the variations in the ratio of unskilled unemployed to unskilled vacancies and where, therefore, $(U/V)_o$ might have been expected to have a more significant explanatory role.

Much the same conclusions can be drawn when we consider Birmingham males and when we extend the analysis to investigate the effect of adding $(U/V)_o$ as a further independent variable to equation (1b). There is simply no evidence that the ratio of unemployed to vacancies for any occupational group assists us in explaining the behaviour of quit rates for that occupational group. Can we then conclude that the occupational context is irrelevant to mobility and, hence, to turnover; that individuals in deciding whether to leave their current employment are not concerned with the employment conditions facing the occupational group in which they are presently employed? An affirmative answer to these questions would imply that the occupational mobility accompanying employer changes was very considerable. Though occupational mobility does appear to be fairly frequent, a substantial fraction of all employees remain in the same occupational group when changing jobs,[1] so that one might expect the quit rate for any occupational group to be sensitive to changes in employment conditions for that group. The lack of

[1] See pp. 282–6 below.

relationship between quit rates and $(U/V)_o$ leads us to the view that the latter variable is a poor indicator of the employment conditions facing a particular group—at best it is an ordinal and not a cardinal measure of the balance between labour demand and supply. In part, this is likely to be due to the fact that we can, with the available data, only measure employment conditions for an occupational group by means of a ratio. We shall see shortly that this process is seldom satisfactory.

6. EMPIRICAL RESULTS

Having discarded $(U/V)_o$, we are left with two regression equations, (1a) and (1b). The first tests whether quit rates fluctuate with changes in unemployment and vacancies in the market as a whole, and the second whether quit rates respond to changes in unemployment and vacancies in engineering. We cannot unfortunately choose between these propositions; they must stand or fall together because of the high correlation between all unemployment and engineering unemployment and between all unfilled vacancies and unfilled vacancies in engineering.

Inspection of the residuals from equations (1a) and (1b) suggests the existence of seasonal fluctuations in quit rates. In each market area the error term in the fourth quarter of each year (October–December) was usually negative, i.e. the quit rate estimated by both regression equations was generally above the observed or actual quit rate.[1] A dummy variable was therefore added to pick up this seasonal fluctuation. We can then rewrite equations (1a) and (1b) to give:

$$Q_{es} = a+b_1U_s+b_2V_s+b_3D+\varepsilon \tag{1c}$$

and

$$Q_{es} = a+b_1U_{es}+b_2V_{es}+b_3D+\varepsilon \tag{1d}$$

In order to simplify the following discussion, Table 7.3 shows the full results of the regression analysis for equation (1c) only. Columns (i) and (ii) show the multiple coefficient of determination (R^2) and the Durbin–Watson statistic[2] obtained for (1c), while columns (iii) and (iv) provide the same information for equation (1d). The full results for (1c) rather than (1d) were shown because the unemployment and vacancy data necessary to calculate (1c) are more readily available and because, as the Durbin–Watson statistic

[1] The reason for this behaviour pattern is examined on p. 196 below.

[2] Used as a test for serial correlation.

TABLE 7.3

RESULTS OF REGRESSION EQUATIONS: QUIT RATES, SEPARATION RATES AND EMPLOYMENT CONDITIONS: BIRMINGHAM MALES AND FEMALES, AND GLASGOW MALES

	Quits				Separations	
	Equation (1c)		Equation (1d)		Equation (1c)	Equation (1d)
	(i)	(ii)	(iii)	(iv)	(v)	(vi)
Equation (1c)	R^2	D.W. stat.	R^2	D.W. stat.	R^2	R^2
BIRMINGHAM MALES						
Skilled						
$Q_{em} = +4{\cdot}96 - 1{\cdot}18^{*}U_m - 0{\cdot}21\ V_m - 0{\cdot}46D$ $(0{\cdot}32)\quad(0{\cdot}34)\quad(0{\cdot}31)$	0·53	1·68	0·52	1·65	0·51	0·46
Semi-skilled						
$Q_{em} = +13{\cdot}24 - 3{\cdot}60^{*}U_m - 1{\cdot}14V_m - 2{\cdot}20^{*}D$ $(0{\cdot}81)\quad(0{\cdot}86)\quad(0{\cdot}77)$	0·61	1·29	0·59	1·22*	0·55	0·48
Unskilled						
$Q_{em} = +8{\cdot}51 - 2{\cdot}25^{*}U_m + 2{\cdot}10^{*}V_m - 1{\cdot}45^{*}D$ $(0{\cdot}87)\quad(0{\cdot}93)\quad(0{\cdot}83)$	0·71	1·61	0·67	1·67	0·58	0·54
All males						
$Q_{em} = +9{\cdot}73 - 2{\cdot}57^{*}U_m - 0{\cdot}40V_m - 1{\cdot}38^{*}D$ $(0{\cdot}55)\quad(0{\cdot}58)\quad(0{\cdot}52)$	0·67	1·04*	0·66	1·02*	0·60	0·54

BIRMINGHAM FEMALES						
All females						
$Q_{ef} = +12{\cdot}75 - 4{\cdot}18^* U_f - 0{\cdot}06\ V_f - 2{\cdot}85^* D$ $(1{\cdot}54)\quad(0{\cdot}71)\quad(0{\cdot}61)$	0·67	1·61	0·68	1·91	0·46	0·51
GLASGOW MALES						
Skilled						
$Q_{em} = +3{\cdot}69 - 0{\cdot}32^* U_m + 3{\cdot}84^* V_m - 0{\cdot}27 D$ $(0{\cdot}13)\quad(0{\cdot}52)\quad(0{\cdot}22)$	0·86	2·29	0·90	1·50	0·64	0·69
Semi-skilled						
$Q_{em} = +4{\cdot}17 - 0{\cdot}35\ U_m + 6{\cdot}07^* V_m - 0{\cdot}60\ D$ $(0{\cdot}25)\quad(1{\cdot}01)\quad(0{\cdot}44)$	0·78	1·71	0·75	1·12*	0·60	0·58
Unskilled						
$Q_{em} = +6{\cdot}91 - 0{\cdot}68^* U_m + 2{\cdot}07^* V_m - 1{\cdot}02^* D$ $(0{\cdot}16)\quad(0{\cdot}64)\quad(0{\cdot}28)$	0·80	2·12	0·73	1·40	0·43	0·37
All males						
$Q_{em} = +4{\cdot}31 - 0{\cdot}38^* U_m + 4{\cdot}13^* V_m - 0{\cdot}55^* D$ $(0{\cdot}14)\quad(0{\cdot}57)\quad(0{\cdot}25)$	0·86	1·76	0·84	0·76*	0·62	0·62

Note: * Significant at the 5 per cent level.

shows, there is stronger evidence of positive serial correlation in equation (1d). In all other respects, however, the results obtained for (1c) are, *mutatis mutandis*, the same as those obtained for (1d). It is therefore unnecessary to show the results for both equations in full as they each support the same general conclusions. Where it is necessary, the results for (1d) are discussed in the following text. Table 7.3 also shows in columns (v) and (vi) respectively the R^2 obtained when separation rates are substituted for quit rates as the dependent variables in equations (1c) and (1d).

Let us deal first with the problem of serial correlation. The Durbin–Watson statistic indicates positive serial correlation significant at the 5 per cent level in one of nine cases for equation (1c) and in four of nine cases for equation (1d). For most groups there is some evidence of a bunching of negative and positive residuals following a cyclical pattern. In Glasgow, the value of the quit rate predicted by equations (1c) or (1d) tends to be above the actual or observed value over late 1963 and 1964, while on either side of this period the predicted quit rate tends to lie below the actual quit rate. In Birmingham, bunching of negative and positive residuals is discernible with roughly the same timing.

Serial correlation, of course, suggests the existence of some other factor not taken into account by equations (1c) and (1d). Further experiments were attempted with lagged variables; dummies were fitted to test whether the *direction* of change of unemployment or vacancies was significant,[1] and the rate of redundancies amongst the case-study plants was introduced as a further explanatory variable. Each variable, particularly the latter, reduced positive serial correlation but had no significant effect on R^2. The problem could have been dealt with by fitting dummy variables for those periods with predominantly negative or positive residuals. This does not, however, assist us in explaining the behaviour of quit rates unless we can ascribe some particular characteristics to the periods identified.[2] Because of this, and because serial correlation is not very serious, this procedure was not adopted.

R^2 in Table 7.3 is always significant at the 1 per cent level and is very similar for equations (1c) and (1d) whatever group is considered. This, however, was only to be expected given the high degree

[1] U_e and U_{em} in equations (1c) and (1d) were set to equal 0 when unemployment was falling and 1 when unemployment was rising; and vice-versa for V_e and V_{em}.

[2] For this reason, we have resisted the temptation to remove the serial correlation, as would have been possible, by fitting dummy variables to represent either election and non-election years or Labour and Conservative administrations!

of correlation between U_s and U_{es} and between V_e and V_{es}. It is more interesting to observe that the sign of the regression coefficient for U_s (U_m for males, U_f for females) is always negative as expected, and that in all cases save one the regression coefficient exceeds its standard error and is significant at the 5 per cent level. However, for each group of males considered the regression coefficient for U_m is much greater in Birmingham than in Glasgow. If U_{em} were considered, the same conclusions would emerge.

While the regression coefficient for U_m or U_{em} is lower in the case of Glasgow males, the reverse is true in the case of V_m or V_{em}. For all groups in Glasgow, the regression coefficient is positive and extremely significant. For Birmingham males the coefficient is weak and negative in three cases (two cases in equation (1d)) and is significant and positive, as expected, in only one case. These phenomena of a high regression coefficient for vacancies and a lower coefficient for unemployment in Glasgow, and the reverse situation in Birmingham, are interrelated. The negative correlation between U_m and V_m and between U_{em} and V_{em} is strong in both markets. Thus, if the coefficient for one variable is large and significant, the coefficient for the other variable is likely to be small. It follows that if the variable with the highest coefficient is dropped the coefficient for the other variable will improve—often quite markedly. For example, if V_m is dropped from equation (1c) the coefficients for U_m rise markedly for all groups of Glasgow males. However, as would be expected, the coefficients for U_m show little change for all groups of Birmingham males (save unskilled males). The same applies for Birmingham females. The reverse situation holds when U_m is excluded (U_f for females); the rise in the regression coefficients for V_m and V_f is more marked for Birmingham males and females. The negative signs for V_m in the case of skilled, semi-skilled and all males and for V_f in the case of females become positive, indicating that the 'true' relationship between vacancies and quit rates is positive as expected.[1] The coefficients are significant at the 5 per cent level but explain a much smaller proportion of the variation in quit rates than is the case for corresponding groups of Glasgow males.

The strong negative correlation between U_s and V_s and between U_{es} and V_{es} raises certain problems of interpretation. Despite this, the use of both these variables gives a better explanation of quit rates than the use of either variable independently or their combination in some ratio such as U_s/V_s or U_{es}/V_{es}. Suppose we begin by dropping V_m (V_f for females) from equation (1c). We are then representing

[1] Where two independent variables are highly correlated, the less significant variable can adopt the 'wrong' sign.

employment conditions by the overall level of unemployment for males and females. In other words, we adopt that approach which is widely applied in Phillips curve analysis. In this situation, the fall in R^2 is, for females or for skilled, semi-skilled and all males in Birmingham, not significant, but *is* significant at the $2\frac{1}{2}$ per cent level for the five remaining groups. Reading down the groups in Table 7.3, R^2 becomes 0·52, 0·59, 0·65, 0·67, 0·67, 0·58, 0·49, 0·72 and 0·60 respectively. It is apparent from these results that the explanatory power of U_m is much the same in the Birmingham and Glasgow markets. It is also evident that if a single variable must be chosen to represent the effect of employment conditions on quit rates, then the level of unemployment is preferable to the level of unfilled vacancies. Nonetheless, reliance on a single measure of employment conditions is unsatisfactory. This is even more obvious if U_m is excluded so that V_m (or V_f) is taken to represent employment conditions. In this instance, the fall in R^2 is statistically significant at the $2\frac{1}{2}$ per cent level in all cases save semi-skilled males in Glasgow. The other alternative is to substitute U_m/V_m (or U_f/V_f) for U_m and V_m separately. This is no better, as the sequential F test shows in all cases a fall in R^2 significant at the $2\frac{1}{2}$ per cent level.

It is preferable to use U_m and V_m (or U_f and V_f) as independent variables, for although the variables are negatively related, the correlation is far from perfect. Hence one cannot take one variable as a satisfactory proxy for employment conditions; we will learn more by including both unemployment and vacancies as independent variables. Nor is it sufficient to regard some ratio of unemployment to vacancies as an indicator of employment conditions because it is not then possible to say which variable—unemployment or vacancies—contributes most to the phenomenon under examination.

That this is an important consideration is clearly demonstrated by Table 7.3 and by the above analysis. The R^2 obtained for each group from equations (1c) and (1d) is always significant at the 1 per cent level. Yet it is apparent that both (1c) and (1d) provide the best explanation of quit rates when applied to groups of Glasgow males. The level of unemployment considered by itself (U_s or U_{es}) contributes roughly the same to an explanation of quit rates in both the Birmingham and Glasgow markets. Hence the reason for the better fit in Glasgow must lie elsewhere. It is found, of course, in the fact that the percentage of unfilled vacancies is of little assistance in explaining variations in quit rates in Birmingham with the sole exception of unskilled males. In contrast, there is a strong, statistically significant relationship between quit rates and unfilled vacancies for all groups of males in Glasgow. Indeed, in Glasgow unfilled vacancies contribute more to an explanation of variations in quit rates than does the percentage unemployed.

This clearly implies that the percentage of unfilled vacancies is a better indication of the unsatisfied demand for labour in Glasgow than in Birmingham. Two questions naturally arise. First, is this difference in some way related to the employment conditions ruling in this market; is there something about a slack market which suggests that unfilled vacancies will better reflect the demand of employers for additional recruits? Second, if there are grounds for accepting this proposition, what are the behaviour patterns which produce this result?

The first question is more easily answered than the second. The relationship between quit rates and unfilled vacancies is so strong for all groups of males in Glasgow and so much weaker for all groups in Birmingham, save unskilled males, that it is difficult to dismiss it as a quirk. Moreover, the only exception to the general rule in Birmingham strengthens rather than weakens the case that unfilled vacancies are a better indicator of labour demand in a slack market. This is because the exception applies to unskilled males and, as we have seen, unskilled males were easier to recruit in Birmingham than any of the other groups considered in Table 7.3.

Why should vacancies be a better indicator of changes in labour demand in Glasgow than in Birmingham, and in Birmingham amongst that group—unskilled males—for whom employment opportunities were most scarce? It is not because vacancies notified to the employment exchanges provided an accurate measure of the actual demand for labour in Glasgow. On the contrary, notified vacancy statistics, and hence unfilled vacancies, in Glasgow accounted for only a fraction of employers' labour requirements at any given point in time. Unfortunately, no similar data are available for Birmingham, but interviews with management suggest that in Birmingham, too, vacancies understated true labour requirements.[1] If vacancies were a better indicator of changes in labour demand in Glasgow, it follows that *movements* in vacancies must in some way have reflected *movements* in actual demand; that the proportion of labour requirements notified to the exchange varied less in Glasgow than in Birmingham. This is, however, a very speculative argument which must be treated with some caution. We are comparing only two labour markets, and generalization from this basis is particularly hazardous because of the deficiencies in vacancy data and, more importantly, because we cannot adduce a behaviour pattern which

[1] See p. 350 below.

lends support to the proposition that the relationship between notified vacancies will be more stable in a slack market. The proposition will require more extensive analysis before it can be accepted as valid. What is important is the discovery that, to explain quits satisfactorily, we must use vacancies and unemployment as independent variables rather than express these variables as a ratio or rely on only one of them.

Some other points of interest emerge from Table 7.3. First, the regression coefficient for the dummy variable in equation (1c), set to equal zero in quarters I, II and III, and one in quarter IV of each year, is always negative and in the majority of cases is significant at the 5 per cent level. There is, then, clear evidence of a seasonal low in quit rates for all groups over the period October–December.[1] This is likely to be caused by an unwillingness to change jobs and hence risk unemployment and/or a fall in earnings at this particular time of year. It is noticeable that the seasonal low is more in evidence in Birmingham, and in Birmingham is most significant for females. By and large, the same conclusions apply if equation (1d) is substituted for (1c).

Inspection of the regression coefficients from equation (1c) for male occupational groups supports our earlier suggestion that changes in employment conditions have less effect on the quit rates of skilled males than on those of other occupational groups. For example, while a 1 per cent increase in male unemployment in Birmingham produces a fall of only 1·18 per cent in the skilled quit rate, the equivalent reduction is 3·60 and 2·25 per cent for semi-skilled and unskilled males. In Glasgow also, a given change in U_m has least effect on skilled male quits although the latter are more responsive to changes in V_m than are unskilled males. In both Glasgow and Birmingham, the seasonal low in quit rates in the fourth quarter is least important for skilled males. There is, then, ample evidence for the view that quit rates for skilled men are less likely to display short-run fluctuations than the quit rates for other groups.

Up to this point, we have assumed that a linear relationship exists between quit rates and the independent variables, unemployment and vacancies. This may not be appropriate. Consider how the quit rate might respond to changes in unemployment. As unemployment rises the quit rate falls, but the assumption of linearity will only hold if a given increase in unemployment causes the same fall in the

[1] It is, of course, quite possible that other seasonal lows or highs might have been identified given a different choice of quarters than that adopted in this study.

quit rate irrespective of the level of unemployment. Higher unemployment will discourage quits amongst those who might have left for work reasons, but the quit rate for certain individuals who change jobs because of age, or illness, or other personal reasons will be less responsive to increasing unemployment. Because of this the quit rate may fall at a diminishing rate as unemployment rises. If this is so, a curve, convex to the origin, such as that in Figure A below, will be traced out.[1]

This is not the only possible alternative. Suppose that unemployment is falling from a high level. At first the quit rate may rise sharply as those dissatisfied with their current job seek new alternative work in the more favourable market situation prevailing. If unemployment continues to fall, however, the stock of employees will increasingly be composed of those who are reluctant to change jobs—long-service or older employees who do not want to leave a familiar environment, individuals with strong ties with the work group, etc. Successive falls in unemployment may produce increases in the quit rate, but the relationship may become progressively weaker. In other words, the curve depicting the relationship between quits and unemployment will be concave to the origin as in Figure B below.[2]

Since, on conceptual grounds, a non-linear relationship appears possible it was decided to test for this by assuming a relationship as in Figure A and substituting for equations (1c) and (1d) equations (1e) and (1f) below:

$$\log Q_{es} = \log a + b_1 \log U_s + b_2 \log V_s + b_3 D + \varepsilon \qquad (1e)$$

and

$$\log Q_{es} = \log a + b_1 \log U_{es} + b_2 \log V_{es} + b_3 D + \varepsilon \qquad (1f)$$

The regression equations for (1e) and (1f) were calculated for each group included in Table 7.3 above. We shall discuss the results by comparing R^2 from (1e) with that shown in column (i), Table 7.3 for (1c). Reading down the column, R^2 becomes 0·54, 0·74, 0·75, 0·79, 0·69, 0·88, 0·79, 0·80 and 0·88 when (1e) is substituted for (1c). In all cases, therefore, R^2 from (1e) is greater than that from (1c), except for unskilled males in Glasgow where R^2 from both equations is identical.[3] There is, then, some indication of non-linearity, but there

[1] *Figure A* [2] *Figure B*

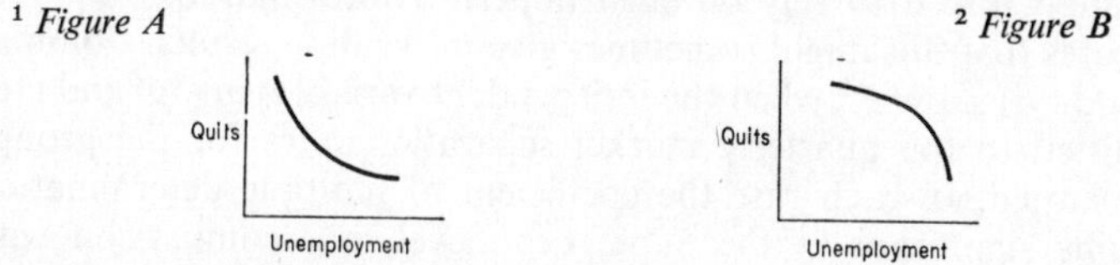

[3] In view of this we can rule out a relationship between quits and unemployment such as traced out by Figure B.

does appear to be a difference in behaviour as between the Birmingham and the Glasgow markets. In Glasgow, the improvement in R^2 obtained from using equation (1e) is extremely small for all male groups and is never statistically significant. For Birmingham males, however, the adoption of equation (1e) yields an improvement in R^2 which is statistically significant at the 1 per cent level for both semi-skilled males and all males. Precisely the same results are obtained if equation (1f) is substituted for equation (1d).

This suggests that in a slack market the relationship between quits and employment conditions can be approximated by a linear function. It is always possible that, even in slack markets, a curvilinear relationship would appear were unemployment higher, and vacancies correspondingly much lower, than was the case in Glasgow. However, even if this were so, it would be of little practical interest, for, given the fact that in Glasgow unemployment was substantially above the national average, non-linearity would only emerge at levels of unemployment well in excess of those normally experienced in Britain. It might also be noted that non-linearity may not emerge even in an extremely tight labour market like Birmingham. Equation (1e) provides superior results to (1c) only for semi-skilled and all males, and the latter result must be heavily affected by the former, for the semi-skilled are the single most important occupational group. For skilled and unskilled males in Birmingham, and for females, there is no evidence of marked non-linearity. For the moment, therefore, we have to accept the possibility that non-linearity *might* emerge at low levels of unemployment. Verification (or rejection) of this suggestion will have to await further evidence.

Lastly, let us consider whether the independent variables in equations (1c) and (1d) can be used to 'explain' separation rates. It has already been indicated that persons who do not leave for voluntary reasons are unlikely to be influenced by employment conditions, as are voluntary leavers. Hence, our hypothesis is simply that separation rates change in response to employment conditions because a high proportion of separations are voluntary quits. Data on separation rates are more readily to hand than data on voluntary quits, and we are often forced to rely on such imperfect information. Table 7.3 indicates that this might sometimes give misleading results. Columns (v) and (vi) show R^2 when the independent variables in (1c) and (1d) are fitted to the quarterly market separation rates for the groups enumerated. In each case the coefficient of multiple determination remains significant at the 5 per cent level, but comparison with columns (i) and (iii) shows that R^2 falls in all cases when separation rates are substituted for quit rates as the dependent variable. The fall in R^2 is substantial in the case of Birmingham females and all groups of Glasgow males, and is largest amongst unskilled males in the latter area. This is partly due to the fact that for the latter groups voluntary quits account for a smaller proportion of separations than for Birmingham males. This difference is not, however, very substantial, and a more important influence is the higher redundancy rate amongst Birmingham females and Glasgow males.[1]

As might be expected, redundancies move anti-cyclically, thus damping down the response of the separation rate to changes in employment conditions. A higher redundancy rate will then reduce the sensitivity of separation rates to changes in employment conditions even if quit rates are the same between groups. If quit rates are lower in the area with higher redundancies, as appears likely, and as is the case in Glasgow compared to Birmingham, then redundancies will account for an even higher proportion of all separations. The relationship between separation rates and quit rates will, then, be especially weak. This can best be illustrated by comparing quit rates and redundancy rates for unskilled males in Glasgow and Birmingham. The unskilled quit rate in Glasgow is substantially below the level in Birmingham, while the reverse applies to the redundancy rate. Hence, while, for unskilled males, R^2 for equations (1c) and (1d) is higher for Glasgow than for Birmingham when quit rates are considered, it is much lower for the former group when separation rates are substituted for quit rates as the dependent variable. This illustrates that the relationship between quits and separations is not stable for all groups. It follows that it is dangerous to draw inferences from a study of the separation rate alone, for that rate covers different groups whose behaviour is conditioned by different sets of factors.

7. CONCLUSIONS

(i) Quit rates and separation rates were, in general, higher in Birmingham than in any other area. In part this reflected differences in employment conditions as measured by unemployment and vacancies, but the characteristics of the labour market, particularly its size and compactness, have an influence on quits independent of employment conditions as conventionally measured. In consequence,

[1] See p. 375 below. The redundancy rate is crucial because redundant leavers are the only category of *non*-voluntary leavers whose leaving rate shows substantial cyclical fluctuations. The leaving rate for other groups—discharges through unsuitability and misconduct, retired, etc.—is much more stable over time.

quit rates tended to be particularly high in the Birmingham and Glasgow conurbations.

(ii) In Birmingham, quit rates (and separation rates) were inversely related to skill, being higher for semi-skilled than skilled and highest for unskilled. There was little sign of such a relationship in the other labour markets, and this held at the level of the individual plant. It appears that quit rates for skilled men are least responsive to differences in employment conditions between markets and to changes in employment conditions within the same labour market.

(iii) In each labour market area and in the case of most occupational groups, quits accounted for some 70 per cent of all separations. There was, however, some variability between groups and between markets and in the relationship between these rates in the same market through time. Thus, although the course of quits generally dictated turning-points in separations, they accounted for a varying proportion of separations over the cycle. In each market and in each group the proportion of quits to separations was inversely related to the level of unemployment. Any 'explanation' of quits must, therefore, apply with diminished force to separations.

(iv) Quit rates were not influenced by the rate of change (or the direction of change) of employment conditions. In Birmingham there was some evidence of a lagged relationship between unemployment and quits, but the evidence was weak, and general considerations suggest that quit rates for manual workers can be expected to respond to employment conditions *currently* ruling in the market.

(v) Collinearity between the independent variables prevents us distinguishing whether quits responded primarily to changes in unemployment and vacancies in the market as a whole (U_s and V_s where s represents sex), or to unemployment and vacancies in the engineering industry (U_{es} and V_{es}). It is apparent, however, that the ratio of the number unemployed to the number of unfilled vacancies by occupational groups, $(U/V)_o$, adds little to the explanation of occupational quit rates. This probably indicates that the employment conditions facing a particular skill group are not satisfactorily represented through a ratio, rather than the fact that such employment conditions are irrelevant to quits.

(vi) Quit rates were more responsive to changes in the rate of unemployment defined to *exclude* those temporarily stopped than to unemployment defined to *include* this category. When the former definition is adopted, there is for all groups considered a negative and statistically significant relationship between unemployment and quit rates. This applies both to U_s and U_{es}. The explanatory power of these variables differed little between Birmingham and Glasgow. In contrast, while the 'true' relationship between quits and unfilled vacancies was positive in all cases, the strength of the relationship was much more apparent in Glasgow. In consequence, the adoption of unemployment *or* vacancies as a measure of employment conditions (or, if one prefers it, excess demand) is unsatisfactory, as is the use of a ratio derived from these variables. It is not clear why unfilled vacancies are more significant in explaining quit rates in Glasgow, but it *may* arise because of a more stable relationship between vacancies notified (and unfilled) and employers' actual demands for labour, in a slack market.

(vii) For all groups, there was a seasonal low in quits in the fourth quarter (October–December) of each year. In Glasgow the assumption of linearity yielded results very similar to those obtained if a curvilinear relationship is postulated between quits and the independent variables. In Birmingham, there was some evidence of non-linearity between the dependent variable, quits, and the independent variables, unemployment and vacancies, which suggests that if the 'true' relationship is non-linear it will only emerge as such at low levels of unemployment (and, of course, high levels of unfilled vacancies). For the moment, however, the possibility of a non-linear relationship in tight markets remains only a suggestion which requires more extensive empirical investigation.

(viii) If the separation rate is substituted for the quit rate as the dependent variable, then poorer 'fits' are obtained from each regression equation. This is most evident in the case of Glasgow males and Birmingham females. In the former case, the fall in R^2 when separation rates are considered is due to the relatively high redundancy rates and low quit rates for Glasgow males.

CHAPTER 8

PERSONAL CHARACTERISTICS AND LABOUR TURNOVER

1. INTRODUCTION

In terms of our original classification, the factors influencing labour turnover have been considered under two headings—environmental influences and personal characteristics. Environmental influences are simply the labour market circumstances in which job choices are made; that is, the job opportunities ruling in the labour market, and the wage and non-wage conditions offered by employers. The effect of such factors on job choice and labour turnover was, of course, the subject matter of the preceding two chapters, but there personal characteristics did enter into the discussion, as when we considered occupational groups and the effect of length of service on the voluntary quit rate. This chapter further investigates the effect of personal characteristics, and in particular the relationship between age, occupation, length of service, labour turnover, and the voluntary quit rate.

The broad subject area of this chapter has been dug over by many previous researchers, but most research in this field, particularly in Britain, has dealt either with the characteristics of all leavers, undistinguished by reason for leaving, or with the characteristics of redundant leavers. These studies have found common ground in the following points: labour wastage is inversely related to length of service and to age, being particularly high amongst young workers with short periods of service, and particularly low amongst older employees with long periods of service; it is, at least amongst manual workers, inversely related to skill and to the length of training undertaken; and it is greater for females than for males and for married females than for single women.

The degree of unanimity is impressive, but certain problems remain. Labour market investigations have been based on plants with differing technologies and with different occupational structures, operating in substantially different labour market conditions. Together with variations in methodology, these factors make it difficult to establish how far the characteristics of redundant leavers differ from those of total separations in a given labour market environment. Still less is it possible to compare the personal characteristics of persons leaving for voluntary reasons, or because of unsuitability, or misconduct, etc., to those of total separations. It is, therefore, worth while considering whether the effect of personal characteristics on wastage (length of service, skill, etc.) is also modified by the employment conditions ruling in the external market.

The following sections investigate two main issues. In Section 2, using data relating to those leaving the case-study plants over the whole of the study period, we investigate how the personal characteristics of different categories of leavers differ within a labour market area. It is also possible to use these results to make a preliminary exploration of our second main field of interest—whether employment conditions ruling in the market modify the relationship between personal characteristics and labour wastage. This can be accomplished by contrasting the results obtained for the separate labour market areas. However, while this procedure yields certain interesting clues, it is not wholly satisfactory, partly because of the difficulty of relating the number of leavers in any category to an appropriate stock concept, and partly because we cannot see how the personal characteristics of leavers might change as labour market conditions change in a given labour market area. For these reasons, the analysis is taken a stage further in Section 3 where a more detailed study is made of the personal characteristics of voluntary leavers. Since it is difficult to separate out the influence of length of service and age on quits, further examination of these factors is undertaken in Section 4, while Section 5 emphasizes the sensitivity of quit rates and separation rates amongst short-service employees to differences in labour market conditions, and examines differences in the 'stability' indices for the different market areas. A brief summary of the conclusions is contained in Section 6.

Before proceeding to our analysis, it is useful to outline briefly why the results obtained may have some importance. They are relevant to labour market theory, for it is possible to construct from that theory a set of predictions as to how labour wastage, and particularly the voluntary quit rate, will vary by sex, occupation, age and length of service. These predictions can then be tested against the results obtained. However, the main value of the results lies in their relevance for policy decisions. At the level of the individual plant, it is evidently useful to be able to identify those groups with high wastage rates, for this has implications for selection and recruitment policies and for predicting future labour wastage and, hence, future labour needs. For public policy, it is essential to know the characteristics of leavers, for this may have an important bearing on the ease or difficulty with which the labour becoming available can be redeployed to other uses. This is most relevant in the case of redundant workers, who often come on to the labour market in unfavourable circumstances. If redundant leavers have other disadvantages, for example an unfavourable age structure, this may affect the ability of the employment exchanges to provide them with alternative work and may indicate the need to reconsider such instruments of manpower policy as the Redundancy Payments Act. This is only one example of why it is important to understand in greater detail the relationships between personal characteristics, employment conditions and labour turnover.

2. Reason for Leaving, Skill, Age and Length of Service

The analysis of this section is based on a count of all separations over 1959–66 distinguishing leavers by reason for leaving and by skill, age and length of service with the plant at the date of leaving. The approach is, therefore, straightforward enough, but two data problems do arise. The first is those Birmingham leavers whose reason for leaving was not known and who were dealt with in Chapters 6 and 7 by assuming that they had the same distribution by reason for leaving as those whose reason for leaving was known. Here no such adjustment is attempted because we are dealing with proportions and not with rates, and proportions are *not* changed by the assumption that not-known leavers had the same distribution as known leavers in terms of a given characteristic.

Nonetheless the validity of our results does depend, to some extent, on known and not-known leavers displaying similar behaviour patterns, and this in turn hinges on whether these groups have similar personal characteristics. We have already seen that the occupational distribution of known and not-known leavers was similar, and the data below show that these groups also had like characteristics with regard to length of service and age.

The only notable differences, amongst Birmingham males, are that those whose reason for leaving was not known had a high proportion of leavers with more than five years' service and, surprisingly in view of this, a higher proportion in the youngest age group. Similar comments apply in the case of Birmingham females, but here, too, the contrasts are not great enough to throw serious doubt on an analysis which ignores those whose reason for leaving was not known.

The second data problem arises when reason for leaving was known but age or length of service at the date of leaving was not known. For reasons similar to those outlined above, no adjustment has been made for those whose age or length of service was not known. In the case of length of service, this qualification is hardly important, as the number whose length of service at the date of leaving was not known never amounts to even 1 per cent of all leavers in any area. However, the number whose age was not known at the date of leaving generally amounts to some 10 per cent of leavers in each area, reaching a peak of 20 per cent in the case of Birmingham females.

Percentage by length of service (Birmingham males)

	Up to 4 wks	4 wks up to 12 wks	12 wks up to 1 yr	1 yr up to 2 yrs	2 yrs up to 5 yrs	5 yrs+
Reason for leaving known	19·1	21·4	31·2	11·3	10·7	6·3
Reason for leaving not known	18·9	17·5	26·6	9·6	9·8	17·6

Percentage by age (Birmingham males)

	⩽20 yrs	21–24 yrs	25–34 yrs	35–44 yrs	45 yrs
Reason for leaving known	11·5	20·4	28·9	19·5	19·7
Reason for leaving not known	16·4	16·5	30·2	19·6	17·1

(i) *Reason for leaving and occupational group*

We begin our analysis with Figures 8.1 and 8.2. Figure 8.1 shows, for males and females whose reason for leaving was known, the proportions by reasons for leaving. The number of leavers on which the proportions are calculated is also shown for each group.[1] Similar information for male leavers distinguished by skill is shown in Table 8.2.

In all cases save unskilled males in North Lanarkshire, voluntary quits accounted for the greater part of all leavers. Thus, although labour market conditions varied from one area to the next, no area produced such unfavourable employment conditions as to make voluntary leavers a minority group, and the variations between the employment conditions in the separate markets, although fairly large

[1] Throughout this chapter we ignore leavers from plants in New Town. The reasons for this are given on p. 170 above. Details are provided for Small Town leavers only where there are sufficient numbers to warrant their inclusion.

by British standards, do not appear to have been sufficient to produce substantial differences in the proportion of voluntary quits. It is true

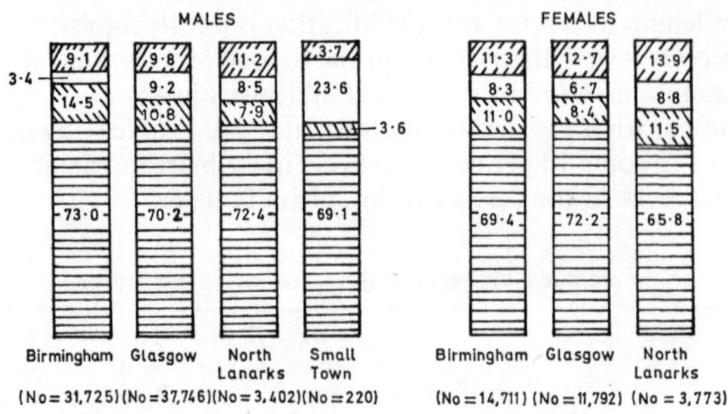

Figure 8.1. Total separations by reason for leaving

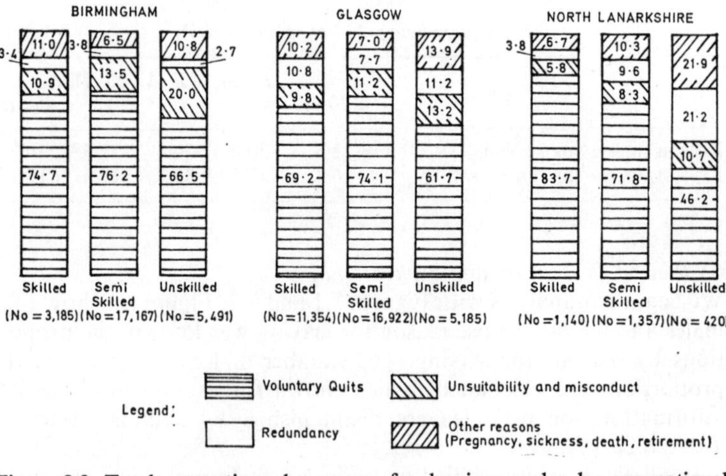

Figure 8.2. Total separations by reason for leaving: males by occupational group

that, amongst males, voluntary quits accounted for the largest proportion of total separations in Birmingham and that this also held for all occupational groups except skilled males in Birmingham and North Lanarkshire, but the differences between the areas were small in most cases, and amongst females the proportion of voluntary quits was higher in Glasgow than in Birmingham.

Comparing males with different skills, we find that the proportion of voluntary quits was always lowest for unskilled males in a given market area. This applied even in Birmingham where the voluntary quit *rate* was highest for the unskilled,[1] indicating that unskilled males were more likely to leave for other reasons. Voluntary quits were a smaller proportion of female leavers than male leavers in Birmingham and North Lanarkshire, and this contrast is heightened if we allow for the effect of skill. Thus, in each labour market area, voluntary quits accounted for a lower proportion of all female separations, who were almost exclusively semi-skilled, than of male semi-skilled separations.

Turning to other categories of leavers, we find that for almost all groups the proportion of separations due to unsuitability or misconduct was substantially higher in Birmingham than in the other areas. As the great bulk of such leavers were dismissed for unsuitability after a probationary period of between one and three months, this might indicate that the selection methods applied to new recruits were particularly inefficient in the case of Birmingham plants. However, as selection procedures were rudimentary in most areas, a more plausible explanation would be that Birmingham plants were forced to lower their hiring standards in face of recruitment difficulties caused by the low levels of unemployment in the area. This would seem to be consistent with the high proportion of unskilled male separations due to unsuitability and misconduct, a feature which was particularly marked in Birmingham.

Amongst females, the *proportion* dismissed through unsuitability or misconduct shows little variation from area to area, and this also applies to those discharged because of redundancy. In some respects, however, this is misleading, for the *rate* of redundancies and the *rate* of dismissals for misconduct and unsuitability were greater for Birmingham females.[2] The similarity in the proportions leaving for these reasons was, then, largely due to the fact that Birmingham females had a higher voluntary quit rate than Glasgow and North Lanarkshire females.[3] For males, the proportion of redundancies was particularly low in Birmingham and especially high in Small Town, while Glasgow and North Lanarkshire occupied intermediate positions. There is no distinct pattern by occupational groups in the former area, but in Glasgow, and particularly in North Lanarkshire, a high proportion of unskilled male leavers were discharged through redundancy. Once again, we cannot make direct inferences about the

[1] See Table 7.1, p. 169 above.
[2] See Table 14.3, p. 375 below.
[3] See Table 7.1, p. 169 above.

rate of redundancies because quit rates varied between skill groups and between areas. Thus, while the overall rate of male redundancies was lowest in Birmingham, both here and in North Lanarkshire the rate was higher amongst the semi-skilled and unskilled than amongst skilled males. In Glasgow, the rate of redundancies differed very little between skill groups, but in each group the rate was higher than in Birmingham and in North Lanarkshire.[1]

(ii) *Reason for leaving and length of service*

Figure 8.3 shows the proportion of total separations,[2] voluntary quits and redundancies by length of service categories. The corresponding information for those leaving through unsuitability and misconduct and for 'other reasons' (pregnancy, sickness, death or retirement) has not been shown because the results are very much those which would be expected. That is to say, a high proportion of those dismissed for unsuitability and misconduct had completed only a short period of service with the plant, while there was a marked concentration of long-service workers amongst those leaving for 'other reasons'.

In each market area the proportion of leavers with at least five years' service was always higher for males than for females, whether one considers total separations, voluntary quits or redundancies. This was predictable in view of the high proportion of females who are secondary workers and who consequently are likely to enter and withdraw from the labour force on more than one occasion throughout their working lives. Hence, relatively few females had a continuous period of employment with one plant exceeding five years. Rather surprisingly, the proportion of female leavers in the shortest length of service category was not consistently higher than that for males. Once again we have to distinguish carefully between proportions and rates. The *rate* of total separations, voluntary quits and redundancies was higher for females than the corresponding rate for males in each market area.[3] We shall also see shortly that in each length of service category the voluntary quit rate was higher for females than males. Thus the reason why short-service categories did not account for a relatively high proportion of female quits was simply that females in *all* length of service categories were more likely to leave the plant than males. The same comments apply to total separations.

[1] See Table 14.3, p. 375 below.

[2] The numbers of total separations on which the analysis of Figure 8.3 is based exceed those shown in Figure 8.1 because the proportions of leavers whose length of service was not known was always smaller than the proportion of leavers whose reason for leaving was not known.

[3] See Tables 7.1 and 7.2, p. 169 above and Table 14.3, p. 375 below.

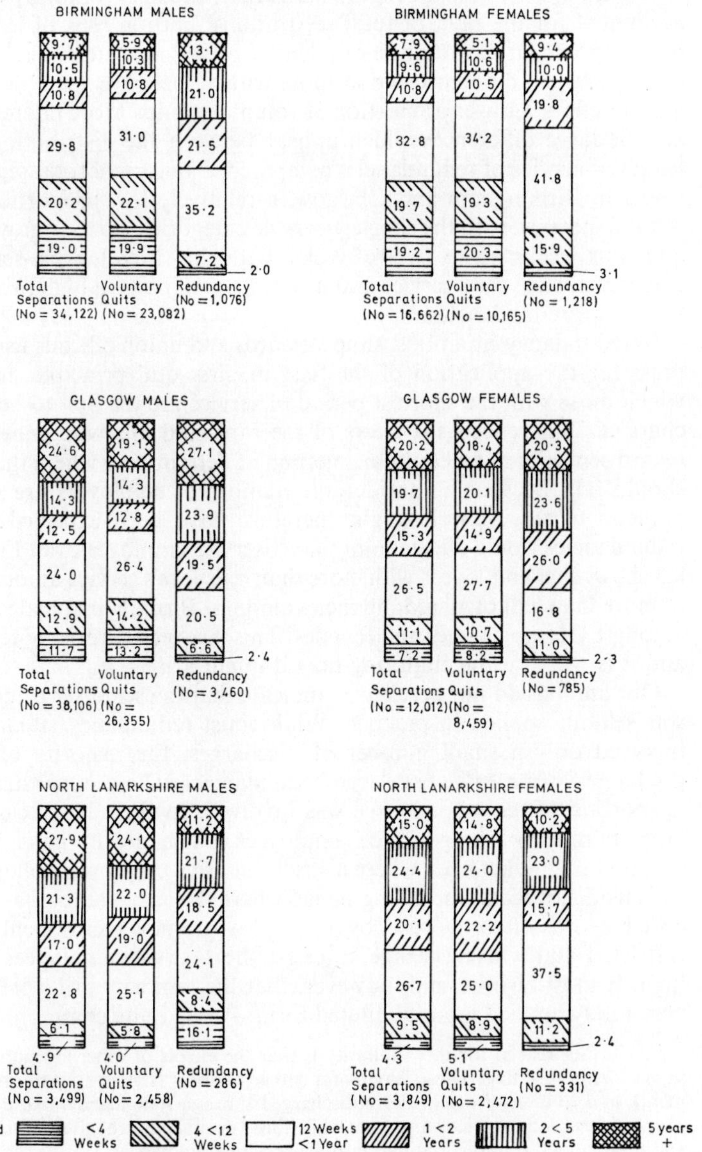

Figure 8.3. Reason for leaving by length of service

A general finding from Figure 8.3 is that, because voluntary quits accounted for the bulk of total separations, both groups of leavers had very similar distributions by length of service categories.[1] The only consistent difference is that those with at least five years' service accounted for a lower proportion of voluntary quits. More interesting are the large differences which appear between the distribution by length of service of redundancies compared to quits and total separations. In Birmingham and Glasgow, a relatively low proportion of redundancies were in the shortest-service categories, up to four weeks and four weeks up to twelve weeks, with the three longest-service categories, one year's service and upwards, making up a high proportion of all redundancies.[2]

In redundancy situations, shop stewards and union officials usually press for the application of the 'last in, first out' principle, under which those with the shortest period of service are the first to be discharged. This reflects the views of the rank and file, who generally regard seniority as an equitable method of determining where the axe should fall. Because a considerable number of employers are sympathetic to this view it might therefore have been expected that redundancies would affect mainly short-service employees, but Figure 8.3 shows that employees with more than one year's service amounted to more than half of all redundancies amongst Birmingham males and amongst Glasgow males and females. This is a very surprising result, and it is worth while enquiring how it might arise.

Our findings do not necessarily indicate that the seniority principle was seldom applied in practice. While most redundancy situations involved only a small number of discharges, the majority of the discharges came in a few large redundancies where a substantial proportion of the labour force was involved. In these latter cases a large number of long-service employees are likely to have been declared redundant even where a strict seniority policy was followed. This tendency would be strengthened where the redundancy was preceded, as was often the case, by a period where new recruitment was restricted and where wastage amongst short-service employees was high. It must also be said, however, that the seniority principle was very rarely applied in an undiluted form. Plants quite commonly de-

[1] A further reason for this similarity is that the effects of other categories of leavers were offsetting. Those leaving for 'other reasons' (sickness, death, retirement), and to a lesser extent those discharged through redundancy, had a high proportion of long-service employees while those dismissed through unsuitability and misconduct were predominantly short-service employees.

[2] The reader should note that most of these redundancies occurred *before* the 1965 Redundancy Payments Act linked redundancy payments to length of service.

clared married women redundant before single women or men, particularly where the married women had few family responsibilities; or men and women over or near retiring age before workers in younger age groups; or 'inefficient' employees before 'efficient' employees. Even where seniority was used to choose between those who should leave and those who should stay, this was usually applied on a job or department rather than on a plant-wide basis. It was unusual to find 'bumping' on the basis of seniority between persons of different skills, or even between persons of the same skill in different departments, with the result that long-service employees in a particular job or department would be dismissed before less senior employees engaged on other tasks. The above issues are discussed in greater detail in Chapter 14. For the moment, it is sufficient to observe that, whatever the causes, redundant workers contain a high proportion of long-service employees and, as would be expected from this, a high proportion of employees in the older age groups.

The distribution of leavers by length of service was substantially different from one area to the next. For male voluntary quits, and for male total separations, the proportion in each of the three shortest-service categories, with under one year's tenure, was always highest for Birmingham and lowest for North Lanarkshire, with Glasgow occupying an intermediate position. Conversely, the proportion in each of the three service categories above one year was always greatest in North Lanarkshire and lowest in Birmingham. As we know that the quit rate and separation rate were highest in Birmingham and lowest in North Lanarkshire, it follows that the greatest difference between the areas must have occurred amongst those in the short-service categories. This conclusion is consistent with the hypothesis put forward earlier, that labour market conditions influence not only the overall quit rate but also the incidence of that quit rate between groups distinguished by length of service. Not only does the quit rate vary by market area with the employment opportunities available, but in areas with favourable employment conditions a higher proportion of all quits and separations is composed of persons who leave plants within a short period of recruitment. It is not therefore surprising to find, as we did in Chapter 6, that differences in quit rates between plants in Birmingham were to a much greater extent due to differences in the past quarter's recruitment rates than was the case in Glasgow. Figure 8.3 shows that persons with less than twelve weeks' service accounted for 4 in 10 of male voluntary quits in Birmingham, for less than 3 in 10 of voluntary quits in Glasgow and for only 1 in 10 of voluntary quits in North Lanarkshire. At the other end of the scale, only 5·9 per cent of male voluntary leavers in

Birmingham had more than five year's service, compared to 19·1 per cent in Glasgow and 24·1 per cent in North Lanarkshire.[1]

(iii) *Reason for leaving and age*

Finally in this section, we can analyse the age distribution of employees distinguished by reason for leaving. This is done in Figure 8.4 which shows reason for leaving by age groups for males and females in Birmingham, Glasgow and North Lanarkshire.[2]

To some extent the results of Figure 8.4 could be predicted from what we already know about the length of service distribution of leavers, since the age of an employee and his length of service must increase together, so that short-service employees are likely to be concentrated amongst the younger age groups and vice versa. In some ways, therefore, the most interesting question is whether age *or* length of service is the important factor in determining quits. As this issue cannot be satisfactorily resolved by the methods of analysis employed in this section, we must postpone the discussion to Section 4. For the moment, we shall briefly consider the age distribution of leavers bearing in mind the probability that age and length of service are likely to be positively related.

Just as the length of service distribution of voluntary quits and total separations was similar within the same labour market area, so these groups of leavers show relatively little variation by age. The most notable difference, which again is an echo of the length of service distribution, is that those in lower age groups, less than 21 years and 21–24 years, accounted for a higher proportion of voluntary quits than total separations in all areas, while those aged at least 45 years always amounted to a higher proportion of total separations. With redundant leavers we find, as would be expected given our previous analysis by length of service, that, relative to quits, a low proportion were 24 years and less and a high proportion were 35 years or more.

A more important and interesting difference between the length of service and age distribution of leavers emerges when comparisons are

[1] For females, much the same comments apply. Once again the proportion of voluntary quits composed of short-service employees was highest in Birmingham and lowest in North Lanarkshire. The contrast between the latter area and Glasgow was not, however, as sharp as in the case for males, and females with at least five years' service accounted for a higher proportion of quits and separations in Glasgow.

[2] The number of observations is less than in Figure 8.3, for, as explained previously, the proportion of leavers whose age at leaving was not known generally averaged some 10 per cent of all separations while length of service was known in almost all cases.

Figure 8.4. Reasons for leaving by age

made between market areas. Once again the results for males are straightforward enough. Males aged 24 years or less were a higher proportion of total separations and voluntary quits in Birmingham than in Glasgow or North Lanarkshire. Remembering that the total male quit rate was higher in Birmingham, and following a similar logic to that employed earlier, we can therefore argue that the differences in quit rates between areas were particularly marked in the case of younger age groups. Hence in Birmingham the male quit rate was especially high for young, short-service employees. With females the picture is quite different, and it must be more closely examined. While, for females as for males, leavers with short periods of service were more important, and those with long periods of service less important, in the case of Birmingham females, the opposite was true for age. In other words, leavers in the older age groups (35 years and above) were a *higher* proportion of voluntary quits and total separations in Birmingham, while leavers in the younger age groups (24 years or less) were relatively *less* important. This is an unexpected result given that age and length of service are likely to be positively related, and the clear contrast in the results for males and females requires some explanation.

Our suggested explanation is as follows. In Birmingham, married women were more likely than they were in Glasgow or North Lanarkshire to continue in paid employment after marriage, and those who did leave for domestic reasons were more likely to take up employment again at a later date. This mainly reflected the greater employment opportunities available in Birmingham, although differences in social attitudes and conventions may also have been influential. Differences in female participation rates were greatest in the older age groups,[1] so that in Birmingham a high proportion of the female labour force, and hence a higher proportion of leavers, were aged 35 years or over. This does not, of course, reflect a tendency on the part of females in the older age groups to have lower quit rates in Birmingham. On the contrary, female quit rates were, with one minor exception, always highest for all age groups in Birmingham.[2] However, while females were *more* likely to leave a given plant, they were *less* likely to leave the labour force. Hence, those in the higher age groups formed a larger part of the female labour force in Birmingham and a large part of voluntary quits and total separations.

The differences in the age composition of those females leaving through unsuitability and misconduct or for other reasons can be

[1] This can be seen from Table 3.7, p. 59 above.
[2] See Table 8.2, p. 219 below.

explained in a similar fashion. Because women in Birmingham were more likely to continue in employment or return to employment after marriage, those in the older age groups accounted for an exceptionally high proportion of those leaving for 'other reasons', in which retirements were the most important single category. The explanation for the differences in the age composition of those dismissed for unsuitability or misconduct is more speculative, but it may be that the large proportions in the older age groups in Birmingham were due to females returning to paid employment after some time out of the labour force. If, as seems possible, such females found it difficult to acquire new skills or to adapt to the requirements of industrial life, managements may have 'weeded out' a high proportion of such recruits.

3. QUIT RATES, PERSONAL CHARACTERISTICS AND EMPLOYMENT CONDITIONS

The previous section reached certain tentative conclusions regarding the effect of employment conditions on the personal characteristics of leavers through an investigation of the differences which

	Birmingham	Glasgow	North Lanarkshire
Males			
Unemployment low	1.4.65–31.3.66 (0·7)	1.1.61–31.12.61 (4·3)	1.1.65–31.12.65 (2·7)
Unemployment high	1.4.62–31.3.63 (2·4)	1.10.62–30.9.63 (7·1)	1.1.59–31.12.59 (7·3)
Females			
Unemployment low	1.4.65–31.3.66 (0·3)	1.7.65–31.6.66 (1·1)	1.7.65–30.6.66 (3·5)
Unemployment high	1.10.61–30.9.62 (1·2)	1.1.63–31.12.63 (2·7)	1.1.59–31.12.59 (7·4)

emerged between the separate labour market areas. It was not possible, however, to examine the influence of changes in employment conditions within each market, and the analysis had to proceed in a rather indirect manner, as the data were presented in the form of proportions rather than rates. In view of this, the current section concentrates on the rate of voluntary quits distinguished by age, length of service and by period. In each market considered—Birmingham, Glasgow and North Lanarkshire—two separate yearly periods were isolated, one at the peak of a labour market cycle (the 'unemployment low') and one at a trough (the 'unemployment high'). These represented the four consecutive quarters over 1959–66 which yielded the lowest and the highest unweighted averages of quarterly unemployment rates. The periods selected are shown above, with the unweighted averages of the four quarterly rates for the wholly unemployed in parentheses.

At the trough of the labour market cycle in Birmingham, unemployment rates were still usually below the lowest levels experienced in Glasgow and North Lanarkshire at the peak of their labour market cycles. Yet, while absolute changes in unemployment levels were greatest in Glasgow and North Lanarkshire, in relative terms these changes were largest in Birmingham, so that although the markets operated at very different levels of unemployment, within each market unemployment 'highs' and 'lows' were identifiable.

In Table 8.1 we show annual quit rates by length of service categories which were calculated in the following fashion. A count was taken of all those employed by the plants in each labour market area at the beginning of the period considered. Two further groups were then distinguished from this stock—those who left voluntarily over the period, and those who remained with the plant over the period. Voluntary leavers were assigned to a length of service category according to their length of service at the time of leaving, and this figure was expressed as a percentage of the *non-leavers* in the same length of service category at the beginning of the period. For example, in the period of the unemployment high, 236 Birmingham males with less than twelve weeks' service at 1.4.62 had remained with the same case-study plant until 31.3.63. Over that yearly period, 111 males in the original stock had left having completed less than twelve weeks' service, giving the annual quit rate shown of 47·0 per cent.[1] The reader should observe that the quit rates have been calculated by reference only to those persons who were *known* to have left for voluntary reasons. The result is that in this and the following section the quit rates presented are below the 'true' quit rates. To place the results for the different market areas on an exactly comparable basis is impossible, but greater comparability will be obtained if the quit

[1] It should be noted that this definition of the quit rate differs from that used in Chapters 6 and 7. In the latter case, all those in the stock were included in the denominator whether or not they left in the succeeding quarter. Here we adopt the rather unusual practice of excluding leavers from the denominator, a procedure which will bias quit rates upwards. This method is employed because for technical reasons the length of service, age etc. of leavers and non-leavers had to be determined at different dates (the date of leaving and the date at the beginning of the yearly period respectively) and because it yields results which are quite serviceable provided the above reservation is borne in mind.

rates by length of service categories shown in Table 8.1 are adjusted upwards by one-tenth for Birmingham males and females. When dealing with quit rates by age groups shown in Table 8.2, the same adjustment can be applied in the case of Birmingham males, but with Birmingham females the rates shown should be increased by one-fifth to produce greater comparability with other areas.[1]

In all three markets there was a strong inverse relationship between length of service and quit rates for both males and females. This held

TABLE 8.1

ANNUAL QUIT RATES BY LENGTH OF SERVICE: 'UNEMPLOYMENT HIGH' AND 'UNEMPLOYMENT LOW'

Length of service	Unemployment high			Unemployment low		
	Birmingham	Glasgow	North Lanarkshire	Birmingham	Glasgow	North Lanarkshire
A. Males						
Up to 12 wks	47·0	32·9	18·8	98·6	50·6	42·6
12 wks up to 1 yr	32·1	29·1	8·7	43·5	34·6	25·4
1 yr up to 2 yrs	13·4	15·6	8·2	21·1	23·0	23·8
2 yrs up to 5 yrs	6·8	14·1	4·7	13·2	14·1	13·1
5 yrs+	2·4	6·3	2·3	3·1	4·2	8·7
B. Females						
Up to 12 wks	158·1	50·0	19·2	127·2	53·7	14·6
12 wks up to 1 yr	40·5	37·0	20·5	70·9	49·1	28·5
1 yr up to 2 yrs	22·4	34·7	19·6	41·8	36·7	29·3
2 yrs up to 5 yrs	15·6	22·2	10·8	12·4	14·6	27·7
5 yrs+	13·8	9·2	13·1	4·8	12·0	13·4

in periods when unemployment was relatively high and when unemployment was relatively low. Hence the quit rate was high for short-service employees and low for long-service employees regardless of employment opportunities. Again, in all length of service categories the quit rate tended to be higher for females than for males, and this held in each market and whether unemployment was high or low.

When we consider how quit rates by length of service categories may be affected by employment conditions, the most obvious point which emerges is that in Birmingham quit rates were exceptionally

[1] This is because when dealing with Birmingham females we have, in addition to a number of leavers whose reason for leaving was not known, a relatively high proportion of leavers whose age was not known at the date of leaving (see p. 205 above).

high relative to other market areas amongst short-service male and female employees. Even if no allowance is made for the slight under-counting of voluntary quits in Birmingham, these differences were substantial, and always in the same direction. Yet it is quite common to find, at both unemployment lows and highs, higher quit rates in Glasgow and North Lanarkshire for employees with at least two years' length of service.[1] One explanation for this could be that in the tight Birmingham labour market relatively few employees survived the 'induction crisis'. Those who remained were persons who found the conditions of work offered by the plant particularly agreeable or persons who displayed a high degree of inertia because they had a narrow view of the labour market, or because they were difficult to employ. If this were not so they could have 'voted with their feet' because alternative jobs were easy to obtain. The quit rate falls off sharply as one reaches employees with at least one year's service, for many of those with a high propensity for mobility left within the first year. On the other hand, many of those who would not have survived the induction crisis in Birmingham remained on in the plant in Glasgow and North Lanarkshire and left thereafter when employment opportunities arose. Quit rates amongst employees with at least one year's service were, then, relatively higher in Glasgow and North Lanarkshire and, in some cases, even absolutely greater than those in Birmingham.

While a distinction emerges between markets with different employment conditions, the effect of differences in employment conditions within the markets is not so clear. For Birmingham and Glasgow males the transition from an unemployment high to an unemployment low produced a sharp rise in the quit rate of employees with less than twelve weeks' service relative to the rise in quit rates for other groups. For males in North Lanarkshire, and for females in each area, the position was usually reversed, quit rates rising relatively slowly for those with less than twelve weeks' service. Nonetheless, whatever happens in relative terms, the absolute rise in quit rates was usually greatest for those in the two shortest length of service categories.

The relationship between quits and age, and the influence of employment conditions, can be investigated in a similar fashion. In other words, from a given stock the number of voluntary leavers in an age group over a yearly period is expressed as a percentage of the number of non-leavers in that age group at the beginning of the yearly period. The results are shown in Table 8.2.

[1] This is true even if we allow for 'not-knowns' by adjusting quit rates in Birmingham upwards by one-tenth.

There appears to be an inverse relationship between age and quits, the quit rate tending to fall as age rises. This does not always hold. For males the quit rate was usually higher for those aged 21–24 years than for those under 21 years, especially in Glasgow and North Lanarkshire, and this was due to the system of apprenticeship training. A high proportion of the male labour force under 21 years of age consisted of apprentices, the bulk of whom remained with a plant until their apprenticeship was completed. This depressed the quit

TABLE 8.2
ANNUAL QUIT RATES BY AGE: 'UNEMPLOYMENT HIGH' AND 'UNEMPLOYMENT LOW'

Age (years)	Unemployment high			Unemployment low		
	Birmingham	Glasgow	North Lanarkshire	Birmingham	Glasgow	North Lanarkshire
A. Males						
⩽20	21·4	10·3	3·4	32·4	14·1	5·5
21–24	22·8	30·2	14·0	31·7	50·8	37·5
25–34	11·6	19·8	7·5	23·5	27·3	23·3
35–44	6·7	9·9	3·6	11·5	11·1	13·1
45+	4·3	4·8	1·7	5·2	4·6	4·5
B. Females						
⩽20	47·9	18·1	12·7	86·3	65·7	26·2
21–24	71·6	52·8	28·5	56·8	52·4	35·1
25–34	27·4	19·2	15·8	73·8	33·0	28·1
35–44	17·5	8·9	5·3	25·9	11·2	12·2
45+	5·7	10·6	8·3	8·8	7·7	5·2

rate amongst this age group, particularly in Glasgow and North Lanarkshire where apprentice training was more widespread and the wastage rate amongst apprentices was very low. Subsequently, when apprentice training had been completed the quit rate rose sharply and thereafter fell with increasing age. Different factors apply to females. The high quit rates for those aged 21–24 years in the unemployment high period probably reflected quits by married women who left the plant and the labour force for domestic reasons. The quit rate for this group was not very sensitive to differences in employment conditions and differed very little over the unemployment high and low. On the other hand, the quit rate for those aged less than 21 years rose sharply over the unemployment low to a level much above that for those aged 21–24 years. Hence, although the quit rates are high for both groups they respond to quite different influences, so that one is stable over the cycle while the other is much more sensitive to changes in employment conditions.

The quit rate tended to be higher for females than for males in each age group and in all markets whatever the employment conditions, but the difference became progressively less important as age increased. Thus, while females of any age group tended to have higher quit rates than males, the contrast was least important both absolutely and relatively for groups aged 35–44 years or 45 years and above.[1] As far as the cyclical sensitivity of quit rates distinguished by age is concerned, it does seem that within each market quit rates changed least for males and females aged 45 years plus, for males aged less than 21 years and for females aged 21–24 years. The factors which give rise to the first conclusion are self-evident. The cyclical insensitivity of quit rates for males aged less than 21 years was due to the apprenticeship system, and that for females aged 21–24 years to the fact that a high proportion of such quits were due to domestic rather than to employment reasons.

4. QUIT RATES, LENGTH OF SERVICE AND AGE

Thus far we have proceeded by the rather unsatisfactory method of considering age and length of service as two independent factors which influence quit rates. However, age and length of service are likely to be positively related, so that from our previous analysis it is impossible to ascertain which factor is more important. We can observe that the quit rate falls with length of service and with age, but we cannot say whether this is due to increasing tenure or increasing age. This problem can be approached by applying chi-square tests to the data employed in the preceding section, and it will simplify the analysis if we discuss only the results obtained for Birmingham and Glasgow over the unemployment low. Table 8.3 shows the number of voluntary leavers expressed as a percentage of the number of non-leavers in each age and length of service category.

If age and length of service had *no* effect on quit rates, then the rate of voluntary quits would always be equal to the overall quit rate shown in the bottom right-hand corner of each section of the table. In other words, for all Birmingham males the quit rate over the unemployment low would be 12·9 per cent regardless of the length of service and age of the group considered. The application of this 'quit rate' to the number of non-leavers by age and length of service category yields the 'expected' number of leavers in each category on the assumption that age and length of service have no effect. The

[1] Once again this reflects changes in domestic responsibilities over the female life-cycle. The narrowing gap between male and female quit rates with increasing age is due to the fact that as females grow older they are much less likely to leave the labour force for primarily domestic reasons.

TABLE 8.3

NUMBER OF VOLUNTARY LEAVERS AS A PERCENTAGE OF THE NUMBER OF NON-LEAVERS BY AGE AND LENGTH OF SERVICE CATEGORIES: BIRMINGHAM AND GLASGOW, 'UNEMPLOYMENT LOW'

Age (years)	Length of service: Up to 12 wks	12 wks up to 1 yr	1 yr up to 2 yrs	2 yrs up to 5 yrs	5 yrs+	Total (age groups)
A. Birmingham males						
≤20	101·7	40·8	19·6	15·3	0·0	32·4
21–24	116·3	59·1	37·5	17·7	9·4	31·7
25–34	100·8	48·2	32·4	16·8	9·1	23·5
35–44	106·6	47·0	16·5	11·4	2·9	11·5
45+	75·0	28·2	11·2	9·9	1·5	5·2
Total (length of service)	98·6	43·5	21·1	13·2	3·1	12·9
B. Birmingham females						
≤20	155·6	94·6	68·4	39·5	0·0	86·3
21–24	476·9	50·0	17·9	5·7	64·7	56·8
25–34	150·9	160·4	156·0	11·1	6·9	73·8
35–44	88·1	54·3	26·1	28·2	4·4	25·9
45+	33·9	40·4	34·9	3·1	3·8	8·8
Total (length of service)	127·2	70·9	41·8	12·4	4·8	26·0
C. Glasgow males						
≤20	39·6	20·2	11·4	7·8	0·0	14·1
21–24	61·9	69·2	34·9	45·4	37·8	50·8
25–34	63·3	39·2	39·2	23·1	8·1	27·3
35–44	63·3	34·1	13·8	9·4	3·8	11·1
45+	24·7	17·3	17·7	6·8	1·7	4·6
Total (length of service)	50·6	34·6	23·0	14·1	4·2	14·1
D. Glasgow females						
≤20	55·9	71·4	87·9	31·3	85·7	65·7
21–24	73·3	118·2	39·1	35·4	31·6	52·4
25–34	59·3	39·0	30·7	21·1	32·3	33·0
35–44	53·1	20·8	9·6	6·2	7·3	11·2
45+	15·4	13·6	40·6	0·8	7·0	7·7
Total (length of service)	53·7	49·1	36·7	14·6	12·0	22·6

application of the χ^2 test confirms what casual inspection of the percentages in the above table would suggest, that the observed distribution of quits differs significantly from the expected distribution. In each case, χ^2 is significant at the 1 per cent level and this also holds if we apply the above procedure to the unemployment high.

This still does not tell us *which* variable was more important in determining quits, although a glance at Table 8.3 would suggest length of service had the greater influence. In each group, save Glasgow females, the decline in quit rates is more substantial and more regular as one moves through length of service groups (given by the bottom row) than when one moves through age groups (given by the right-hand column). For example, in the case of Birmingham males, the quit rate declined from 98·6 per cent for those with less than twelve weeks' service to 3·1 per cent for those with at least five years' service. On the other hand, the decline in quits with increasing age was much less spectacular, from 32·4 per cent amongst those aged 20 years or less to 5·2 per cent for those aged at least 45 years.

In each age category, quits tended to decline with increasing length of service in all market areas. This rule applies to all age groups in the case of Birmingham males and is relatively rarely upset in the other three cases considered. There is also an age-specific quit rate, the rate in a given service category tending to fall as age increases, but exceptions to this rule were more common. In each length of service category males aged 21–24 usually had the highest quit rate, and those aged 25–34 and 35–44 tended to have higher quit rates than those aged 20 years or less. This can, of course, be explained by the prevalence of apprentice training; but even if allowance is made for this, differences across the rows seem more substantial than those down the columns. Moreover, with females, where apprentice training does not apply, the inverse relationship between age and quit rates does not always hold. Thus, quit rates were often highest for those aged 21–24 years, and the ordering of quit rates between this group, those aged 20 years or less and those aged 25–34 years, is subject to substantial changes.

A more systematic manner of demonstrating the above argument is to isolate the effect of length of service and age by standardizing for each factor in turn. Let us illustrate this by reference to Birmingham males in Table 8.3. For this group the overall quit rate for those aged 20 years or less was 32·4 per cent. If length of service had no effect on quit rates then the quit rate for those aged 20 years or less would be 32·4 per cent, regardless of the length of time spent with the plant. The length of service distribution of voluntary leavers would then be the same as that of non-leavers. Using the length of service distribution of non-leavers and applying the standard rate of 32·4 per cent, one can generate the expected distribution of voluntary leavers by length of service and compare this to the actual distribution. The χ^2 obtained by comparing the actual and expected distributions of voluntary leavers by length of service is always significant at the 1 per cent level. In other words, we can reject the null hypothesis that the discrepancies between the distributions could have arisen by chance.

TABLE 8.4

χ^2S WHEN (A) AGE IS CONSTANT, (B) LENGTH OF SERVICE IS CONSTANT: BIRMINGHAM MALES, 'UNEMPLOYMENT LOW'

Age constant		Length of service constant	
Age (years)	χ^2	Length of service	χ^2
⩽20	117·7	12 wks	7·4
21–24	180·7	12 wks up to 1 yr	20·6
25–34	521·1	1 yr up to 2 yrs	50·0
35–44	1128·9	2 yrs up to 5 yrs	18·4
45+	1145·6	5 yrs+	188·4

When length of service is held constant and when the actual and expected age distribution of leavers in each length of service category is compared, we also have to reject the null hypothesis that discrepancies between the age distributions could have arisen by chance. Age, then, is an important factor but the χ^2s resulting from this process are generally much smaller than those obtained when age is held constant and length of service is varied. This can be illustrated by Table 8.4 which shows, for Birmingham males over the unemployment low, the χ^2s obtained when first age and then length of service is held constant.

All the χ^2s shown are significant at the 1 per cent level. However, while quits are inversely related to length of service *and* to age, it is clear that the former factor is the more powerful. When age is held constant and the actual and expected distributions by length of service are compared, the resulting χ^2s are always higher than those obtained when length of service is held constant. In each age group the actual number of quits is much greater than 'expected' in length of service categories up to 12 weeks and 12 weeks up to 1 year. The reverse appears for those with 2 years up to 5 years and 5 years or more service, the low quit rates being especially marked in the latter group. More detailed examination also reveals that the effect of length of service depends on the age group considered, as the older the age group the greater the difference between the actual and expected distribution of quits by length of service.

These conclusions are confirmed if the same analysis is applied to Glasgow males and to females in both Birmingham and Glasgow, nor are they much affected by considering the effect of service and age over unemployment highs. In all cases both age and length of service are inversely related to quits, and of these two factors length of service is the more important. Again, there is a tendency for the inverse relationship between quits and length of service to strengthen as the age of the group considered increases. While this applies to both males and females it is more noticeable in the former case. For females the progressively stronger relationship between quits and length of service as age increases is disturbed by the fact that length of service has most impact on the quit rate of those aged 25–34 years and those aged 45 years and above. There is also some evidence which suggests that the effect of age on quit rates does become stronger as length of service increases, but this relationship is less powerful than that found between length of service and quits as age increases.

5. SHORT-SERVICE EMPLOYEES AND STABILITY INDICES

We have already produced some evidence to the effect that quit rates were particularly high amongst short-service employees in Birmingham and that greater mobility amongst this group contributed substantially to the differences in overall quit rates, and hence separation rates, found between the labour market areas. The intention of this section is to demonstrate this point more fully by considering, first, quits and separations amongst employees joining and leaving a plant within a three-monthly period and, second, indices of labour stability which show the proportion of employees remaining with a plant over a yearly period.

The number of persons joining *and* leaving case-study plants voluntarily *within* each quarter was expressed as a percentage of the total stock of employees at the beginning of the quarter.[1] The quarterly rates obtained were then summed and divided by the number of quarters (32) to give unweighted averages for the whole

[1] In other words, the quit rate was calculated by a method similar in principle to that used in Chapters 6 and 7. In the case of Birmingham males and females, adjustments have been made for leavers whose reason for leaving was not known according to the procedure outlined on p. 169 above.

period. The same procedure was also applied to total separations. The resultant figures, for what we shall term 'short-service employees', are shown in Tables 8.5 and 8.6.

Quit rates and separation rates[1] for short-service employees were highest in Birmingham and lowest in North Lanarkshire. The differences between the markets were very substantial and are consistent with what would be expected given our knowledge of labour

TABLE 8.5

AVERAGE QUARTERLY MARKET QUIT RATES FOR SHORT-SERVICE EMPLOYEES

Area	Males				Females
	Skilled	Semi-skilled	Unskilled	All	
Birmingham	0·5	1·8	2·1	1·4	2·3
Glasgow	0·6	0·9	0·9	0·7	0·7
North Lanarkshire	0·1	0·1	0·2	0·1	0·3

TABLE 8.6

AVERAGE QUARTERLY MARKET SEPARATION RATES FOR SHORT-SERVICE EMPLOYEES

Area	Males				Females
	Skilled	Semi-skilled	Unskilled	All	
Birmingham	0·7	2·2	3·2	1·8	3·0
Glasgow	0·9	1·1	1·2	0·9	1·0
North Lanarkshire	0·1	0·1	0·7	0·2	0·5

market conditions. The quit rate, and hence the separation rate, for short-service employees were lower the worse were employment opportunities, and were particularly sensitive to differences in labour market conditions. Comparing Table 8.5 with Table 7.1 on p. 169,

[1] Strictly speaking, the term 'rate for short-service employees' is incorrect, for we are expressing, for those who joined and left within a quarter, the number of voluntary leavers and the number of separations as a percentage of the total stock of employees rather than as a percentage of all those employees joining during the period. The correct terminology would be 'the number of short-service employees as a percentage of the total stock of employees', but as this is so clumsy the phrase 'rate' is preferred.

we find that employees joining and leaving voluntarily within a quarter accounted for 1 in 4 of all voluntary quits in Birmingham in the case of males and females. In Glasgow, less than 1 in 6 of male quits and only some 1 in 10 of female quits were short-service employees, and the ratios in North Lanarkshire were 1 in 20 for males and 1 in 14 for females. Hence, one of the chief reasons for differences in overall quit rates between areas is that, as a market tightens, the quit rate rises particularly quickly for short-service employees.

Distinguishing males by occupational groups, we find the same pattern emerges. In each case, for skilled, semi-skilled and unskilled males, the proportion of all quits due to quits by short-service employees was highest in Birmingham and lowest in North Lanarkshire. The other interesting feature is that quits by short-service employees were a relatively small proportion of all quits for skilled employees and tended to be a particularly high proportion of unskilled quits. Within markets there was little difference between total males and females in this respect. The proportion of all quits due to quits by short-service employees was the same for males and females in Birmingham, was lower for females than males in Glasgow and higher for females than males in North Lanarkshire. As the bulk of separations were due to voluntary quits the above conclusions, *mutatis mutandis*, emerge from a comparison of Table 8.6 with Table 7.2, on p. 169 above.

The importance of high leaving rates amongst short-service employees can also be demonstrated by comparing stability indices for the separate labour market areas. These indices were constructed in the following fashion. The total number of persons employed by each plant at the beginning of each year was obtained and, of these, the number still employed by that plant a year later. The sum of the latter totals for all relevant plants, as a percentage of the sum of the former totals, yields the stability index for the market.[1] The lower the index the lower the proportion of employees remaining one year with a plant and vice versa. The results for the different market areas are shown in Table 8.7.

In each labour market area the stability index was lower for females than for males but, as many previous labour market studies have suggested, the bulk of the labour force, male or female, does not change employers very frequently. Even with females those remaining

[1] Let N_j be the number employed at the beginning of the year by the jth plant, S_j is the number from the original stock (N_j) still employed by the jth plant at the end of the year. Then the stability index for all plants in the market is given by

$$\left(\sum_{=1}^{n} S_j \div \sum_{j=1}^{n} N_j\right) . 100$$

TABLE 8.7

ANNUAL STABILITY INDICES: 1959–66

Year	Males				Females			
	Birmingham	Glasgow	North Lanarkshire	Small Town	Birmingham	Glasgow	North Lanarkshire	Small Town
1959	86·3	88·8	92·8	91·7	78·2	81·4	82·3	94·7
1960	81·9	87·6	89·5	94·0	71·6	77·6	80·1	89·5
1961	82·7	83·5	94·2	91·1	75·3	73·4	84·0	90·9
1962	87·6	80·5	90·1	83·7	77·8	71·0	78·0	50·0
1963	87·7	82·9	94·2	94·3	79·9	71·3	81·7	88·2
1964	87·3	86·7	91·6	93·7	75·2	79·4	81·3	89·6
1965	83·8	82·1	85·3	92·1	73·5	79·4	73·8	82·4
1966	83·5	81·7	90·2	95·1	72·6	75·9	75·6	88·5

with a plant over at least one year seldom amounted to less than 75 per cent of the labour force. For males, this percentage never fell as low as 80 per cent. Hence where problems of 'excessive' labour turnover arose they were mainly due to high quit rates and high separation rates during and immediately after the induction crisis.

This point can be illustrated more graphically by contrasting the results of Table 8.7 with those of Table 7.2 on p. 169 above. Table 7.2 showed very substantial differences in average market separation rates over the period 1959–66. For example, the market separation rate for males in Birmingham was approximately three times as high as the equivalent rates in North Lanarkshire and Small Town. This does not mean that the stability index in Birmingham was only one-third of that in North Lanarkshire and Small Town. Certainly the stability index was usually lower in Birmingham, indicating that a smaller proportion of the labour force in that area remained with a plant over a yearly period, but the differences between the areas were not very large. Again, for males, the stability index in Glasgow was, in five of eight years, *below* that for Birmingham although the market separation rate was, on average, higher in the latter area. The same conclusions apply to females. The contrasts between the stability indices derived for different areas were much less marked than those found when female separation rates were compared. In short, while the proportion of the labour force remaining with a plant over a yearly period tended to be lower in a 'tight' labour market, this was not the chief cause of differences in separation rates between markets. These varied substantially from area to area mainly because the rate at which employees joined and left plants within a yearly period was particularly sensitive to labour market conditions.

6. CONCLUSIONS

(i) The proportion of total separations due to voluntary quits did not show much variation between the different labour markets although, in the case of males, it was generally greatest in Birmingham. Within each market, the ratio of voluntary quits to total separations was always lowest for unskilled males. Dismissals for unsuitability and misconduct (the former reason being most important) were a relatively high proportion of total separations in Birmingham and of unskilled separations within each area. This reflected the tendency to lower hiring standards in tight labour markets.

(ii) Persons who had completed at least five years' service with a plant were a higher proportion of male than female leavers, whichever category of leavers is considered—total separations, voluntary quits or redundancies. Rather surprisingly, the reverse conclusion does not always apply, i.e. leavers with short periods of service did *not* always account for a higher proportion of female leavers. That this result occurred, despite higher quit *rates* and separation *rates* for females with less than one year's service, was due to the fact that females in *all* length of service categories had higher quit rates and separation rates than males. Comparing results for the different labour market areas, we find that short-service employees comprised a relatively high proportion of all types of male and female leavers in Birmingham.

(iii) Redundant employees contained, in both Birmingham and Glasgow, a high proportion of long-service and older workers and a low proportion of short-service and younger workers relative to voluntary quits and total separations. In each market area, voluntary quits relative to total separations had a high proportion in the younger age groups and vice versa. With males, those in the younger age groups were a larger proportion of quits and separations in the tight Birmingham labour market, but the reverse conclusion applied for females. This reversal was due to the higher participation rates in the older female age groups which prevailed in Birmingham. In like manner, we can explain differences between Birmingham females and females in other areas in the age composition of those dismissed for unsuitability and misconduct, and those leaving for 'other reasons'.

(iv) There was a strong inverse relationship between quit rates and length of service. In each length of service category female quit rates tended to be higher than male quit rates. The most noticeable difference between the markets was that quit rates were exceptionally high for both Birmingham males and females with less than one year's service. Quit rates were also inversely related to age although, because of apprenticeship training, male quit rates were usually higher for those aged 21–24 years than for those less than 21 years. Female quit rates in each age group tended to be higher than the corresponding rates for males, but this difference became less substantial as age increased.

(v) While age and length of service each had an independent effect on quit rates the latter was the more powerful influence. For both males and females in each age group, quit rates diminished as length of service increased. This relationship, while always significant, was most important for employees in older age groups.

(vi) Persons joining and leaving plants within a quarterly period were a higher proportion of all voluntary quits and all separations in Birmingham. In other words, the quit rate and the separation rate for short-service employees was particularly sensitive to differences in labour market conditions, being especially high in tight labour markets. This was also true of all employees with less than one year's service. In consequence, differences in the annual stability indices for the separate markets were much less noticeable than differences between market separation rates. The chief problem a plant faced in retaining labour in a tight market was, therefore, to reduce the high rate of quits and separations amongst those employees recruited in the recent past.

PART IV
LABOUR RECRUITMENT AND MOBILITY

CHAPTER 9
RECRUITMENT AND MOBILITY BY AREA AND INDUSTRY

1. INTRODUCTION

In this chapter and the next we shall examine the sources from which the case-study plants drew their labour force. In particular, we want to know the geographical areas, industries and occupations in which new recruits were employed prior to their engagement by the case-study plants. The patterns of mobility so revealed will indicate how managements have adjusted their recruitment policies in response to different labour market conditions. Given the specialized skills and experience required of a large part of the labour force in engineering, plants are likely to have some preference for labour with previous experience of engineering employment. However, the engineering plants upon which this study is based were operating in labour markets with very different levels and types of economic activity. While adequate supplies of engineering labour may have been available in Glasgow, this was much less true of the other labour market areas. The ability of plants to secure an adequate supply of labour in these different conditions will, therefore, have depended on whether they could vary their recruitment policies to suit the relevant labour market. The extent to which such adaptation took place in the five market areas is studied in this and the following chapter.

Labour market studies have indicated that the principal constraint on a plant's labour supply is determined by the reluctance of labour to undertake geographical mobility involving a change of residence. All job-changing between plants involves some geographical mobility, but it is clear from earlier research that most of these employer-changes take place within the same local labour market and that relatively few require movement between different labour market areas. Indeed, movement between geographically distinct labour markets appears to be the least common form of mobility associated with a change of employer: manual workers are much more likely to change their industry or occupation than the area in which they work. It follows that a plant's potential labour supply is primarily determined by the resources of the local labour market in which it is itself situated. Yet we have little detailed knowledge about the spatial dimensions of local labour markets in Britain, and we begin with an examination of certain aspects of their structure. The first concerns the pattern of geographical mobility itself, the extent to which plants obtained their labour from other units within the locality, and the extent to which labour was drawn from plants situated beyond the immediate market area. A comparison of this information with travel-to-work journeys, allows us to obtain estimates of the amount of residence-changing which accompanied employer shifts, before we examine travel-to-work journeys in their own right.

If the case-study plants were operating in clearly defined local labour markets, then the industrial and occupational characteristics of the local labour supply will have been important so long as managers had specific preferences regarding the previous industrial or occupational experience of recruits. If, however, labour could be hired from one industry or occupation as readily as from another, then the plant will have been able to adjust much more easily to different or changing employment conditions in its local market. Following on from our study of geographical mobility and travel-to-work journeys, therefore, we investigate the industrial pattern of recruitment. In particular, were recruits drawn mainly from a particular type of industry or from a wide range of industries, and did different patterns of behaviour emerge in the various labour market areas?

The present chapter is limited to these two forms of 'simple' mobility, geographical and industrial.[1] Section 2 examines the evidence arising from previous empirical research, in order to establish whether a consistent pattern of behaviour emerges and to outline the hypotheses which have been put forward to explain such behaviour. The section concludes by indicating, on the basis of this evidence, what we might expect to find in the present study. Section 3 analyses the extent of geographical mobility into and within our labour market areas, while travel-to-work journeys are the subject of Section 4. Section 5 deals with industrial mobility, and Section 6 attempts a brief summary of the conclusions reached.

[1] The third aspect, occupational mobility, is left until the next chapter, as is 'complex' mobility. The latter occurs when a job-change involves some combination of geographical, industrial and occupational mobility.

2. EXISTING EVIDENCE

A common finding of studies of geographical mobility is that short-distance mobility is more common than long. For example, the United States Bureau of the Census has estimated[1] that, while 12·4 per cent of the male labour force moved to another house in the same county between March 1966 and March 1967, only 3·3 per cent moved to another county within the same state and only 3·7 per cent to another state. A similar situation prevails in the U.K. Of those who changed house for work reasons between 1953 and 1963, 39·8 per cent moved less than 10 miles, while only 23·4 per cent moved more than 100 miles.[2] From our point of view the most interesting findings on the personal characteristics of migrants are that males have a greater propensity to undertake long-distance moves than females, and that mobility declines steadily after about the age of 30, but increases with the level of educational attainment and with the skill of the occupational group examined.[3]

This evidence relates to employer-changes involving a change in the employee's residence. We shall, however, also be discussing the travel-to-work journeys of employees. These are important, since a substantial body of evidence indicates that many employees undertake long-distance mobility reluctantly or not at all. For example, Jefferys found that only one-tenth of those recruited to plants in Dagenham and Battersea had moved in from beyond the London area, the great majority of the new employees having come from other plants within the immediate region.[4] In most cases, therefore, management must rely on obtaining new recruits from within the local labour market area, and the extent of that area will be largely determined by travel-to-work patterns. If employees are unwilling to undertake long journeys to work, and are reluctant to undertake long-distance mobility, this will limit the elasticity of labour supply to the plant. The employer may have to do without additional labour, or may have to pay a high price to overcome the inconvenience and other costs of commuting and/or changing residence. Commuting patterns can be expected to differ substantially from one market to the next, depending on population density, transport facilities, etc. They may also vary within the same market according to the type of labour involved, as the available evidence indicates that personal

[1] U.S. Dept. of Commerce: Bureau of the Census, 'Mobility of the Population of the United States, March 1966 to March 1967', *Current Population Reports: Population Characteristics*, Series P20, No. 171, April 1968, p. 26.

[2] Government Social Survey, p. 17.

[3] U.S. Dept. of Commerce, *loc. cit.;* Government Social Survey, p. 9.

[4] Jefferys, *op. cit.*, pp. 110–11.

characteristics have much the same impact on travel-to-work patterns as on long-distance mobility: thus, male workers are more prepared to undertake long journeys to work than females, and skilled workers than unskilled.[1]

In view of these results, we would expect to find that most of our case-study plants recruited the bulk of their labour from other establishments within a fairly short distance of their own location. Nonetheless some differences are likely to emerge between the separate labour market areas. We might expect plants in Birmingham to have attempted and been able to attract a relatively high proportion of recruits from other labour market areas. In contrast, long-distance in-migration appears likely to have been less common in Glasgow, given the easier labour market conditions in that area; in North Lanarkshire, too, employers may have been able to meet their labour requirements quite easily from within the immediate labour market. New Town presents a rather special case where long-distance mobility into the area has been considerable as the community has built up.

We can formulate further hypotheses which can be outlined briefly. Geographical mobility between plants will be greater for males than for females. Males will also have longer travel-to-work journeys. The length of these journeys will increase with skill and will also be dependent on the characteristics of the labour market area, and on the size of the establishment. Large plants will, other things being equal, have to draw their labour from a wider area than small plants, a factor which in turn may be reflected in wage differentials. Each of these propositions is examined in Sections 3 and 4 below.

The evidence relating to movement between industries shows that a substantial amount of industry-changing takes place as workers move between employers. Palmer's study in the United States and that of the Government Social Survey, based on the industrial attachment of workers at the beginning and end of a given period, show that about one-third of the workers had changed their industry of employment.[2] If the count is based on actual job-changes *throughout* the study periods, the amount of recorded industrial movement is increased. Using this method, Palmer concluded that three-quarters of the workers who changed their employer also changed their industry, while Bancroft and Garfinkle have shown that half of the employer-changes in the United States in 1961 involved a change of industry.[3]

[1] See, for example, J. H. Thompson, 'Commuting Patterns of Manufacturing Employees', *Industrial and Labor Relations Review*, Vol. 10, 1956.

[2] Palmer, *op. cit.*, p. 106; Government Social Survey, p. 76.

[3] Palmer, *op. cit.*, p. 78; Bancroft and Garfinkle, *op. cit.*

It is not clear that this substantial amount of industrial mobility exhibits any systematic pattern. The main question is whether the flow of labour between industries is random in character, or whether certain groups or pairs of industries experience an above-average or below-average interchange of labour. If the pattern were random, new recruits to a particular industry would be drawn from each other industry in the same proportion as the latter's share of total employment in the relevant labour market. If the proportion drawn from one industry is materially above or below its share of employment, this suggests an unusually strong or weak link between the two industries. The findings of earlier studies are not conclusive in relation to inter-industry linkages. For example, Reynolds, using two different approaches, arrived at the seemingly contradictory conclusions that 'the pattern of inter-industry employment is close to what one would expect on a random basis' and that industrial 'movement is more intense between certain pairs of industries than others'.[1]

If evidence is found of unusually strong or weak links between two or more industries, the explanation of this phenomenon is likely to be found in their respective occupational structures. If little interchange of labour takes place, this probably reflects different and incompatible skill requirements, rather than any impediment to industrial mobility as such. On the other hand, a close affinity, reflected in considerable movement, may be expected between industries with similar occupational structures, so that all that is necessary to change industry is to change employer—in which case it is the factors which determine the latter which are important. As Parnes points out, 'if industrial attachment is only a reflection of attachment to employer or occupation, and if the direction of movement between industries is merely a function of their occupational composition, then the concept of industrial flexibility in labor supply is superfluous. So long as workers are able and willing to change employers and to make specified occupational shifts, their current industrial affiliation becomes immaterial.'[2]

The high incidence of industrial mobility shown by earlier studies leads us to expect that the engineering plants in the present investigation will have drawn a substantial proportion of their new recruits from industries other than engineering. The earlier enquiries do not, however, provide any convincing evidence of close pairings between engineering and other industries; if anything, they suggest that we may well find inter-industry movement to be random. Nonetheless, it is likely that the proportion recruited from non-engineering units will vary between the markets, for the extent to which plants can exercise

[1] Reynolds, *op. cit.*, pp. 32 and 35 respectively. [2] Parnes, *op. cit.*, p. 96.

their preference for labour with experience in their own industry will depend largely on the availability of such labour in the local labour market. Thus in Glasgow, where unemployment has been high and employment in engineering declining, plants should have found it relatively easy to obtain a high proportion of new recruits with previous experience of engineering employment. One would expect the proportion of recruits from engineering to be lower in the other market areas. In Birmingham, the plants will probably have been obliged to recruit more workers from non-engineering organizations, given the scarcity of skilled and semi-skilled engineering workers. In the Other Scottish Areas, particularly New Town and Small Town, the case-study plants have built up a labour force in market areas with little existing engineering industry. Consequently, we would expect to find that the growth in employment has been achieved by drawing heavily upon the largely non-engineering activities in their respective areas, although the pattern in New Town will have been modified by in-migration to the area.

In sum, the case-study plants have been operating in labour markets with quite different characteristics, which will have had a marked effect on their ability to recruit the necessary supplies of labour. The first point to establish is the geographical limits of these labour markets, in other words, whether plants are, in fact, operating in *local* markets for manual labour, as distinct from regional or national markets. If the former proves to be the case, the important question for management then becomes the industrial and occupational flexibility of the labour force presently living within daily travel-to-work distance. If we find that labour has been drawn from a wide range of other industries, this will imply that management has a good deal of flexibility in its response to different labour market conditions. It is true that the recruitment of non-engineering labour may have necessitated greater expenditure on in-plant training, but a high degree of recruitment from other industries will nonetheless indicate that a plant's labour demands can be adapted to the supply conditions prevailing in a particular local market. Such a conclusion would be of considerable interest to those concerned with the development of manpower and regional policy. We return to this theme in Section 5, after discussing geographical mobility and travel-to-work journeys.

3. THE GEOGRAPHICAL PATTERN OF RECRUITMENT

In this and the following chapter, our analysis is based upon those persons recruited by the case-study plants for whom the necessary information relating to location, industry and occupation with the previous employer could be obtained from personnel records.[1] One difficulty is that such information was not available for a large proportion of recruits, so that the sample on which the following analysis is based may be atypical of the total inflow to the case-study plants. Fortunately we can obtain some indirect check on this possibility in the following manner. The total male[2] inflow is divided into two categories according to whether the occupational group with the previous employer was 'known' or 'not known'. Then, for each group in turn, the job taken up on joining the case-study plant is analysed by occupational groups. The percentage distribution of known and not-known recruits by occupation taken up with the case-study plant is shown in Table 9.1 along with the total number in each group.

The results of Table 9.1 are important for the following reason. We shall see from our subsequent analysis that a recruit's occupation with his previous employer will affect travel-to-work patterns, industrial mobility and the occupation taken up with the case-study plant. Thus if known and not-known recruits had the same occupational distribution on joining the case-study plants, this may be taken to reflect the fact that they had much the same occupational distribution with their previous employer; and if this is so, the known recruits are likely to be typical of the total inflow with regard to geographical and industrial mobility, travel-to-work patterns, etc.

In Birmingham and North Lanarkshire there was considerable similarity in the occupations taken up by known and not-known recruits on joining the case-study plants. In these two areas, therefore, we can be fairly confident that the known recruits on whom our subsequent analysis is based were reasonably representative of the total population of recruits. However, the results presented below for Glasgow will have to be treated more cautiously, as it is likely that the occupational structure of known recruits was biased in favour of the skilled and against the semi-skilled. 38 per cent of those whose previous occupation was known took up skilled jobs on joining the plant compared to only 25 per cent of those whose previous occupation was not known. In the case of the semi-skilled, the position was reversed, the percentages being 27 and 48 respectively.

[1] For reasons stated elsewhere (see p. 44 above and Appendix), we do *not*, in this and the following chapter, weight our results to take account of the fact that only a sample of personnel records was taken from the larger case-study plants.

[2] We ignore the female inflow because, as almost all females were semi-skilled, the following analysis has no relevance.

TABLE 9.1

PERCENTAGE OF RECRUITS IN EACH OCCUPATIONAL GROUP ON JOINING THE PLANT: PREVIOUS OCCUPATION KNOWN AND NOT KNOWN; MALES

Occupational group on joining the plant	Birmingham		Glasgow		North Lanarkshire	
	Previous occupation		Previous occupation		Previous occupation	
	Known	Not known	Known	Not known	Known	Not known
Skilled	11·4	8·6	38·1	25·3	23·2	29·9
Semi-skilled	53·7	52·8	27·0	48·0	36·8	35·5
Unskilled	18·9	17·7	18·6	14·2	21·4	18·3
All others	16·0	21·0	16·3	12·5	18·6	16·3
Total numbers	13,653	26,044	38,834	32,574	5,399	1,142

This apparent bias in the occupational structure of known recruits in Glasgow affects the interpretation of the results obtained only in a limited fashion, and it is only when we are dealing with the mobility pattern for *all* males in Glasgow that there is some danger of obtaining misleading results from basing our analysis on known recruits. Since labour mobility is affected by skill, the inclusion of too high a proportion of skilled employees amongst known recruits is likely to yield results different from those which would be obtained if the total population of recruits could be analysed. Hence, while dealing with *all* males for Glasgow, careful interpretation is necessary, and we have, wherever possible, provided an analysis by occupational groups as well as one for total males.

Two sets of information are used to analyse mobility between plants and the travel-to-work journeys within the local labour market areas. The precise nature of the data and the sampling procedures used are explained more fully in the Appendix but may briefly be recounted here. For those employees recruited during the period 1959–66, and who were *still employed at the end of the period*,[1] the address of the last previous employer was ascertained wherever possible. A comparison of this address with the location of the case-study plant gave the distance between the two establishments. This was then coded to one of five ranges: under 1 mile, 1 up to 2 miles, 2 up to 5 miles, 5 up to 20 miles and 20 miles and over.[2]

Travel-to-work journeys were examined by a rather different technique. To obtain the most up-to-date information, a sample was drawn from those employed at the end of our study period.[3] The home addresses of these employees were then compared with that of the plant to give travel-to-work journeys over the same set of distance ranges as are used to measure mobility between plants. The comparison of these two sets of data, although based on different populations, will give us some indication whether mobility between plants is likely to have required a change in residence or merely a change in an employee's travel-to-work journey.

We begin our discussion by considering the 'catchment' areas from

[1] When investigating the previous industry and occupation of recruits the analysis was always based on *all* those recruited over 1959–66 for whom the necessary data were available. However, ascertaining previous location was a much more arduous task and one which in view of the numbers involved could not be completed for all known recruits. This explains why this particular analysis was restricted to those employed at the end of the period.

[2] In presenting this information in the following pages, the first two ranges have been amalgamated into a single 'under 2 miles' category.

[3] The reasons for restricting the sample are similar to those described in footnote 1 above. For further details of the sampling procedure see the Appendix.

TABLE 9.2
DISTANCE OF RECRUITS' PREVIOUS EMPLOYER FROM CASE-STUDY PLANT

Distance range (miles)	Percentage								
	Birmingham		Glasgow		North Lanarkshire		New Town		Small Town
	Males	Females	Males	Females	Males	Females	Males	Females	Males
Under 2	28·7	55·1	39·0	47·0	22·4	27·7	24·6	39·0	72·7
2 up to 5	47·8	32·1	33·1	31·8	25·2	38·0	8·6	12·9	0·6
5 up to 20	17·2	10·6	20·7	17·8	46·3	30·5	38·4	33·5	7·1
20 and over	6·3	2·2	7·3	3·4	6·1	3·8	28·4	14·6	19·5
Total numbers	2,509	557	2,808	321	2,475	1,572	476	720	323

Note: The thirty-three females in the sample in Small Town were all recruited from plants less than 2 miles distant.

which the case-study plants drew their recruits. Table 9.2 shows, for males and females in each labour market area, the percentage of our sample who moved over given distances when leaving their previous employer to join a case-study plant. The total number of workers, upon which the percentages are based, is shown in the bottom row.

North Lanarkshire and New Town both obtained more than half of their male recruits from plants more than 5 miles away, although the comparable percentage for females was somewhat smaller. We remarked earlier that New Town was something of a special case, as the development of industry in the area was dependent upon workers being attracted from beyond the immediate vicinity. In the event, less than one-quarter of the male recruits came from plants less than 2 miles away, while 38·4 per cent came from those between 5 and 20 miles away. This covers an area containing many small and medium-sized settlements connected with a declining coal-mining industry, which, as we shall see, was a valuable source of labour for the New Town plants. More than one-quarter of the male recruits were drawn from beyond 20 miles away, and much of this mobility must have involved a change of residence as well as of employer.[1]

In North Lanarkshire the considerable recruitment of labour from establishments more than 5 miles away largely reflects the siting of the case-study plants. Four of the five plants studied were deliberately established at a point which, although itself fairly isolated, allowed recruitment from a number of existing population settlements several miles away. The figures in Table 9.2 show the effect of this policy, with more than half of the men, and about one-third of the women, coming from plants beyond the immediate locality. The experience of both these areas, although somewhat exceptional, nevertheless indicates the quite considerable extent to which plants are able to secure labour from a relatively long distance, provided that conditions are favourable.

In Birmingham, 76·5 per cent of the male recruits were drawn from other establishments within a 5-mile radius of the case-study plants, in Glasgow 72·1 per cent, and in Small Town 73·3 per cent. Most of the remaining recruits in the two cities came from between 5 and 20 miles away, but an unexpected point to emerge from the table is that the proportion drawn from beyond 20 miles was much the same in Birmingham and Glasgow, despite the very different employment conditions which prevailed in these market areas. We expected that the Birmingham plants would have recruited from establishments over a relatively wide radius, while those in Glasgow would have

[1] See pp. 244–5 below.

been able to meet most of their requirements from within the immediate area. We should however note that, because the level of labour turnover differed between the two areas, this result does not necessarily imply a similar *rate* of in-migration to Birmingham and Glasgow.

In Small Town, most labour was recruited from other employers within a very short distance of the case-study plant. The fact that no less than 72 per cent were recruited from establishments less than 2 miles from the case-study plant, and that recruitment in the intermediate ranges was small, indicates the isolation of this town. Nonetheless, a much higher proportion of the recruits to the Small Town labour force was drawn from establishments more than 20 miles away than in any other area save New Town. One-fifth of the sample came from plants beyond a 20-mile radius, and half of this group was recruited over distances exceeding 100 miles. While generalization is dangerous in view of the small numbers involved, it is probable that such recruits represented return migration on the part of employees who had previously moved away from the area. This suggests that the establishment of new plants in such an area might be able to check the rate of population drift, and also indicates that, even in more remote areas, plants need not be wholly dependent on local labour supplies; they may in certain circumstances be able to benefit by the return to the area of some workers who had previously left it.

Table 9.2 clearly illustrates that the catchment area for female employees is much more restricted than that for males. In each area a substantially higher proportion of females than males were taken from plants less than 2 miles distant. For example, in Birmingham, 55·1 per cent of females were recruited from plants less than 2 miles away, against only 28·7 per cent of the males. The same general conclusion, though with some modification in detail, emerges from a study of the other areas. To an even greater extent than males, female employees when changing employers tend to take up new employment at a plant within 5 miles of their old establishment. The corollary of this is that only a small fraction of female employees moved over distances greater than 20 miles when changing employers. Overall, then, we find that most job changes took place between employers within a 5-mile radius, with women tending more than men to move between plants relatively close to one another. Only a minority shifted over longer distances to work at the case-study plants, although plants in every market secured some workers from units more than 20 miles away.

It is possible that the area over which labour is recruited may vary with the size of the plant concerned. Small plants may be able to recruit most of their limited requirements from their immediate vicinity, as their demand for labour is likely to be small relative to the available supply. As plant size increases, however, management may be forced to recruit over a wider area. To investigate this possibility, plants were allocated to one of the three size ranges shown in Table 9.3 according to the number of male manual employees at December 31, 1966. For each size range, the percentage of recruits moving between plants over the specified ranges was calculated.

TABLE 9.3

DISTANCE OF RECRUITS' PREVIOUS EMPLOYER FROM CASE-STUDY PLANT, BY SIZE OF PLANT: BIRMINGHAM AND GLASGOW MALES

Distance range (miles)	Percentage					
	Birmingham			Glasgow		
	Number of employees			Number of employees		
	1–499	500–999	1000+	1–499	500–999	1000+
Under 2	46·8	21·4	21·9	44·3	38·3	28·7
2 up to 5	34·9	55·1	50·5	30·4	39·9	26·2
5 up to 20	11·6	16·6	22·2	19·3	14·1	35·7
20 and over	6·6	6·9	5·4	6·0	7·7	9·3
Total numbers	707	916	886	1,210	1,041	557

In both markets the small plants with 1–499 male manual employees obtained a higher percentage of recruits from nearby units than did the larger units. One in two of all recruits to small plants in Birmingham came from establishments less than 2 miles distant, and the ratio was almost as high in Glasgow, compared with a much lower proportion for plants with over 1,000 employees. The opposite side of the coin is that medium-distance mobility was more substantial into the large than into the small plants. While long-distance movement (from over 20 miles) into large plants was not much different from that into small plants, it is apparent that the former did recruit a relatively high proportion of employees over distances between 2–5 and 5–20 miles. This implies higher recruitment costs and, as we shall see shortly, longer and more costly travel-to-work journeys on the part of employees working in larger units.[1]

[1] It could, of course, be objected that such a general conclusion should not be based on a study of only two market areas where special factors might obtrude. For example, the above results might arise simply because of differences in the

4. TRAVEL-TO-WORK JOURNEYS AND RESIDENTIAL MOBILITY

If labour were completely immobile between separate markets, then the absolute limit to the elasticity of the labour supply would be set by the size of the working population within the relevant market. Of course, labour is not completely immobile, and plants in all our labour markets drew some new recruits from fairly distant areas. Yet we have also seen that such long-distance mobility was the exception rather than the rule, so that in all areas plants obtained the bulk of new recruits from nearby units. This will generally involve a change in the employee's travel-to-work journey, and the flexibility of employees in this respect is another factor which, like their propensity to undertake long-distance mobility involving residential movement, is important in determining the elasticity of labour supply to the individual plant. If employees are unwilling to undertake long journeys to work then our labour markets, particularly in the conurbations, will consist of a number of loosely related sub-markets. Employers will then find their catchment areas limited by the costs associated with journey to work as well as by the costs involved in geographical mobility.

Details of travel-to-work journeys are shown for each labour market and for males and females in Table 9.4. Although the sample on which Table 9.4 is based differs from that used to construct Table 9.2,[1] we can nonetheless use the results of these two tables to provide a general indication of the extent to which employees on changing plants were obliged to change their residence or merely their travel-to-work journeys. We will take this as our starting-point before proceeding to a discussion of travel-to-work journeys as such.

Considering for a moment only males, we begin with the New Town plants who are *known* to have attracted a relatively high proportion of their labour force from establishments more than 20 miles away. Table 9.2 showed that 28·4 per cent of the male recruits were previously employed more than 20 miles away from the case-study plants, while from Table 9·4 we can see that only 1·9 per cent of male employees had a travel-to-work journey over that distance. There is a similar disparity, though much less marked, at the 5–20 mile range, the respective figures being 38·4 and 22·9 per cent. It seems reasonable to assume from this that most of those previously employed more

spatial distribution of large and small plants throughout the market. To test this, an analysis was carried out for a group of Glasgow plants on the same trading estate. The same pattern was found, *viz.* the small plants drew a higher proportion of recruits from other units in their immediate vicinity.

[1] See p. 239 above.

than 20 miles away changed residence on joining the case-study plants, and some house-changing must also have accompanied moves between plants in the 5–20 mile range. If we lump these two groups together, then we find that 66·8 per cent of male recruits to New Town plants were previously employed at units more than 5 miles away, while only 24·8 per cent of male employees had travel-to-work journeys over that distance. While we cannot assume that the difference between these totals accurately measures changes in residence, it does seem reasonable to conclude that a considerable proportion of all male recruits to New Town plants must have changed residence on joining the case-study plants.

New Town is, of course, a somewhat special case. The nearest parallel is Small Town. 26·6 per cent of male recruits were previously employed more than 5 miles from the case-study plants, but only 4 per cent of male employees had a travel-to-work journey over this range, so that here too a relatively large proportion of residence-changing will have accompanied employer-shifts. In Birmingham and Glasgow, the percentage of males drawn from plants more than 5 miles away was not very different from the percentage with travel-to-work journeys over that distance, so that in both conurbations few new recruits seem likely to have changed their residence as a consequence of changing their job. Hence job-changing in the conurbations probably involved less residential mobility, but here, as in the Other Scottish Areas, only a very small proportion of employees had travel-to-work journeys exceeding 20 miles.

Using the same technique of comparing travel-to-work journeys and distance of recruitment, we would reach the conclusion that female recruits appear less likely than male recruits to have changed residence when changing their employer, except in Glasgow, where both sexes appear to have much the same tendency to move house. It is not surprising that, in general, women are less prone than men to change residence when changing their employer. For a married woman with domestic responsibilities, the acceptance of a job is likely to be heavily dependent upon its proximity to her present place of residence. This will probably have been selected for its accessibility to the *husband's* place of work, as the husband's wage will usually be the chief source of family income. Where the married woman changes her employer, this is likely to involve an adjustment of her travel-to-work journey without a concomitant change of residence. The conjunction of a change in employer and in residence is only likely to occur for the married female where a change of employer on the part of her husband necessitates a change in family residence. Employers seeking female rather than male labour will therefore be more heavily dependent on recruiting their labour forces from within the local labour market, especially if, as seems probable, single females are less willing to undertake long-distance mobility for work reasons than single males.[1] The employer seeking female labour may, of course, be able to obtain those previously employed in other labour market areas, but such in-migration will be largely dependent, certainly in the case of married females, on the availability of male jobs. In an area like New Town, therefore, the balance between the provision of male and female jobs is of critical importance.

The foregoing discussion indicates that, for both males and females, there have probably been substantial differences between the five market areas in the extent to which residence changes were associated with changes of employer. The greatest amount of residence-changing seems to have occurred for those joining the New Town plants. To some extent this was only to be expected given the nature of the area, but it does demonstrate that in certain circumstances geographical mobility may be considerable and that we have to bear in mind the interrelationships between local labour markets. Rather more unexpected is the extent to which workers moved into Small Town from other parts of the country, in order to take jobs with the case-study plants. This is likely to reflect return migration, and it demonstrates that even in remote areas plants are unlikely to be completely dependent for labour supplies on the local market. In the two conurbations the proportion of the recruits moving house when joining the case-study plants appears to have been a good deal lower than in New Town and Small Town. To some extent this is likely to be due to the fact that employment in the Birmingham and Glasgow case-study plants did not expand at the rate experienced in the Other Scottish Areas, but the supply of labour to individual concerns is bound to be more elastic in the conurbations given the ability of employees to change employers by adjusting their travel-to-work journeys.

It is worth while examining travel-to-work journeys in their own right, as it is apparent from Table 9.4 that there were considerable differences in travel-to-work patterns in the various markets. In North Lanarkshire one-third of the males had a travel-to-work journey exceeding five miles, but this reflects, as we have seen, the deliberate policy of siting the case-study plants in a location where they could draw labour from a number of different towns, so that it is difficult to generalize from the experience of the North Lanarkshire case-study plants. More important is the contrast between the conurbations and the smaller labour market areas, the characteristics

[1] For a list of some studies which have investigated the influence of sex on geographical mobility, see Hunter and Reid, *op. cit.*, p. 47.

TABLE 9.4
DISTANCE OF TRAVEL-TO-WORK JOURNEYS

Distance range (miles)	Percentage								
	Birmingham		Glasgow		North Lanarkshire		New Town		Small Town[1]
	Males	Females	Males	Females	Males	Females	Males	Females	Males
Under 2	42·1	68·5	42·0	62·6	29·3	34·1	58·9	69·9	92·5
2 up to 5	42·9	25·6	39·2	26·5	36·3	42·1	16·3	13·9	3·5
5 up to 20	14·3	5·6	17·7	10·6	34·4	23·4	22·9	16·2	4·0
20 and over	0·7	0·3	1·1	0·4	0·1	0·4	1·9	0·0	0·0
Total numbers	2,306	746	2,663	283	1,222	560	375	345	199

Note: 1. The eleven females in the Small Town sample all had travel-to-work journeys of less 2 than miles.

of travel-to-work journeys within the conurbations, and the difference between male and female work journeys within each labour market.

Travel-to-work journeys of less than 2 miles were more common in New Town or Small Town than in any other labour market area, both for males and females, but it is more interesting that even in the conurbations a high proportion of the labour force lived close to their place of work. 42 per cent of males and over 60 per cent of females in both Birmingham and Glasgow lived within 2 miles of their employment, and more than 80 per cent of the males and about 90 per cent of the females lived within 5 miles of the plant. Table 9.2 showed that in both Glasgow and Birmingham about three-quarters of the male entrants, and an even higher proportion of females, had been recruited from plants less than 5 miles away. This, together with the high percentage of employees with travel-to-work journeys of less than 5 miles, indicates that it is perfectly sensible to treat the conurbations as distinct labour markets. Moreover, within the conurbations it is apparent that most manual employees seek work close to their homes. The limits to the job search are set by the costs associated with long travel-to-work journeys and with residential mobility. In other words, the conurbations do not constitute single labour markets, with the interchange of labour taking place with complete freedom over the whole of their area. Rather there exists within each conurbation a series of sub-markets. Residential mobility and travel-to-work journeys across the boundaries of these sub-markets can and do occur, as does in-migration from other areas of the country to the conurbations as a whole. Nonetheless, such movement is restricted, so that it is possible to view the conurbations as a series of loosely connected sub-markets with weaker links to other labour market areas.

From the point of view of management, the relatively restricted employee movement within the conurbations means that a plant's potential labour supply is more limited than would be the case if mobility took place easily over the whole of the conurbation. It is therefore possible that imbalances may arise between the demand for and supply of labour in different parts of the city; the expansion of plants in one part may mean that the labour supply available in that area is fully used, while there are simultaneous surpluses of labour in another part of the city. Of course, the walls between the sub-markets are more porous than those between clearly distinct local labour markets, and it will be possible for plants in the conurbation to extend their catchment area in conditions of shortage. Yet given the costs involved in residential mobility and the evident preference of employees for short travel-to-work journeys, such a policy must involve additional recruitment and other costs.

A common feature of each labour market area is that women, and particularly married women, chose work closer to their homes than did men. In every market area, a higher percentage of women than men had a travel-to-work journey of less than 2 miles—69 per cent as against 42 per cent in Birmingham, 63 per cent against 42 per cent in Glasgow and 70 per cent against 59 per cent in New Town. Also, travel-to-work journeys were shorter for married than for unmarried women in each of these markets. For example, in Glasgow the percentage of married women with travel-to-work journeys of less than 2 miles was 74 per cent compared to 48 per cent for unmarried females. These figures, and the similar results obtained for the other areas, reflect the tendency for married females with domestic responsibilities to seek work which minimizes the time spent on travel-to-work journeys.

This has implications for the physical planning of new or existing communities. A widely accepted objective of manpower policies in economies where unemployment is low is that of encouraging greater labour force participation on the part of married women. This involves overcoming those obstacles to participation peculiar to married women, particularly those with young families. There are several possible approaches to this problem, but our evidence suggests that the location of plants relative to residential areas will be of fundamental importance in determining the proportion of married women in employment. Thus, a recent study of participation rates in Birmingham found 'the highest rates in council estates immediately adjoining concentrations of light industry. The lowest rates obtained on peripheral post-war estates far distant from centres of employment.'[1] In other words, management decisions or town-planning policies which locate housing away from industrial developments will seriously discourage married women from taking a job. Conversely, the careful siting of suitable factories in or close to housing schemes could greatly increase the labour force participation of married women.

Turning from our analysis of the effect of sex on travel-to-work journeys, we can now consider whether the skill of the employee has an appreciable effect on the travel-to-work journeys of male manual workers. The necessary data are assembled in Table 9.5 which shows the distance of travel-to-work journeys for skill groups in Birmingham and Glasgow.

[1] D. Eversley and K. Gales, 'Married Women: Britain's biggest reservoir of labour', *Progress*, No. 3, 1969, p. 145.

The unskilled tended to live nearer to their place of work than the skilled, with the semi-skilled occupying an intermediate position. In both conurbations only about one-third of the skilled workers lived less than 2 miles from the plant, compared with more than half of the unskilled. One reason for this difference is probably that the recruitment policy adopted by a plant varies with the skill-content of the job vacancy considered. Plants fill the great bulk of unskilled vacancies through the informal channels of casual callers and friendship and kinship networks. These methods are likely to favour those living close to the unit, and while informal channels are important for

TABLE 9.5

DISTANCE OF TRAVEL-TO-WORK JOURNEYS, BY OCCUPATIONAL GROUP: BIRMINGHAM AND GLASGOW MALES

Distance range (miles)	Percentage					
	Birmingham			Glasgow		
	Skilled	Semi-skilled	Unskilled	Skilled	Semi-skilled	Unskilled
Under 2	33·5	42·6	53·9	37·8	42·1	52·1
2 up to 5	47·6	44·4	38·0	41·6	37·4	37·1
5 up to 20	18·4	13·0	7·4	20·5	19·9	10·4
20 and over	0·4	0·2	0·8	0·2	0·5	0·3
Total numbers[1]	462	906	258	977	733	385

Note: 1. The total numbers on which this table is based are lower than those for Table 9.4 as we exclude all semi-skilled or unskilled non-production jobs.

all groups of manual employees, greater emphasis is placed on more active recruitment policies, such as advertising, in the case of the skilled. Thus information on skilled job vacancies is likely to be disseminated more widely through the market. The method of recruitment is, however, likely to be in most cases only the proximate cause of different travel-to-work journeys by the skilled and unskilled. More fundamentally, resort to advertising in the case of the skilled reflects their comparative scarcity in the market and the fact that skilled employees, whose earnings are generally higher than the unskilled, are more likely to find it worth while to undertake long travel-to-work journeys. Finally, skilled employees, with higher living standards, will more often buy or rent a house in more pleasant surroundings away from the industrial areas. Hence only some 10 per cent of the unskilled had travel-to-work journeys exceeding 5 miles compared to approximately 20 per cent of the skilled. However, while this implies that the boundaries of the submarkets within the conurbations were more tightly drawn for unskilled workers, too much should not be made of this distinction, since the great majority of male manual workers, irrespective of occupation, had travel-to-work journeys of less than 5 miles.

Finally, in our discussion of the limits to plant recruitment set by spatial considerations, we can consider, as a counterpart of Table 9.3, the travel-to-work journeys of employees by size of plant. In

TABLE 9.6

DISTANCE OF TRAVEL-TO-WORK JOURNEYS, BY SIZE OF PLANT: BIRMINGHAM AND GLASGOW MALES

Distance range (miles)	Percentage					
	Birmingham			Glasgow		
	Number of employees			Number of employees		
	0–499	500–999	1000+	0–499	500–999	1000+
Under 2	51·2	43·9	26·8	44·6	44·5	30·9
2 up to 5	39·9	43·3	47·2	39·5	39·8	37·3
5 up to 20	8·8	11·3	24·9	14·4	15·7	30·7
20 and over	0·1	1·5	1·1	1·6	0·1	1·1
Total numbers	1,078	520	708	1,391	797	475

Table 9.6, the plants are allocated to size ranges on the same basis as in Table 9.3, while the results are obtained by rearranging the data of Table 9.4.

Short travel-to-work journeys, of less than 2 miles, were more common in the smaller and medium-sized plants than in those employing 1,000 or more male workers, especially in Birmingham. Conversely, the largest establishments had a relatively high proportion of employees with travel-to-work journeys of more than 5 miles compared to those units with less than 500 men: for example, in Glasgow 16·0 per cent of those employed in the smallest plants had a travel-to-work journey over this distance compared to 31·8 per cent for the largest establishments. This finding is in line with what one would expect on *a priori* grounds and with our previous findings regarding work shifts between employers analysed by size of plant. As we saw, smaller units tended to obtain a relatively high proportion of new recruits from other establishments within their immediate neighbourhood. The implication we drew from this, which is confirmed by Table 9.6, was that in smaller units employees would tend to have relatively short travel-to-work journeys. In consequence, small units are likely to use more informal and less costly methods of recruitment and, other things being equal, to pay lower wages because they do not have to compensate employees for long travel-to-work journeys to the extent necessary in larger units. This is unlikely to be the only reason for the positive association between plant size and wage levels, but it is one factor which must be borne in mind when considering labour market behaviour.

5. THE INDUSTRIAL PATTERN OF RECRUITMENT

Our discussion of the industries from which recruits to the case-study plants came is based on those individuals for whom information on previous employment could be obtained. In most cases, the recruit's previous industry was ascertained through checking the name of his previous employer against trade and other directories, and his job with his previous employer was recorded where possible and coded in the manner described in the Appendix, thus enabling us to examine the interrelationships between occupational and industrial mobility. The proportion of not-knowns is lower in this case than when dealing with work shifts and travel-to-work patterns,[1] but it is still high. This prevents us from 'weighting' the results[2] and, of course, leaves us with the problem discussed previously in this chapter, of the characteristics of 'known' and 'not-known' recruits. However, while these qualifications must be borne in mind, particularly the latter, it is possible with the data at our disposal to carry through a fairly detailed analysis of the industries from which the case-study plants obtained their labour.

The measurement of any type of mobility raises problems of classification. It is always possible to 'increase' or 'decrease' the amount of industrial mobility by defining industries more narrowly or more broadly. Thus if we had taken a change between any of the 153 Minimum List Headings of the Standard Industrial Classification as a change in industry, industrial mobility would have been 'greater' than if we had used Main Order Heading Classifications. It is, however, possible to become so hypnotized by the problems of measurement that all analysis is stultified, and some decision has to be taken. For the purpose of the following analysis, industries have been classified by the 24 Main Order Headings of the 1958 Standard

[1] The reason for this is explained on p. 239, n. 1 above.

[2] See p. 237 above.

Industrial Classification, because we are interested in establishing broad patterns of movement, and because comparison by Minimum List Heading would suggest a specious accuracy which our data do not possess. Basing our classification upon Main Order Headings does produce a slight problem, in that Order VII includes both marine engineering and shipbuilding. Elsewhere in this volume we have taken the engineering industry to include the former but not the latter sector. An exception is, however, made in this instance in order to base the entire analysis on Main Order Headings, and the engineering industry should here be taken to mean the four Main Order Headings of Engineering and Electrical Goods, Vehicles, Other Metal Goods, and Shipbuilding and Marine Engineering.[1]

One last point of classification should be noted. Hereafter we ignore all movement between the four Main Orders of the engineering industry. We have decided on this course for two reasons. First, the system of classification is itself not entirely satisfactory, involving an element of arbitrariness in decisions to allocate multi-product plants to one Order or another. Second, since an analysis for each of the four Orders showed little difference between the Orders in the industries from which they drew labour, it considerably facilitates presentation of the results to treat the four Orders as one industry.

We begin our analysis with Table 9.7 which is based on the total number of males recruited by the case-study plants in each labour market area over 1959–66, for whom the industry of their previous employer was known. The percentage of this total recruited from each of the industries in the table is shown, as is the percentage of insured male employees engaged in these industries in each labour market area.[2] All industries have been included which employed more than 3 per cent of insured male employees in at least one of the labour market areas.[3]

Inspection of the first column for each labour market area in Table 9.7 reveals the spread of industries from which new recruits were obtained, while a comparison of the first and second columns for each labour market area allows us to relate the pattern of recruitment by industries to the existing industrial distribution of insured

[1] The decision to include shipbuilding only affects the analysis for Glasgow and Small Town. It can be defended on the additional grounds that much of the labour employed in shipbuilding yards has skills similar to those required by the other activities classified as engineering.

[2] As we are dealing with recruitment over 1959–66, the percentage of insured employees in each industry was calculated for each year over this period. The unweighted average of these percentages is shown in Table 9.7.

[3] It may be noted that an identical list of industries is obtained if the criterion used is 3 per cent of the recruits in at least one labour market area.

TABLE 9.7

PERCENTAGE OF RECRUITS AND OF INSURED EMPLOYEES, BY INDUSTRIES: MALES

Industry of previous employer	Percentage									
	Birmingham		Glasgow		North Lanarkshire		New Town		Small Town	
	Recruits (1)	Insured population (2)	Recruits (1)	Insured population (2)	Recruits (1)	Insured population (2)	Recruits (1)	Insured population (2)	Recruits (1)	Insured population (2)
Primary	0·4	0·2	0·2	0·2	0·6	1·1	1·3	3·1	18·4	30·2
Extractive	1·6	*	0·4	0·7	5·1	3·5	21·1	34·4	0·5	0·1
Food, drink and tobacco	4·2	3·6	2·3	4·8	2·0	1·6	2·1	2·2	9·9	8·7
Metal manufacture	12·0	5·5	5·4	3·0	18·0	17·5	7·0	1·3	2·6	0·2
Engineering	47·6	45·0	64·8	28·3	34·5	26·7	23·0	5·4	14·5	14·2
Paper and printing	0·7	1·9	0·6	3·0	1·6	1·1	3·5	4·7	0·7	0·3
Construction	6·3	8·7	5·9	11·5	9·4	14·0	8·4	12·2	13·7	12·0
Transport	4·8	4·6	4·8	10·5	5·5	6·6	5·7	6·7	10·6	6·2
Distribution	5·0	8·1	3·9	11·2	5·6	6·4	4·6	7·4	7·8	10·9
Professional and scientific services	1·1	4·5	0·4	5·8	2·0	3·7	0·4	3·7	1·2	2·1
Public administration	3·4	2·4	3·8	4·2	4·3	5·0	9·5	3·4	9·0	4·7
All other industries	13·1	15·5	7·6	16·8	11·5	12·8	13·5	15·5	11·1	10·4
Total numbers	13,080	—	9,320	—	3,730	—	1,072	—	424	—

Note: * Signifies less than 0·05 per cent.

employees. As the ratios of the figures in columns (1) and (2) approach unity, the process of recruitment becomes more random, in that recruits are drawn from other industries according to their share of total employment in the market rather than because strong or weak linkages exist between industries. A marked deviation from unity is *prima facie* evidence that industrial attachment does shape mobility; a ratio substantially greater than unity indicates strong linkages between the industry concerned and engineering, while the reverse would be indicated by a ratio substantially below unity.

In each area except Small Town the biggest single industry from which recruits to the case-study plants were drawn was engineering itself. For example, 47·6 per cent of those recruited to the Birmingham case-study plants had previously been employed in engineering establishments, and this figure rose to 64·8 per cent in Glasgow. Recruitment from other engineering units was less important in the Other Scottish Areas, but to some extent this was only to be expected, given the relatively low percentage of the labour force engaged in engineering, particularly in New Town and Small Town.

Another important source of recruits has been the metal-manufacturing industry. In Birmingham and North Lanarkshire this came second to engineering as a source of labour supply (providing 12·0 per cent and 18·0 per cent of the inflow respectively) and in Glasgow it was the third most important source of new recruits. Indeed, it is noteworthy that in each labour market area the proportion of recruits drawn from metal manufacture was greater than that industry's share of local employment. This suggests a more than usually strong linkage between the two industries, which, as we shall show later,[1] is probably due to certain common elements in their occupational structure. No other industry displayed any marked linkage with engineering and, for most of the industries shown, the ratio of recruits to employment share was slightly below unity.

Nonetheless, the overall impression gained from Table 9.7 is that the case-study plants drew extensively on a wide range of other industries for their recruits. Although in Birmingham, Glasgow and North Lanarkshire more than half of the recruits came from engineering and metal manufacture, this still left a considerable proportion to be drawn from other industries, and the degree of dependence on non-engineering labour was much more marked in New Town and Small Town. A further point of interest is the frequency with which the proportion of recruits drawn from an industry corresponded quite closely to that industry's share of employment in the labour market concerned. In other words, the pattern of recruitment often

[1] See p. 263 below.

appeared to be of an essentially random character, corresponding quite closely to the industrial distribution of employment in the local area.

This was not true of New Town, but then a close match between the industrial attachment of recruits and of employment in the local area was not to be expected given the rather unique situation of New Town, to which a high proportion of new recruits were attracted from outside the local labour market. Thus, although a very small proportion of the insured population in the area was engaged in the engineering industry, about a quarter of the recruits to the case-study plants were drawn from that industry, due to the in-migration of engineering workers from other parts of the country. Nevertheless the plants have had to rely on non-engineering labour for three-quarters of their labour requirements, and the declining coal-mining industry was, next to engineering, the most important single source of labour, accounting for one in five of all new recruits.

The Small Town plants were in a similar position to those in New Town, in that they were established in an area with virtually no engineering employment other than that which they themselves provided. This clearly limited the extent to which either of the two case-study plants in Small Town could recruit labour from the engineering industry, so that, as in New Town, the plants were forced to rely on recruits drawn from a wide range of industries. The largest single source of labour in Small Town was the primary sector of agriculture, fishing and forestry; 30·2 per cent of the insured labour force were employed in primary industry and 18·4 per cent of those recruited to the case-study plants came from this sector. The relationship between recruitment from, and employment in, the basic sector of the local economy was, then, much the same in both Small Town and New Town.

The experience of these plants in New Town and Small Town is significant in that, despite the lack of any significant supplies of indigenous engineering labour, the establishments concerned still secured the labour necessary for their growth. Discussions with plant managers confirmed that the plants in these areas were confident of their ability to secure any additional labour necessary for future expansion. Managers were well satisfied with the quality of past recruitment and with the general labour supply position, despite an extremely rapid expansion in engineering employment over 1959–66 and the severely restricted base from which this expansion began. The plants were able to expand by adapting their recruitment policies to the supply conditions prevailing in the relevant local labour markets, for example, by recruiting heavily from the coal industry in New

Town, and from the primary sector in Small Town. It is true that in each case the industry's share of recruitment was below its share of local employment, probably reflecting its unfavourable age and occupational structure. Nonetheless, despite these obstacles, the engineering units *have* recruited heavily from the basic industries in the local market. This must reflect both that the plants could not acquire experienced engineering labour locally, and that they felt that labour from coal-mining and primary industry could be trained economically for factory jobs. Indeed, this argument must have a much wider application in view of the fact that the engineering plants in New Town and Small Town recruited respectively some 75 per cent and 85 per cent of their labour from non-engineering employments.

The experience of New Town and Small Town suggests that in certain circumstances the supply of labour to any particular industry may be much more flexible than a casual inspection of the existing employment distribution might suggest. Of course, plants in New Town and Small Town employed a relatively high proportion of semi-skilled labour and a relatively low proportion of skilled labour[1] and, being new units, they were able to install a technology and a method of working which, together with extensive induction training, enabled them to draw heavily on non-engineering labour. Yet this should not detract from the conclusion that careful selection and a flexible management response may enable plants to flourish in what appears to be a hostile labour market environment. The implications for regional policy are self-evident.

Turning more directly to labour recruited from other engineering units and excluding New Town as a special case, we find that in three of the four remaining areas, Birmingham, North Lanarkshire and Small Town, the percentage of recruits obtained from other engineering units corresponded closely to that industry's share of the insured population in the area concerned. Engineering accounted for 47·6 per cent of recruitment in Birmingham and for 45·0 per cent of the insured labour force in that area. The comparable figures for North Lanarkshire were 34·5 per cent and 26·7 per cent and for Small Town, 14·5 per cent and 14·2 per cent. The exceptional position in Glasgow stands out clearly, where 64·8 per cent of recruitment was from other engineering units, while the industry accounted for only 28·3 per cent of the insured population.

To some extent this contrast between Glasgow and the other areas is due to the bias in the Glasgow data in favour of skilled recruits,[2]

[1] See Table 3.5, p. 53 above.

[2] See p. 237 above.

for it is reasonable to suppose that a high proportion of such labour would be drawn from other engineering plants. Yet this is by no means the whole story, since Glasgow plants recruited heavily from other engineering units at most skill levels.[1] The main explanation of this difference in behaviour seems to lie in the high level of unemployment in Glasgow and the rapid reduction in engineering employment. This meant an abundance of labour available with engineering experience, and the case-study plants naturally exercised a preference for this labour over that from other industries. On the other hand, severe recruitment difficulties in Birmingham and the rapid expansion of engineering employment in North Lanarkshire and Small Town compelled plants in these areas to recruit labour over a wide range of industries. In both areas, therefore, the pattern of recruitment was very close to the prevailing pattern of employment and remarkably close to what we would expect on a random basis. This suggests that linkages between industries in terms of labour supply will emerge most clearly where the labour market is particularly slack, so that employers are able to exercise their preference for employees with the appropriate experience and skills.

Before proceeding to discuss the occupational characteristics of male recruits by industry, we should examine briefly the pattern of industrial mobility displayed by female recruits to the case-study plants. This is done with the aid of Table 9.8 which has been constructed on the same principles as Table 9.7 for males. Small Town is excluded because of the small number of females recruited over 1959–66, and the list of industries shown in Table 9.8 does not correspond exactly with that in Table 9.7, since it includes only those industries which employed 3 per cent or more of insured *female* employees in at least one labour market area.

In Birmingham, Glasgow and North Lanarkshire, the most important single source of female recruits to the case-study plants was the engineering industry itself; 47·6 per cent of female recruits were drawn from engineering in Birmingham, 27·8 per cent in Glasgow and 32·8 per cent in North Lanarkshire. Engineering also accounted for almost one-fifth of the recruits in New Town, although a slightly greater proportion came from the distributive trades in that market. Distribution was, in fact, the most important, or second most important, source of female labour for the case-study plants in each area, a situation which derived from that industry's position as a major employer of female labour.

It is important to observe that for females there was not the same degree of correspondence between the share of recruits coming from

[1] See Table 9.10, p. 263 below.

TABLE 9.8

PERCENTAGE OF RECRUITS AND OF INSURED EMPLOYEES, BY INDUSTRIES: FEMALES

Industry of previous employer	Percentage							
	Birmingham		Glasgow		North Lanarkshire		New Town	
	Recruits (1)	Insured population (2)	Recruits (1)	Insured population (2)	Recruits (1)	Insured population (2)	Recruits (1)	Insured population (2)
Food, drink and tobacco	7·8	4·9	11·7	7·4	5·4	4·6	7·8	4·0
Engineering	47·6	32·2	27·8	8·5	32·8	21·6	19·2	3·7
Textiles	2·3	0·9	7·5	2·9	8·1	5·2	11·7	8·6
Clothing and footwear	0·5	1·3	9·3	6·9	8·2	7·4	2·2	1·6
Paper and printing	0·8	2·2	1·9	3·7	3·9	1·2	6·5	4·8
Transport	2·6	1·9	4·7	3·6	3·6	4·0	3·7	5·0
Distribution	10·6	15·4	17·9	25·2	22·6	17·4	22·9	24·5
Insurance, banking and finance	0·1	3·2	0.2	3·0	0·5	1·1	0·6	0·9
Professional and scientific services	2·9	15·8	4·0	20·4	1·9	18·2	4·2	19·4
Public administration	1·4	1·6	0·8	1·8	1·0	3·4	5·2	2·4
All other industries	23·4	20·6	14·2	16·6	12·0	15·9	16·0	25·1
Total numbers	5,054	—	1,448	—	3,457	—	1,094	—

engineering and that industry's share of total employment as was observed in the case of males. If we express the percentage of recruits from engineering as a ratio of the percentage of the insured labour force in engineering, then the results obtained for males in Birmingham, Glasgow, North Lanarkshire and New Town are 1·1, 2·3, 1·3 and 4·3. For females, the corresponding figures are 1·5, 3·3, 1·5 and 5·2. In other words, in the case of females, linkages between engineering plants appear to have been stronger, and this suggests that a smaller proportion of females than males leave engineering employment when changing employers. This is frankly a mystifying result as, given the higher skill-content of the male labour force, we would expect that the tendency for industry changes to accompany employer changes would be less, and not more, common for males. Nonetheless, the evidence appears clear enough.

Inspection of Table 9.8 shows that the recruitment of female labour did not simply mirror the industrial structure of the insured labour force. For example, the proportion of engineering recruits drawn from professional and scientific services, which accounted for a substantial part of female employment, was extremely small in every market area and a great deal less than its share of employment. The linkage between the two industries is extremely weak, presumably because of occupational and perhaps social differences, differences which are reflected on a somewhat smaller scale if one repeats a similar analysis for Insurance, Banking and Finance. Recruitment is not, then, a random process to which previous industrial attachment is irrelevant, but the walls between *manufacturing* industries do not appear to be very high so that there is little impediment to inter-industry mobility. It does appear, however, that females may behave rather differently from males and that in both cases the pattern of recruitment will, to a large extent, be determined by the employment conditions in the local labour market areas as well as by the industrial structure of the area.

The most interesting point to emerge from our discussion of industrial recruitment is the extent to which the case-study plants depended upon non-engineering establishments for their labour supplies. It does not, of course, follow that labour drawn from engineering or non-engineering establishments was of equal initial suitability to the case-study plants. Given the specialized nature of much engineering work, one would expect a higher proportion of those recruited from engineering plants to have had the necessary skills and experience than those coming from elsewhere, and it may well be that the more plants were obliged to recruit from other industries, the more internal training they will have had to undertake.

TABLE 9.9

PERCENTAGE OF RECRUITS FROM ENGINEERING AND ALL OTHER INDUSTRIES, BY OCCUPATIONAL GROUP WITH PREVIOUS EMPLOYER: MALES

Occupational group with previous employer	Percentage									
	Birmingham		Glasgow		North Lanarkshire		New Town		Small Town	
	Engineering	All others	Engineering	All others	Engineering	All others	Engineering	All others	Engineering	All others
Skilled	22·1	7·6	57·1	21·5	49·7	18·8	46·1	16·7	60·4	17·1
Semi-skilled	46·8	16·1	30·5	10·9	32·2	14·0	38·7	7·7	17·9	2·6
Unskilled	11·1	16·2	5·9	16·2	5·1	14·9	3·9	14·0	14·3	9·1
Non-production	6·5	12·1	2·7	13·3	4·1	10·6	4·3	9·4	1·8	5·7
Miscellaneous	13·5	48·0	3·8	38·1	8·8	41·8	7·0	52·3	5·6	65·5
Total numbers	4,936	5,477	2,902	1,364	1,207	2,314	230	778	56	351

We can examine the previous industry and occupation of male recruits through Table 9.9. The data were initially analysed for each industry included in Table 9.7, although in Table 9.9 the only distinction is between engineering and all other non-engineering industries. The total number recruited from these two broad industrial categories in each market area is shown at the foot of the table. These entrants were allocated to one of the occupational groups shown, on the basis of their job with their previous employer, and the percentage in each occupational group was calculated.

The occupational groups are the same as those used throughout this study. In other words, 'skilled' and 'semi-skilled' recruits are employees whose previous occupation required a type of skill specific to, or closely allied to, the engineering industry. The category unskilled workers is self-explanatory, and non-production workers are semi-skilled and unskilled manual workers not directly engaged on production. The 'miscellaneous' group requires some further explanation. It consists of a small number of persons previously employed in clerical and staff work, a larger category of workers whose skill could not be ascertained accurately (e.g. welders, inspectors) and, most important, a wide range of employees regarded as skilled or semi-skilled by their previous employers whose skills were not transferable to engineering (e.g. face workers in the coal industry; tractor drivers; bakery workers and so forth).

Most of those recruited from the engineering industry did possess some experience or training relevant for either skilled or semi-skilled engineering work, while those drawn from industries other than engineering were generally either unskilled or had some non-engineering skill. For example, in Birmingham almost 70 per cent of those coming from the engineering industry had been engaged on jobs with close affinities to skilled or semi-skilled engineering work. This was true of only about 24 per cent of those from other industries.[1] The proportions differed between market areas, but the overall pattern remained the same.

For almost every non-engineering industry the percentage of recruits previously engaged in skilled or semi-skilled work relevant to the engineering industry was small. The only exception occurred with metal manufacturing, where the proportion of recruits with the

[1] It will be noted that, in Birmingham, 13.5 per cent of the recruits from engineering plants were in the miscellaneous occupational category. This was a good deal higher than any other area, and was largely accounted for by the recruitment of welders, inspectors, etc. Whether properly referred to as skilled or semi-skilled, these are evidently occupations allied to engineering, so that the observation that 70 per cent of those from engineering plants were in skilled or semi-skilled occupations understates the true position.

appropriate experience or training was similar to that shown for the engineering industry itself. This explains our finding in Table 9.7, that in each market area the proportion of male recruits drawn from metal manufacture was in excess of that industry's share of employment in the local area, i.e. that there appeared to be an unusually strong linkage between these two industries. The result is due to certain similarities in occupational structure because some of the tasks carried out by the respective work forces are common to both industries.

The data used in Table 9.9 can also be looked at in another way, in order to show the proportion of the recruits in a particular occupational group who were previously employed in engineering or non-engineering industries. This is done in Table 9.10 for Birmingham and Glasgow only.

TABLE 9.10

PERCENTAGE OF THOSE IN EACH OCCUPATIONAL GROUP WITH PREVIOUS EMPLOYER RECRUITED FROM ENGINEERING AND FROM ALL OTHER INDUSTRIES: BIRMINGHAM AND GLASGOW MALES

Occupational group with previous employer	Birmingham			Glasgow		
	Engineering	All others	Total numbers	Engineering	All others	Total numbers
Skilled	72·5	27·5	1,506	85·0	15·0	1,951
Semi-skilled	72·3	27·7	3,194	85·6	14·4	1,034
Unskilled	38·1	61·9	1,434	43·6	56·4	392
Non-production	32·7	67·4	981	30·0	70·0	260
Miscellaneous	20·3	79·7	3,298	17·5	82·5	629

As our previous analysis would suggest, we find that in both markets the great majority of those previously in skilled or semi-skilled work were recruited from engineering employment. For example, in Birmingham, 72·5 per cent of those recruited from skilled jobs came from the engineering industry, while in Glasgow the comparable figure was 85·0 per cent. Conversely, only four-tenths of the unskilled and three-tenths of the non-production workers were recruited from engineering plants; the majority of such workers came from non-engineering establishments. While the overall pattern in Birmingham and Glasgow was similar, it is noticeable that a smaller proportion of each occupational group (save non-production and miscellaneous workers) was recruited from engineering plants in

Birmingham than was the case in Glasgow, and this despite the fact that the engineering industry accounted for a much larger proportion of insured male employees in the former area.

These patterns of behaviour are consistent with the known labour market conditions prevailing in these different areas over the period 1959–66. In Glasgow, the level of engineering unemployment was high relative to Birmingham, but, even more important, employment in engineering was falling rapidly in Glasgow while it was stable in Birmingham. Glasgow was, therefore, a 'buyers' market' and engineering plants were easily able to recruit labour with previous experience of the engineering industry. The result was to produce a largely closed system for the skilled and the semi-skilled occupations where the great bulk of new recruits were drawn from other engineering establishments. For unskilled workers, the proportion recruited from engineering in Glasgow was still high relative to other areas, though the bulk of those recruited from unskilled jobs even in Glasgow came from non-engineering industries. This reflects the greater mobility of unskilled workers between industries due to the fact that previous experience of the engineering industry confers a lesser advantage to the potential recruit.

In Birmingham, with the level of unemployment extremely low through most of the study period, employers were forced to obtain recruits from a wider field. While other engineering plants remained the most important source of skilled and semi-skilled labour, one-quarter of those in skilled and semi-skilled occupations had previously been employed outside the engineering industry. This process of recruitment from non-engineering plants was carried even further in the Other Scottish Areas. In North Lanarkshire, New Town and Small Town, employment in engineering expanded extremely rapidly over 1959–66. In the two latter areas the expansion took place in labour markets with relatively few reserves of engineering labour and with low levels of turnover in established engineering plants. Engineering plants seeking additional labour had therefore to recruit over a wide range of industries, so that in each occupational group the proportion drawn from engineering plants was lower than the respective proportions in Birmingham and Glasgow.

It is clear that, in a market with substantial reserves of engineering labour, employers are able to exercise a preference for labour with previous experience of engineering employment. The most important effect of this is that an extremely high proportion of skilled and semi-skilled labour is drawn from other units in the engineering industry. Occupational linkages are less important in the case of unskilled and non-production workers where recruitment takes place over a wide field. As a market tightens and particularly where engineering employment is expanding rapidly, employers are forced to recruit more heavily from non-engineering plants. This produced recruitment patterns in Birmingham, and particularly in the Other Scottish Areas, very different from those found in Glasgow. The widening of the area of recruitment in terms of the industries from which individuals are drawn is, therefore, a very important part of the adjustment process to tighter labour markets.

6. CONCLUSIONS

(i) As earlier studies have suggested, most job changes involved only a limited amount of geographical mobility. Save in New Town where rather special factors were at work, only a small fraction of employees were recruited from establishments more than 20 miles from the case-study plants. Movements over this range were least common for female employees, reflecting their propensity to seek employment close to their existing residence. The tendency to recruit employees from other establishments in the vicinity of the case-study plants was most apparent in the Birmingham and Glasgow conurbations.

(ii) A comparison of travel-to-work journeys with distance moved between plants on changing employer suggests that the great bulk of employer changes involved an adjustment of travel-to-work journeys rather than a shift in residence. The incidence of residence changing did, however, vary substantially from one market to another. It was high in Small Town and especially in New Town, but was much lower in the Birmingham and Glasgow conurbations, where employer shifts could more easily be accomplished by adjustments to travel-to-work journeys.

(iii) In New Town and Small Town, a very high proportion of employees lived within 2 miles of their place of work, reflecting the small size of the two communities. Yet even in the conurbations, more than 40 per cent of male employees had a travel-to-work journey of less than 2 miles. This suggests that the elasticity of labour supply to a plant will depend to a considerable extent on the existing population within a fairly short travel-to-work journey of the unit, and that there are a number of loosely linked sub-markets within the conurbations. This is particularly so in the case of female employees who had a noticeable tendency in both Birmingham and Glasgow to seek work within 2 miles of their residence.

(iv) The length of travel-to-work journeys was directly related to skill in both Birmingham and Glasgow. In other words, the proportion of employees with a travel-to-work journey of 2 miles or more was greatest for the skilled and least for the unskilled, with the semi-skilled occupying an intermediate position. Examination of the distance moved between employers and of travel-to-work journeys suggests that in both cases the catchment area of recruitment was widest for plants with more than 1,000 employees and narrowest for those with less than 500 employees. This indicates that part of any advantage which large units may enjoy in terms of higher wages must be dissipated in compensating employees for relatively long travel-to-work journeys.

(v) As regards the industrial pattern of recruitment, the biggest single source of male labour in every area was the engineering industry itself. Engineering was also the most important single source of female labour in all areas save New Town, and for males and females the proportion of recruits drawn from engineering always exceeded that industry's share of employment in the respective areas. This was also true of male recruitment from metal manufacture. Yet, although other industries tended to provide a share of recruits a little below their share of employment, the overall impression is that the plants drew on a wide range of industries for their recruits. This was especially true of plants in New Town and Small Town, where a large proportion of recruits were drawn from coal-mining and primary industry respectively. An exception to the general rule does, however, occur in Glasgow, where plants were able to exercise their preference for labour with prior experience of engineering due to the declining level of employment in that industry and the high level of unemployment in the market. This suggests that the pattern of industrial recruitment will be affected by the employment conditions prevailing in the market as well as by the industrial structure of the area. There is also an important distinction to be drawn between male and female employees, as the latter seem to have a greater tendency to remain in the engineering industry when changing employers.

(vi) Most males recruited from other engineering establishments had training or skills relevant to the case-study plants, in contrast to males from non-engineering units. The occupational characteristics of those recruited from metal manufacturing corresponded closely to those recruited from engineering, explaining the strong linkages between these two industries noted in conclusion (v) above. It was evident, however, that a lower proportion of those in each occupational group (except 'non-production' and 'miscellaneous') were recruited from engineering plants in Birmingham than in Glasgow. This suggests that the tighter market conditions in Birmingham forced plants to extend the amount of recruitment from outside the industry even for skilled and semi-skilled labour. This is an important means of adjustment to tighter labour market conditions, and suggests that the labour force is more flexible than a superficial reading of employment and unemployment statistics might convey.

CHAPTER 10

OCCUPATIONAL AND COMPLEX MOBILITY

1. INTRODUCTION

We have seen that, while plants did secure most of their recruits from other units in the same local labour market, they did vary their pattern of recruitment both spatially and, even more so, industrially, to meet the labour supply conditions prevailing in the relevant local labour market. We must now consider whether the same flexibility of response was shown in relation to the occupational pattern of recruitment. The plants were not just recruiting labour into the engineering industry, but into a series of specific jobs, embodying a variety of types and levels of skill. Some of these, particularly unskilled or non-production jobs, required aptitudes which are not specific to one industry, but other tasks undertaken by skilled and semi-skilled workers in engineering require skill and experience not shared by many employees outside that industry. Hence, if plants were obliged to go beyond engineering for labour, they are likely to have found it difficult to fill vacancies by the recruitment of workers who had already acquired the necessary expertise for the jobs in hand. Outside Glasgow, therefore, and particularly in the Other Scottish Areas, we might expect considerable occupational mobility to have occurred as plants were forced to take on labour without previous experience of engineering employment. The main part of the following discussion is, therefore, devoted to an examination of the extent of occupational mobility in the different labour market areas.

We also deal with 'complex' mobility, which arises when some combination of geographical, industrial and occupational mobility occurs. So far we have discussed these aspects of mobility in isolation from one another. However, a worker's job change is as likely to be complex as simple in character, and it is therefore appropriate to conclude the discussion of the sources of recruitment by examining the relative frequency of both simple and complex moves as workers were recruited to the case-study plants. The next section outlines the evidence relating to occupational mobility and considers what implications it has for the present investigation. Sections 3 and 4 look at the empirical data on occupational mobility arising from this study; first, from the point of view of the employer and, second, from the point of view of the employee. Turning from occupational mobility as such, we take up our second main area of enquiry in Section 5 which discusses the frequency of the various types of simple and complex mobility. The conclusions derived from this analysis are summarized in Section 6.

2. EXISTING EVIDENCE

The biggest problems which arise when discussing the results of earlier studies of occupational mobility concern the method of classification used. Although such problems loom large in any investigation of geographical and industrial mobility, they seem to be particularly acute where the occupational aspects of mobility are involved. One has not only to distinguish one individual job from another but also to be able, for purposes of analysis, to group these individual jobs together to represent different levels or types of skill. In both these respects, earlier studies have adopted different conventions so that precise comparison of their results is not possible. Nonetheless, they do provide some valuable insights into the broad pattern of occupational mobility as workers move between plants, and those conclusions which are of particular interest to the present study may usefully be outlined at this point.

A common finding is that, however measured, occupational mobility is less common than industrial mobility. For example, only 60 per cent of male workers in the Six Cities Study had changed occupation during the study period, against 72 per cent who had changed industry.[1] Reynolds also concluded that 'the number of occupational shifts . . . is consistently well below the number of industry shifts',[2] and a similar result was reached by Jefferys in her study of Battersea and Dagenham.[3] It has also been shown that the occupational group to which a worker belongs has a strong influence on his propensity to change occupation. Jefferys and Palmer[4] suggested that occupational change is more common amongst labourers and semi-skilled workers than amongst skilled craftsmen, and this is confirmed by the finding that 77 per cent of skilled manual workers in Britain in 1953 were still in that occupational group ten years later compared to 69 per cent of semi-skilled and unskilled workers.[5] In addition, mobility tends to take place more frequently

[1] Palmer, *op. cit.*, p. 77. [2] Reynolds, *op. cit.*, p. 27.

[3] Jefferys, *op. cit.*, pp. 58–9.

[4] Jefferys, *op. cit.*, p. 94; Palmer, *op. cit.*, p. 126.

[5] Government Social Survey, *op. cit.*, p. 51

between 'adjacent' occupational groups than between those which are widely separate in terms of the type and level of skill required. For example, Reynolds observed more movement between operative (semi-skilled) and skilled jobs than between the unskilled and skilled. He also noted, as have other investigations, that there was comparatively little mobility between manual and non-manual types of work.[1]

Three factors seem to be important in explaining the volume and the type of occupational mobility which occurs: the predominant technology, and therefore the occupational structure, in the labour market concerned; the availability of particular types of labour in that market; and the policies of the relevant trade unions towards occupational mobility. If the predominant technology in a particular market is such that the available labour force is composed of groups with quite distinct levels and types of skill, then occupational mobility is likely to be rare. This will be most apparent when the production process requires a labour force made up primarily of a number of specific and highly skilled crafts. Those already in these craft jobs will have built up a considerable personal investment in their craft, and a move to another occupation, either at the same or at a lower level of skill, will involve the sacrifice of all or part of this investment. Unless there has been some substantial shift of relative earnings it will not normally be economic for them to move out of their original job: hence the findings that mobility is less usual amongst skilled workers. A highly structured labour force also inhibits mobility because the training which serves to keep the craftsman from moving to another job can also serve as a barrier to the entry of others to the trade. The more lengthy and costly the training, the more this will discourage unskilled and semi-skilled workers from trying to move into the occupations concerned, thus limiting the amount of upward occupational mobility in the market.

On the other hand, the technology commonly used by plants in a particular market can be such as to facilitate occupational mobility, if the production process requires a labour force with a high proportion of employees at relatively low levels of skill. In this case, the necessary skill can be acquired quickly and inexpensively, enabling workers to move rapidly between jobs. Occupational mobility will also be encouraged if the job structure is such that the training and experience obtained on one job assist the employee to master the tasks required in the performance of another. In other words, the smaller the discontinuities between the skills required by different jobs the greater will be the degree of occupational mobility.

[1] Reynolds, *op. cit.*, p. 136; Hunter and Reid, *op. cit.*, pp. 83–5 also discuss this point, and provide further references which support Reynolds's conclusion.

The nature of and the relationship between jobs do not of themselves determine whether occupational mobility takes place, for account must be taken of the second factor listed above, namely employment conditions in the local labour market. We can assume that managements prefer to recruit workers who already possess the skill and experience necessary for the job, so as to minimize the costs of training and familiarization. If such labour is readily available, then plants will by and large be able to meet their manpower requirements without inducing any occupational mobility. However, if certain types of labour are in scarce supply, then managements will have an incentive to adjust their recruitment policies by drawing on a wider range of potential recruits. Consequently, occupational mobility will tend to be encouraged when markets are tight, whereas in slack markets there will be little incentive to draw labour from other occupations, as trained and experienced labour of the required type is more easily available.[1]

The third factor likely to influence the amount of mobility is the attitude of trade unions and the work force towards occupational mobility. A union may, for example, try to preserve certain types of work for its own members, or for those with a particular type of training or qualification. This policy has been followed in printing, in steel, and in some sections of the engineering industry. Ostensibly, such regulations or agreements limit the ability of management to choose the type of labour which it considers most suitable for the job concerned, but the practical effect of these conventions is more difficult to establish. In many cases a plant may not wish to recruit workers to skilled jobs unless they have undergone the appropriate craft training, since it may consider that only apprentice-trained labour has the level of skill necessary to carry through certain tasks, so that the impact of any institutional restriction is small. In other cases, however, trade union policies may impede to some extent the amount of occupational mobility which would otherwise take place.

The evidence of earlier studies enables us to make a number of predictions which can be investigated by means of the present enquiry. One is that most of those recruited to skilled work in all our areas will have been in the same occupational group with their previous employer, while the other occupational groups (semi-skilled, unskilled and non-production) will have drawn a higher proportion of workers from jobs with disparate skills. As regards the experience of particular markets, each of the three factors discussed above would

[1] Of course, while *upward* occupational mobility is likely to be greater in tight markets, *enforced downward* mobility may be more substantial in slack markets.

suggest a lower degree of occupational mobility in Glasgow. Apart from the traditional craft-based technology and labour force structures, the principal engineering trade union has in its Glasgow district a firm policy of restricting certain skilled jobs to those workers who have served a formal engineering apprenticeship. This will limit the movement of semi-skilled men to skilled work and, by the same token, the opportunities for the unskilled workers to progress to semi-skilled work. A more important obstacle to occupational mobility in Glasgow is likely to have been the slack state of the labour market: management experienced relatively little difficulty in securing even skilled labour over most of the study period, and, in consequence, will have had less incentive to try to draw labour from other occupations. Indeed, it is possible that the depressed state of the market may have caused some downward mobility with, for example, craftsmen accepting semi-skilled jobs because of an inability to obtain skilled work.

We would expect to find a very different pattern in Birmingham and the Other Scottish Areas. Birmingham experienced very tight labour market conditions over most of the study period, particularly for skilled labour. Consequently, plant managements will have had a strong incentive to seek to recruit semi-skilled workers to skilled jobs, which, in turn, will have created opportunities for the unskilled. In this, they will have been helped by more favourable trade union attitudes and perhaps also by the prevailing technology and occupational structure. Furthermore, the fact that the Birmingham plants recruited more heavily from non-engineering establishments than those in Glasgow also leads us to expect rather more occupational mobility.

Recruitment from non-engineering industries was, of course, most pronounced in the Other Scottish Areas, particularly in New Town and Small Town. The occupational mobility made necessary by the lack of trained engineering labour in the markets may have been assisted by a rather undeveloped trade union structure, and also by the nature of the plants themselves. Being for the most part only recently established, and being aware of the absence of trained engineering labour, they were able to design their plant from the outset so as to make most use of untrained labour, much of it without any factory experience.

3. SOURCES OF SKILLED AND SEMI-SKILLED RECRUITS

As the great bulk of the females recruited to the case-study plants had been in semi-skilled jobs with their previous employer and moved to semi-skilled jobs in the case-study plant, very little occupational mobility actually took place, and we confine the following analysis to male workers. Furthermore we are dealing only with those males recruited to the case-study plants over 1959–66 whose job with both the previous employer and with the case-study plant was known. The recruit's job on joining the case-study plant was allocated to one of the 86 job codes.[1] Similarly, his job with his previous employer was allocated to a series of 95 job codes—the 86 already referred to plus a further nine codes used to classify various forms of non-labour force activity (school, armed forces, etc.) and labour force activity outside the engineering industry. These data were used to prepare a matrix showing the job a recruit held with his previous employer against the job which he took up on joining the case-study plant. This matrix, suitably condensed, forms the basis of our analysis of mobility both within and between occupational groups.

One area of possible difficulty must be mentioned. The analysis is based upon plant personnel records, and the terminology used in labelling jobs may have differed between plants. Considerable efforts were made to devise a list of job specifications appropriate to the engineering industry, and this was discussed with the case-study plants before data collection began, in order to take account of variations in the method of classifying jobs. This procedure helped to ensure that persons doing similar work in the separate case-study plants were assigned to the same job code. In this way, inter-plant comparisons were facilitated, as was analysis of aggregate data from the case-study plants as a whole.

In the case of jobs held with a previous employer, however, it was not possible to employ the same procedure, so that job and occupational classification depended on information held by the case-study plant as to the nature of the work previously undertaken by the recruit. Here differing terminology may be crucial. The employee joining a case-study plant may describe his previous work in terms of the job description used by his previous employer, and if this differs from the terminology used by the case-study plants, there may be a danger of recording as job mobility shifts between employers which involved no substantial change in work content. This is most likely to occur when comparing shifts between individual job codes. Where mobility is defined as shifts between broad occupational groups, as in most of this chapter, minor differences in terminology will have little effect. Some degree of inaccuracy in our data is inevitable despite attempts to reduce this to the minimum, but there is no reason to suppose any bias which overstates 'upward' occupational mobility

[1] See Appendix.

relative to 'downward' occupational mobility or vice-versa. We can therefore compare the relative magnitude of flows in either direction.

Although the job data were originally classified and analysed according to the detailed job matrix described above, this job matrix has been summarized and presented so as to show the movement which took place between broad occupational groups. These summaries are to be found in the Annexe to this chapter and they are used to construct the tables in the following text. It should be noted that, while the original matrices were based on the total number of records at our disposal which contained the necessary occupational information, we have excluded three groups of workers in constructing the Annexe and the text tables. These categories were those

TABLE 10.1

PERCENTAGE OF RECRUITS EMPLOYED IN THE SAME OCCUPATIONAL GROUP AS WITH THEIR PREVIOUS EMPLOYER, BY OCCUPATIONAL GROUP ON JOINING: MALES

Occupational group on joining	Birmingham	Glasgow	North Lanarkshire	New Town	Small Town
Skilled	65·1	90·5	87·7	82·9	66·2
Semi-skilled	45·1	56·8	34·7	27·3	4·9
Unskilled	49·2	41·8	29·5	33·6	15·2
Non-production	32·5	29·5	24·1	23·1	13·3
Total	47·4	68·9	47·0	42·6	18·7

engaged in non-labour force activity, those whose skill could not be ascertained accurately and those employed as apprentices by their previous employer. As apprentices on changing employers normally moved into the skilled occupational group there was little to be gained from including their movements in the analysis. The 'skill not specified' category was excluded because of the small number involved and because the uncertainty surrounding their skill level meant that nothing could usefully be said about occupational mobility.[1] Those classified as being in 'non-labour force' activity prior to joining the plant were excluded for an equally obvious reason: we are here concerned with *occupational* mobility, rather than with movement into or out of the labour force.

The matrices which form the Annexe Tables can be interpreted in two ways. By analysing the figures down a column we can consider, for those recruited to an occupational group by the case-study plants,

[1] The category was made up mainly of welders and inspectors who may have been skilled or semi-skilled.

the sources from which these entrants were drawn in terms of occupational attachment with the previous employer. Alternatively, by considering the figures along a row, we can show for those who were employed in any given occupational group with their previous employer the occupations taken up on joining the case-study plants. The latter aspect allows us to ascertain the propensity of different types of labour to change occupation on changing employer. The former shows the relative importance of each occupational group in meeting the case-study plants' need for particular types of labour. It is this aspect—the sources from which an employer draws his labour—which we now consider, deferring the discussion of occupational mobility until Section 4.

Table 10.1 shows the percentage of all those recruited to an occupational group over 1959–66 by the case-study plants who had worked in the same occupational group with the previous employer.

Two main points emerge from Table 10.1. The first is that the majority of those who took up skilled jobs with the case-study plants had been in skilled jobs with their previous employer. There was a very limited amount of movement between jobs *within* the skilled group, but even here plants met most of their requirements for skilled labour by recruiting workers who already possessed the specific experience required. This was true even of New Town and to a lesser extent of Small Town, where one might have expected plants to have experienced some difficulty in this respect, on account of the relative absence of engineering employment. A combination (presumably) of some recruitment from other case-study plants in the local market, together with an ability to attract skilled labour from other areas, and often over long distances,[1] was apparently sufficient to meet most of the fairly limited requirements of these plants for skilled workers. Nonetheless, it is important to emphasize the point that, in each area, plants recruiting skilled labour were largely dependent upon obtaining workers already in skilled occupations with other employers. The contrast with those recruited to semi-skilled, unskilled and non-production jobs emerges clearly. With a single exception (the semi-skilled in Glasgow) less than half of those recruited to semi-skilled, unskilled and non-production jobs were in the same occupational group with their previous employer. The non-production group, in particular, drew extensively upon other occupational groups in every labour market area.

This difference in experience is nowhere more marked than in the Other Scottish Areas. The plants expanding their labour forces rapidly had to draw recruits from a wide range of occupations. In

[1] See pp. 242 above.

this they were largely successful, as few complaints were made of the quality and quantity of labour available. Yet it is clear that this process of adapting the labour supply to new requirements must possess certain limits, especially where skilled labour is concerned. Here the availability of trained and skilled labour is crucial. While the plants could circumvent some skill shortages by providing additional training, a nucleus of experienced, skilled labour was essential to their expansion.

If we look at the total for all recruits, Glasgow clearly stands apart from the other areas, with an exceptionally high proportion remaining in the same occupational group as they moved to the case-study

TABLE 10.2

SKILLED RECRUITS: PERCENTAGE BY OCCUPATIONAL GROUP WITH PREVIOUS EMPLOYER: MALES

Occupational group with previous employer	Birmingham	Glasgow	North Lanarkshire	New Town	Small Town
Skilled	65·1	90·5	87·7	82·9	66·2
Semi-skilled	17·8	4·4	5·4	5·2	9·1
Unskilled	1·8	0·6	0·4	1·0	1·3
Non-production	3·5	0·7	0·5	0·0	1·3
Miscellaneous	11·8	3·8	5·9	10·8	22·1
Total numbers	1,436	2,110	955	287	77

plants. This is in line with our expectation that the slack labour market in Glasgow combined with the contraction of the engineering sector, would have enabled managements to secure a higher proportion of their recruits from amongst those already possessing relevant training or experience than would be the case in the other markets. Conversely, plants in these other markets were obliged to obtain a greater proportion of their recruits to a particular occupational group from those previously working at a different level of skill. Thus less than half of all recruits in the other four areas, and less than a fifth in Small Town, were previously working in the occupational group for which they were recruited by the case-study plants.

Having indicated certain broad features of the occupational pattern of recruitment, we must now examine the direction of the occupational mobility which took place as workers were hired by the case-study plants. We shall direct our discussion to those recruited into skilled and semi-skilled jobs, since these are the groups for which some specialized engineering training is required, and particularly to skilled employees, where the most serious shortages of labour occurred. Table 10.2, therefore, shows the total number of skilled recruits in the five labour market areas and the percentage of this total by the occupational group in which they were employed by their previous plant.

The situation in Glasgow is largely as expected, in view of the readily available supplies of engineering labour in the city. Rather less expected is the high proportion of skilled recruits in North Lanarkshire, New Town and (to a lesser extent) Small Town, who already had experience of skilled work. The North Lanarkshire figure is explained by the pattern of industrial recruitment, for as we saw in the previous chapter more than half of the recruits in North Lanarkshire came from engineering or metal manufacture. The explanation in New Town probably lies in the extensive migration of labour into the area, and the likelihood that many of those who moved to the New Town plants from other industrial areas were skilled engineering workers. Another point is, of course, that by the end of the study period a certain amount of movement could have developed between the case-study plants in New Town, which would also help to produce the result shown.

The pattern of recruitment to the skilled trades in Birmingham is rather different from that found in the markets discussed so far. Only two-thirds of the skilled entrants were previously employed in that occupational group and, of the remainder, 17·8 per cent came from the semi-skilled and 11·8 per cent from the miscellaneous group. Such a result was predictable in view of the state of the labour market and the policies of the trade unions towards occupational mobility. On both these counts, conditions in Birmingham were favourable to mobility and will have helped to produce the results shown in Table 10.2. There is, however, the third determinant—the influence of technology on the occupational and job structure, and, in turn, upon the amount of mobility. Since the amount of mobility from semi-skilled to skilled work in Birmingham was a good deal greater than that found in Glasgow, we must establish how important differences in technology were in facilitating occupational mobility relative to the influence of market or institutional factors.

The first possibility we shall consider is that the amount of movement from semi-skilled to skilled work varied simply because the two occupational groups were themselves of different significance. The amount of mobility between two groups will, in part, be determined by their relative size, in the sense that other things being equal, the larger the proportion of skilled workers relative to semi-skilled workers in a particular market, the less is it likely that a skilled vacancy will be filled by up-grading a semi-skilled worker, and vice-versa. Table 10.3 extends this argument by considering the skill mix in Birmingham and Glasgow.

The first two lines of Table 10.3 represent the percentage of the labour force of the case-study plants engaged in skilled and semi-skilled work at the end of our period.[1] These figures were used to calculate the third line, which shows that the ratio of semi-skilled to

TABLE 10.3

ACTUAL AND 'EXPECTED' PERCENTAGE OF SKILLED DRAWN FROM SEMI-SKILLED: BIRMINGHAM AND GLASGOW MALES

	Birmingham	Glasgow
(1) Percentage of labour force Skilled	16·0	29·4
(2) Percentage of labour force Semi-skilled	51·0	45·2
(3) Ratio of (2) to (1)	3·2	1·5
(4) 'Expected' percentage of Skilled from Semi-skilled	9·4	—
(5) Actual percentage of Skilled from Semi-skilled	17·8	4·4

skilled employees in Birmingham was about twice that for Glasgow. We argued above that such differences in the relative importance of the occupational groups would lead us to expect corresponding differences in the percentage of skilled drawn from semi-skilled work. We know that 4·4 per cent of recruits to skilled jobs in Glasgow had previously been employed in semi-skilled work[2] and we now wish to estimate what percentage we would expect to find in Birmingham, solely on the basis of differences in occupational structure. This 'expected' figure is shown in line 4,[3] and the actual percentage of skilled labour drawn from the semi-skilled in line 5. We can see that in Birmingham we would have 'expected' 9·4 per cent of the skilled recruits to have been drawn from semi-skilled jobs if the occupational structure had been similar to that in Glasgow. Comparing this to the actual percentages of skilled labour drawn from the semi-skilled in the two markets, it would seem that about one-third of the

[1] See Table 3.5, p. 53 above.

[2] See Table 10.2, p. 276 above.

[3] The 'expected' figure is calculated by multiplying the actual figure for Glasgow (4·4 per cent) by the ratio of the figures in line 3, thus $4{\cdot}4 \times (3{\cdot}2/1{\cdot}5) = 9{\cdot}4$.

variation is 'due to' differences in occupational structure between them.[1] In other words, while differences in the ratio of skilled to semi-skilled jobs were important, they were not the whole explanation for the higher percentage of skilled recruits drawn from semi-skilled jobs in Birmingham.

The second way in which technology can affect the amount of mobility is through its effect on the job structure of particular occupational groups. The relationship between the jobs, in terms of skill and experience required for their performance, can be expected to affect the amount of mobility which occurs. If there are sharp discontinuities in skill between jobs, then movement will be difficult, whereas if an employee with experience in one job can easily transfer to others, then movement between jobs will be a good deal easier. For example, a machine operator will, because of his experience and skill, find it easier to obtain employment as a machine-tool setter than as a toolmaker.

The definition of occupational groups on which Table 10.2 is based inevitably involved some degree of arbitrariness, and this may in certain circumstances have an effect on the results obtained. For most jobs, allocation to an occupational group was clear-cut. For others it could not be established with certainty whether a particular job was properly defined as skilled or semi-skilled. In consequence, mobility between some jobs may in fact involve little change in job content while appearing as a major change in occupation in Table 10.2. This is particularly true of those formerly employed as setter-operators who became machine-tool setters on joining a case-study plant. Experience in this former job facilitated 'promotion' to machine-tool setting.

As far as inter-market comparisons are concerned, little difficulty would be caused if these two jobs were present in roughly similar proportions in each labour market area. Unfortunately, this was not the case. In Glasgow about one-tenth of all skilled employees were machine-tool setters compared to one-fifth in Birmingham. A considerable amount of the upward 'occupational mobility' measured in Table 10.2 in fact arose from the recruitment of those formerly employed as setter-operators or as semi-skilled machinemen to jobs as machine-tool setters. Not surprisingly this was most important in Birmingham, reflecting the nature of the production technology rather than institutional factors or market pressures. Any allowance for this different job-mix and for the differing proportions of semi-skilled to skilled labour must be rather speculative, but as a rough guess it would seem that some half to two-thirds of the greater

[1] Obtained by the calculation (9·4 – 4·4)/(17·8 – 4·4).

mobility into skilled groups in Birmingham was due simply to technological factors. Greater difficulties in securing skilled labour and less severe institutional restrictions may then be less important than a superficial reading of Table 10.2 would suggest.

Another important source of skilled labour in Birmingham was the miscellaneous group, from which 11·7 per cent of the skilled recruits were drawn. Detailed analysis shows that most of this movement was accounted for by those classified as semi-skilled or unskilled workers in other industries. There was virtually no movement from those who had attained skilled status in other sectors. This is understandable because the costs which a skilled worker will normally incur if he leaves his trade will generally serve to discourage such moves from being made. This 'miscellaneous' source was relatively unimportant in Glasgow and North Lanarkshire, but in New Town and Small Town it was much more significant. This is, of course, a reflection of the greater dependence of those areas on recruits from non-engineering sources, particularly those in the extractive and primary sectors.

TABLE 10.4

SEMI-SKILLED RECRUITS: PERCENTAGE BY OCCUPATIONAL GROUP WITH PREVIOUS EMPLOYER: MALES

Occupational group with previous employer	Birmingham	Glasgow	North Lanarkshire	New Town	Small Town
Skilled	12·1	12·8	15·0	12·5	19·1
Semi-skilled	45·1	56·8	34·7	27·3	4·9
Unskilled	10·1	8·3	10·5	12·3	13·4
Non-production	7·6	6·7	9·3	10·0	6·1
Miscellaneous	25·1	15·4	30·5	37·9	56·5
Total numbers	6,663	1,502	1,456	512	246

It was evident from Table 10.1 that the majority of the semi-skilled recruits had been working in a different occupation with their previous employer. Table 10.4 enables us to examine the source of semi-skilled recruits in more detail, as it shows the total number of semi-skilled entrants, and the percentage of this total by the occupational group in which they were employed by their previous plant.

Even in Glasgow where semi-skilled labour was readily available, little more than half of the semi-skilled entrants were directly recruited from workers previously engaged at that skill level. In the other markets the proportion was a good deal lower, so that there was a substantial amount of occupational mobility into semi-skilled jobs. Much of the difference in the extent to which plants were able to recruit semi-skilled labour directly has evidently rested on the nature of the area concerned. In those areas of rapidly expanding employment in engineering, especially in New Town and Small Town, plants were forced to rely on semi-skilled and unskilled workers from non-engineering industries; hence the high percentages of semi-skilled labour recruited from the 'miscellaneous' category. This conclusion is, of course, consistent with the findings of Chapter 9 which showed that, as plants experienced difficulty in recruiting labour from other engineering establishments, they turned increasingly to labour employed in other industries.

The high percentage of semi-skilled jobs filled by workers recruited from the miscellaneous category is not particularly surprising. Many semi-skilled jobs can be mastered after only a few days or weeks of training, and there was no evidence, in any of the markets, of institutional barriers to workers moving into these jobs. Consequently, plants unable to secure labour from semi-skilled employees experienced in engineering work generally had little difficulty in extending their recruitment to take those in other occupational groups. A more interesting and surprising feature of Table 10.4 is the high percentage of semi-skilled recruits who were skilled workers with their previous employer. Recruitment from this source accounted for 12·1 per cent of the semi-skilled recruits in Birmingham, 12·8 per cent in Glasgow and 15·0 per cent in North Lanarkshire. It could be argued that such a pattern was predictable in Glasgow on the grounds that the relative slackness of the market would have meant that even skilled workers sometimes experienced difficulty in obtaining employment. In consequence, some of them would have been obliged to accept jobs at a lower skill level. In the other areas, particularly Birmingham, such an explanation would not, however, be consistent with our knowledge of persistent shortages of skilled labour. It is therefore unlikely that the movement from skilled to semi-skilled in these areas was involuntary, and some other explanation for this unexpected feature of occupational mobility must be sought.

We attempt to provide such an explanation in the following section, when we examine the propensity of occupational groups to change occupation on joining the case-study plants. This cannot be done by means of the results of Table 10.4 which, unless interpreted carefully, may be misleading. In Glasgow, North Lanarkshire and Birmingham the percentage of the semi-skilled employees recruited from skilled occupations was of the same order of magnitude. Yet the *propensity* of skilled employees to take up semi-skilled jobs was much greater in Birmingham. That this high propensity did not result in a high percentage of semi-skilled jobs being filled by skilled employees was due to the fact that the latter group formed a relatively small proportion and the semi-skilled a relatively large proportion of engineering employment in Birmingham.

4. OCCUPATIONAL MOBILITY

We have so far examined occupational mobility from the standpoint of the plant, looking at the sources *from which* specified types of labour were obtained. It is also of interest to approach the same data from the other direction, and to examine for each skill group the occupations to which they moved on joining the case-study plants. We begin such an analysis with Table 10.5, which shows the percentage of those previously employed in each of the occupational groups specified who did *not* change occupation on joining the case-study plants.

TABLE 10.5

PERCENTAGE OF RECRUITS BY OCCUPATIONAL GROUP WITH THEIR PREVIOUS EMPLOYER EMPLOYED IN THE SAME OCCUPATIONAL GROUP BY THE CASE-STUDY PLANTS: MALES

Occupational group with previous employer	Birmingham	Glasgow	North Lanarkshire	New Town
Skilled	50·4	89·6	75·0	76·3
Semi-skilled	77·9	76·8	66·5	74·5
Unskilled	58·8	59·5	46·6	55·1
Non-production	27·4	29·2	30·0	15·0

The results for Glasgow, North Lanarkshire and New Town bear out the findings of earlier studies, that the degree of occupational stability increases with the skill level. In other words, it appears to be generally the case that the higher the degree of skill, the greater the tendency to remain in the same occupation when changing employers. This pattern was, however, broken in Birmingham. While the occupational stability of the semi-skilled, unskilled and non-production groups was roughly comparable with that displayed in other areas, the behaviour of those previously employed as skilled workers was distinctly out of line with our expectations, and with the experience of the other areas. Thus, in Birmingham a high proportion of those previously in skilled jobs moved into other occupational groups on joining the case-study plants.

Given the unexpected nature of this observation, it is necessary to examine the underlying data more closely in search of a possible explanation. We have, therefore, broken down each of the broad skilled and semi-skilled occupational groups into two parts. The first of the skilled sub-groups covers a variety of skilled machining jobs,[1] while the second includes all other skilled engineering jobs. The semi-skilled occupational group was broken down in a similar manner, again distinguishing machining operations from other semi-skilled jobs. The data for the original job matrices were then re-analysed to show movement from skilled to semi-skilled jobs in terms of these new groupings.

The Annexe table shows that 804 persons previously engaged in skilled work took up semi-skilled jobs on joining the case-study plants in Birmingham. Of these, 376 took up jobs as semi-skilled machinists, and a relatively high proportion of such movement was undertaken by those previously employed as skilled machinists. The problem raised is whether such movements represented genuine occupational mobility, or merely a change in job title without a corresponding change in job content, as would arise if the same job was given different titles by a worker's previous employer and by the case-study plant. For example, the former may refer to a job as that of a 'turner' or 'machine-tool setter', which we have classified as skilled jobs, while the latter may call the same jobs 'machine operator' or 'machine-tool setter operator' respectively, which we have treated as semi-skilled. A misleading impression of occupational mobility could also arise if two jobs *are* different, but not so different as to justify their inclusion in different occupational groups. Thus, we have taken machine-tool setters to be skilled employees while setter-operators have been taken as semi-skilled. This seems to us a reasonable distinction, but nonetheless a marginal one, for both jobs could have been treated as skilled. If this had been done, 'occupational mobility' from the skilled to the semi-skilled group in Birmingham would have been less marked than Table 10.5 suggests, for both these jobs were particularly important in that area and there was considerable movement between them.

Examination of the detailed matrices of job-changes show that some part of the 'greater' occupational mobility from skilled to semi-skilled jobs in Birmingham can be explained on the above grounds. Thus, those engaged by their previous employer as 'production turners' and 'machine-tool setters' accounted for a large part of the movement to semi-skilled machining, with 68 of the former and 111 of the latter making this transition. The 'occupational mobility' of

[1] Turners, horizontal and vertical borers, machine-tool setters, etc.

machine-tool setters is particularly suspect, as the semi-skilled category includes the job of setter-operator, a job with evident affinities to machine-tool setting. Differences in terminology are likely, then, to have resulted in some overcounting of occupational mobility. More important than this, however, was the existence in Birmingham of a large number of jobs linked by training and experience and lying on the margin between the skilled and semi-skilled occupational groups. Whether movement between such jobs is properly termed as occupational mobility is less important than the point that the existing technology did encourage movement between jobs.

While considerations such as those discussed above are important in explaining the apparently greater degree of downward mobility on the part of those formerly engaged as skilled workers in Birmingham, this is by no means the whole of the story. Recruits were drawn to semi-skilled machining jobs from a very wide range of skilled tasks, many of which were quite clearly distinct as regards skill level and job content. Moreover, if the bulk of the skilled to semi-skilled job shifts were spurious, we would expect it to operate both ways. In other words, there would be considerable upward mobility from semi-skilled jobs to skilled jobs as well as movement in the opposite direction. This is not borne out by a detailed analysis of job-changes. While some semi-skilled workers did move up to skilled jobs, their numbers were considerably less than the number moving down the occupational ladder.[1] One is, therefore, obliged to conclude that considerable real downward occupational mobility occurred when employees joined the case-study plants in Birmingham, and that this was more substantial than the upward occupational mobility involved in changing employers. This indicates a serious malfunctioning of the labour market, as both unemployment and vacancy data and the recruitment experience of the case-study plants clearly indicated that skilled labour was most difficult to obtain and, given this, one would have expected to find relatively little downward mobility from skilled to semi-skilled jobs.

One possible explanation may lie in the wage structures of the case-study plants. For example, a number of Birmingham case-study plants complained that high earnings in motor manufacturing and component establishments allowed such units to recruit as semi-skilled workers those previously engaged by other employers as skilled employees. To test this possibility, a further analysis of occupational mobility was conducted distinguishing separately the four large case-study plants principally engaged upon the manufacture of vehicles or their components. This revealed that, of the 400

[1] See Annexe.

persons recruited by these plants who were employed in skilled jobs by their previous establishment, no less than 245 (61·3 per cent) were engaged for semi-skilled work. Since in Birmingham as a whole only 43·3 per cent of those formerly on skilled jobs moved to semi-skilled work with the case-study plants,[1] it is evident that the motor-car plants accounted for an above-average amount of downward mobility from skilled to semi-skilled jobs. Such downward mobility was not, however, confined to motor-car units. Indeed, although it was *relatively* more important in this case, in *absolute* terms the motor-car plants accounted for less than one-third of the downward mobility observed. Thus, if downward mobility did reflect the high wages offered to semi-skilled workers in some plants, these high earnings were not confined to the motor-car establishments. The problem was more widespread than this and is of a piece with our earlier findings regarding the wage structures, external and internal, ruling in the engineering industry.

We saw in Table 10.5 that, except in Birmingham, the second most stable occupational group after the skilled was the semi-skilled category. The most interesting question relating to this group is how those who did move to other occupational groups were distributed amongst them. An examination of the data in the Annexe shows that in every labour market area more moved down than up the occupational ladder on changing employers. In Birmingham, for example, 11·1 per cent of those who were semi-skilled with a previous employer became unskilled workers with a case-study plant, while only 6·6 per cent became skilled. In Glasgow the comparable proportions were 11·0 per cent and 8·3 per cent, and in North Lanarkshire 18·8 per cent and 6·9 per cent. The same pattern was also evident in New Town. In other words, even in the Birmingham market and in the other markets which were short of skilled engineering labour, only a very small proportion of the semi-skilled labour force found it possible to move to skilled work as they moved between employers.

In Birmingham, this will again in part be due to the occupational structure of the labour force. There were about three times as many semi-skilled as skilled workers in that market, which helps to explain the relatively high percentage of skilled who were drawn from the semi-skilled. The corollary of this is, of course, that relatively few of the semi-skilled had the opportunity of moving to skilled work, as more of them were competing for fewer jobs. But with chronic shortages of skilled labour and a relatively favourable attitude on the

[1] 1,856 Birmingham recruits were engaged by their previous employer as skilled men. Of these, 804 were recruited to undertake semi-skilled work with the case-study plants. See Annexe.

part of the trade unions towards occupational mobility, one would still have expected a higher proportion of the semi-skilled in Birmingham to be upgraded to skilled jobs as they changed employers. The fact that less than 10 per cent were upgraded could be a further result of the high level of earnings for semi-skilled males which explained some of the movement from skilled to semi-skilled jobs. By the same token, the earnings structure must have discouraged some semi-skilled workers from taking up skilled jobs, even though they had the ability to undertake work at a higher skill level.

The importance of training facilities in determining the amount of occupational mobility can be illustrated by the experience of those who were engaged in unskilled jobs by their previous employer. Table 10.5 showed that about half of such recruits moved to other occupational groups on joining the case-study plants, and from the Annexe we can see that the majority of those who changed their occupational group entered semi-skilled employment. In Birmingham, 58·8 per cent remained in unskilled work, while 33·3 per cent moved to semi-skilled jobs. The experience of the other labour market areas was similar, so that upward occupational mobility was most apparent in the case of the unskilled and was mainly into semi-skilled production work. The reason for this is that semi-skilled jobs can often be performed satisfactorily after relatively short periods of informal, on-the-job training. However, while such methods are adequate in facilitating the upward mobility of the unskilled, they are much less likely to be satisfactory in bringing about movements from semi-skilled to skilled work. The ease of entry to semi-skilled work is also demonstrated by the mobility patterns of those who were in non-production or miscellaneous jobs with their previous employer. These jobs lack the training or experience which would be directly relevant to skilled or semi-skilled engineering work, yet a great many such workers joined the case-study plants as semi-skilled employees. The pattern in each market was that, while many of those in the non-production and miscellaneous groups entered the unskilled category, at least as many were able to move to semi-skilled jobs; very few became skilled. In other words, the relatively informal training methods adopted by the plants[1] were quite adequate to provide the expertise necessary to undertake semi-skilled work. They were not, however, adequate to bridge the gap between skilled and other employments even in areas like Birmingham, Small Town and New Town where institutional restrictions were less severe.

[1] We are, of course, ignoring apprentice training in this argument as it applies only to boys and not to adults engaged in unskilled or semi-skilled tasks.

5. COMPLEX MOBILITY

We have so far considered the various types of mobility independently of one another, by looking in turn at the geographical, industrial and occupational elements of the mobility which took place as workers were recruited to the case-study plants. While some of the mobility will have been of this simple sort, involving the recruit in a change of industry *or* occupation *or* area, it is also to be expected that some mobility will have been complex in character, involving a combination of such changes. We must now describe the nature of the actual employer-shifts made by recruits, and put the earlier discussion into perspective by examining the extent to which the various types of mobility, both simple and complex, were undertaken as workers moved to the case-study plants. In doing so, we are merely presenting the earlier data in a different way, so that the findings are largely foreshadowed by our earlier discussion.

The position is summarized in Table 10.6 for each of the three largest labour market areas. The table shows the nature of employer-shifts undertaken by those workers who joined the case-study plants during the study period, for whom both the location and industry of their previous employer was known, as well as their occupation in that previous employment. In constructing the table, an individual was counted as having changed the area of his work if his previous employer was located at least 2 miles from the case-study plant. The definition of industrial mobility is the same as that used in the previous chapter, namely, movement from any Main Order Heading other than Orders VI to IX, the latter having been taken to represent the engineering industry. The data in Table 10.6 relating to occupational mobility are based on those who moved between occupational groups (as distinct from moves between individual jobs) when joining the case-study plants.[1] There is also a category which shows the percentage of shifts which involved only a change of employer without any change in occupation, industry or area of work.

The first point which can be made is that in the conurbations roughly one-quarter of employer-shifts involved the worker in a change of employer only. In other words, the case-study plants were able to secure approximately one in four of their recruits from other engineering plants less than 2 miles away, and these recruits were

[1] Comparable tables were also produced which defined occupational mobility as a change between individual jobs. This naturally served to increase the proportion of shifts involving a change of occupation, but the most common form of mobility remained a simple shift between industries, and changes in industry were more common than changes in occupation (or job) even where simple *and* complex changes were considered.

employed in the same occupational group as the job for which they were engaged by the case-study plants. This was true of only about one in ten of the recruits in North Lanarkshire. In the latter area the nature of the labour market forced the plants to draw a higher proportion of their labour from plants outwith their immediate vicinity. Thus more than half of the new recruits to the North Lanarkshire

TABLE 10.6

GEOGRAPHICAL, INDUSTRIAL AND OCCUPATIONAL MOBILITY: MALES

Type of employer-shift	Percentage making each type of employer-shift		
	Birmingham	Glasgow	North Lanarkshire
(1) Employer only	22·2	29·8	10·3
(2) Area only	3·9	9·2	17·5
(3) Industry only	34·5	33·9	18·9
(4) Occupation only	8·6	4·1	3·7
(5) Area and industry	7·2	12·6	20·0
(6) Area and occupation	2·0	1·4	5·6
(7) Occupation and industry	18·1	6·8	13·6
(8) Area, occupation and industry	3·6	2·3	10·5
Total employer-shifts[1]	1,038	1,311	1,256

Note: 1. The numbers on which this table is based are lower than those used for the tables relating to simple industrial and occupational mobility. This is partly because we are dealing here with complex mobility, so that information must be available on three aspects of employer-change simultaneously (geographical, industrial and occupational), but mainly because it proved necessary for practical reasons to limit the analysis of geographical mobility to a sample of the population for whom the information was available; see p. 239 above.

plants had previously been employed in units at least 2 miles distant. This compares with 25·5 per cent in Glasgow and 16·7 per cent in Birmingham.

Of the various types of mobility which were undertaken the commonest type, with the (marginal) exception of North Lanarkshire, involved a change of industry only. About one-third of the recruits in the conurbations, and about one-fifth of those in North Lanarkshire, remained in the same area and occupation upon joining the case-study plant, but did undertake a change of industry. Additional groups of workers also changed industry in conjunction with some other form of mobility: for example, 18·1 per cent of the employer-changes in Birmingham involved a change of industry as well as of occupation. This predominance of industrial mobility over other types is in line with the results of most earlier studies,[1] and is a reflection of the discussion in Chapter 9 which showed how the case-study plants, especially in Birmingham and North Lanarkshire, had drawn upon non-engineering sources for their labour supplies.

The rather greater occupational mobility which occurred in Birmingham by comparison with Glasgow is also reflected in Table 10.6. The percentage of employer-changes which involved a change of occupation only was about twice as high in Birmingham as in Glasgow, and the percentage who undertook a change of occupation *and* industry was about three times as high. In the conurbations, too, movement involving a change of area was generally the least common type of shift, although this type of mobility was a major feature of the North Lanarkshire labour market.[2]

We can analyse the data upon which the preceding table was based in another way, so as to show the extent to which the various types of mobility occurred, whether singly or in combination with one another. The procedure adopted is best explained by way of an illustration. Suppose we wish to establish the percentage of employer-shifts in Birmingham which involved geographical mobility of a simple or complex type. To obtain this from Table 10.6, we add together all the percentages relating to employer-shifts in which a change of area was involved, i.e. 3.9 (area only), 7.2 (area and industry), 2.0 (area and occupation) and 3.6 (area, industry and occupation). This shows that 16·7 per cent of employer-shifts involved a change of area. The same process was undertaken for changes of employer, industry and occupation, and the results are shown in Table 10.7. It should be noted that these rows do *not* add up to 100 per cent, since many employer-shifts are complex in character and will therefore be recorded under more than one type of mobility. As it is possible that the type of employer-shift undertaken by a worker varied with his skill, Table 10.7 provides an analysis for skilled, semi-skilled and unskilled recruits, occupational classification being determined by their occupation with the previous employer.

[1] Parnes, *op. cit.*, p. 79.

[2] As explained above, Table 10.6 was necessarily based only upon those recruits for whom geographical, industrial and occupational data relating to the previous employer were available. Since the geographical data obtained were analysed for only a small sample of the recruits (see p. 239 above) this limited the number of observations in Table 10.6 quite severely. The data relating to the industrial and occupational aspects of mobility were therefore analysed separately: this gave a much larger number of observations in each market area (6,044 in Birmingham, 3,659 in Glasgow and 2,220 in North Lanarkshire) and confirmed that simple industrial mobility was the most common form of employer-shift.

TABLE 10.7

GEOGRAPHICAL, INDUSTRIAL AND OCCUPATIONAL MOBILITY, BY OCCUPATIONAL GROUPS: BIRMINGHAM AND GLASGOW MALES

Occupational group with previous employer	Birmingham					Glasgow				
	Number of employer shifts	Percentage involving change of:				Number of employer shifts	Percentage involving change of:			
		Employer only	Area	Industry	Occupation		Employer only	Area	Industry	Occupation
Skilled	332	24·1	13·5	57·8	40·6	786	34·5	24·0	54·3	5·6
Semi-skilled	549	23·7	18·2	63·1	27·6	391	25·6	27·6	52·9	24·0
Unskilled	157	12·7	17·8	76·4	30·6	134	14·9	26·8	70·1	39·4
Total	1,038	22·2	16·7	63·4	32·3	1,311	29·8	25·5	55·6	14·6

Table 10.7 clearly confirms that the most common type of employer-shift in both labour market areas was that involving a change of industry. In Birmingham the next most common type of mobility was that between occupational groups, with a change of area being the least common type of movement, a result which is in line with that of most earlier studies. In Glasgow, however, a movement between areas[1] was more common than a movement between occupational groups. Further light on these differences emerges from an examination of the effect of skill on the type of mobility undertaken. In both markets, skilled workers were the most likely to undertake an employer-shift which involved a change of employer only without a concomitant change in area or industry or occupation. Again, in both markets the unskilled had a much weaker attachment to an industry than the skilled—about three-quarters of the unskilled, compared with little over half the skilled, moved to the case-study plants from some non-engineering industry. There was a difference, however, between the markets in the way in which occupational mobility was affected by the level of skill. Earlier studies have shown that the degree of occupational mobility amongst manual workers increases as the level of skill of the worker declines. From Table 10.7 we can see that this pattern was followed in Glasgow, where very little occupational mobility was undertaken by skilled workers, and a considerable amount by the unskilled, with the semi-skilled in an intermediate position. The overall result was that only one-seventh of the recruits in Glasgow changed occupational group, so that this is seen to be the least common type of mobility. The differences in the Birmingham results are produced by the high degree of occupational mobility displayed by those engaged as skilled men by their previous employers. Of this group, no less than 40 per cent changed occupational group when joining the case-study plant. The effect is that almost one-third of the Birmingham job-changes involved a change of occupation; twice the level experienced in Glasgow.

The overall impression of Tables 10.6 and 10.7 is clearly to confirm the underlying theme of this and the previous chapter, namely that quite a considerable amount of mobility occurred as workers were recruited to the case-study plants. Only a quarter of employer-shifts involved merely a change of employer; all the rest involved the recruits in a change of area, industry or occupation or some combination of these types of mobility. In this respect, it should be recalled that much of the discussion in the present section has been

[1] It should perhaps be recalled that the definition used here for an area change is a movement from an establishment more than 2 miles from the case-study plant.

based upon plants operating in established engineering centres. The North Lanarkshire data in Table 10.6 indicate that away from such centres, in places such as New Town or Small Town, the incidence of the various types of mobility will have been even greater. The importance of these findings, both for industry and for policy-makers, is that they demonstrate how employees and managers can adapt to meet prevailing market conditions. Employees are shown to have been very ready to move from other industries and, to a lesser degree, from other occupations and areas to meet plants' demands for labour, while, for their part, managements have evidently been able to modify their requirements to suit available resources within the local labour market. While this is encouraging, not all the mobility observed appears to have assisted the process of adjustment in response to the demand and supply conditions for labour ruling in the market. In particular, we have seen that downward occupational mobility between skilled and semi-skilled jobs was a more common feature of employer-shifts than upward occupational mobility, and this despite the almost continual shortages of skilled labour experienced by the case-study plants. This may well reflect a wage structure problem but at the very least sets us on our guard against assuming that, because mobility occurred, it was necessarily of the 'right' type.

6. CONCLUSIONS

(i) There was much less occupational mobility in Glasgow than in any of the other markets. In Glasgow, some two-thirds of those hired by the case-study plants were recruited to the same occupational group as that in which they had been engaged by their previous employer; this was true of less than half the recruits in the other areas, and less than one-fifth in Small Town. The majority of those recruited as skilled workers had been in that occupational group with their previous plant, the figure being over 80 per cent in Glasgow, North Lanarkshire and New Town. In contrast, with the exception of semi-skilled males in Glasgow, less than half of those recruited to semi-skilled and unskilled jobs were previously employed at the same level of skill.

(ii) The ability of the Glasgow plants to meet most of their skilled requirements by recruiting labour already possessing the necessary training and experience is in line with our expectations, in view of the depressed state of that labour market. In Birmingham, however, plants recruiting skilled labour drew on a much wider range of occupations, particularly from those previously employed in the semi-skilled or miscellaneous categories. In part, this greater mobility into skilled jobs was due to the tighter labour market conditions in Birmingham and to the fact that institutional restrictions on movement into skilled jobs were less severe in that area, but it appears that the job structure of employment in Birmingham was an important contributory factor to upward mobility. In other words, the prevailing production technology of the Birmingham plants was such as to facilitate movement from semi-skilled to skilled work on a greater scale than in Glasgow.

(iii) In all markets save Glasgow, the majority of persons recruited to semi-skilled jobs were previously employed in a different occupational group. That there was considerable movement into semi-skilled work from other jobs is not unexpected, since the fairly limited training required provided no serious obstacle to entry, and there was no evidence of institutional barriers. Nonetheless, it is worth while noting that in each market area a higher proportion of semi-skilled jobs were filled by persons previously employed in skilled jobs than by persons previously employed as unskilled.

(iv) A comparison of occupational group with previous employer to occupational group on joining the case-study plant indicates that, except in Birmingham, occupational mobility was inversely related to skill level. Thus, ranking the skill groups in terms of increasing occupational mobility, we find the skilled were least mobile followed by the semi-skilled, unskilled and non-production workers in that order. This progression was, however, broken in Birmingham, where skilled employees showed a greater degree of occupational mobility than either the semi-skilled or the unskilled. When problems of definition were taken into account, and due allowance made for the effect of differing technologies and job structures in promoting occupational mobility, the contrast between the markets was less substantial, but remained marked in that a relatively high proportion of skilled men in Birmingham moved to semi-skilled work on changing their employer. Moreover, downward mobility from skilled to semi-skilled jobs was more substantial than upward mobility between these groups on changing employer. This is a somewhat surprising result in view of the acute shortages of skilled labour in Birmingham, but it may be a consequence of the complex wage structures to which attention was drawn in Chapters 4 and 5, which make it possible for semi-skilled workers to earn more than skilled workers within, and certainly between, plants. Hence it is noticeable that downward occupational mobility was greatest where formerly skilled employees were recruited by motor-car plants who were the wage leaders in the market.

(v) Of those who moved out of semi-skilled work on changing their employer, more moved down to unskilled work than moved up to skilled work with the case-study plants. This could, of course, be consistent with the wage structure problem referred to in conclusion (iv) above, but it seems also likely to reflect the fact that current training facilities, which, apart from those for apprentices, were largely informal and 'on-the-job', were inadequate to facilitate much upward mobility into skilled jobs by new recruits.

(iv) An analysis of simple and complex mobility shows that about one-quarter of employer shifts in the conurbations and one-tenth in North Lanarkshire involved only a change of employer with no change of area, industry or occupation. The most common type of mobility was that involving a change of industry only, with about one-third of all employer-shifts in Birmingham and Glasgow involving such a move. Occupational mobility was more frequent than geographical mobility in Birmingham, but the position was reversed in Glasgow, because the ability of employers to recruit directly the skills they required held down the amount of occupational mobility undertaken in that area. In both conurbations skilled employees were most likely to undertake an employer-shift which involved *only* a change of employer, and they displayed greater industrial attachment than the unskilled. As would be expected from conclusions (i) and (iv), however, occupational attachment was strong for skilled employees in Glasgow, but weak in Birmingham.

ANNEXE TABLE 10A

OCCUPATIONAL GROUP WITH PREVIOUS EMPLOYER BY OCCUPATIONAL GROUP ON JOINING THE CASE-STUDY PLANTS (NUMBERS)

Occupational group with previous employer	Occupational group on joining				
	Skilled	Semi-skilled	Unskilled	Non-production	Total
Birmingham					
Skilled	935	804	70	47	1,856
Semi-skilled	255	3,004	427	170	3,856
Unskilled	26	676	1,193	135	2,030
Non-production	50	504	260	307	1,121
Miscellaneous	170	1,675	473	287	2,605
Total	1,436	6,663	2,423	946	11,468
Glasgow					
Skilled	1,910	193	14	14	2,131
Semi-skilled	92	853	122	44	1,111
Unskilled	12	125	251	34	422
Non-production	15	100	89	84	288
Miscellaneous	81	231	125	59	496
Total	2,110	1,502	601	235	4,448
North Lanarkshire					
Skilled	838	219	45	16	1,118
Semi-skilled	52	505	143	59	759
Unskilled	4	153	207	80	444
Non-production	5	135	77	93	310
Miscellaneous	56	444	229	138	687
Total	955	1,456	701	386	3,498
New Town					
Skilled	238	64	5	5	312
Semi-skilled	15	140	27	6	188
Unskilled	3	63	92	9	167
Non-production	0	51	34	15	100
Miscellaneous	31	194	116	30	371
Total	287	512	274	65	1,138
Small Town					
Skilled	51	47	6	4	108
Semi-skilled	7	12	0	1	20
Unskilled	1	33	5	2	41
Non-production	1	15	4	2	22
Miscellaneous	17	139	18	6	180
Total	77	246	33	15	371

CHAPTER 11

INTERNAL MOBILITY

1. INTRODUCTION

The earlier chapters have been concerned with those forms of mobility which occur as workers leave one employer and move to another. Here we shall consider those occupational changes made by workers while continuing to work in the *same* plant. Such internal occupational mobility is of interest in its own right as one part, though a neglected one, of the total process of occupational mobility. In addition, it can be a major element in a plant's employment policy, enabling management to adjust the existing labour force to meet changes in manpower requirements. For example, internal mobility can help to alleviate labour shortages, in that the re-training and transfer of part of the existing labour force may allow the plant to avoid the recruitment difficulties caused by shortages in the external market.[1]

Most of the work on the operation of internal labour markets has been American in origin[2] and is not necessarily an accurate indication of what happens in the very different institutional and economic framework of British industry. Moreover the literature on internal mobility has largely been concerned with the development of analytical models, few of which have been subjected to extensive empirical tests. Thus, while one can hypothesize that internal mobility may vary from industry to industry and from market to market as a result of differences in technology, labour market conditions, and trade union policies, little hard evidence exists on the extent and the nature of the internal mobility which actually takes place.

In Reynolds' New Haven study, two-thirds of all employees had not changed their jobs since joining their current employer; 28 per cent considered they had been promoted to a better job, and 5 per cent that they had been demoted.[3] Palmer's Six Cities study also

[1] In addition the attractiveness of a plant to potential employees may itself be enhanced by the knowledge that it has an active policy of internal promotion.

[2] For example, Reynolds, *op. cit.*, pp. 139–51; H. M. Gitelman, 'Occupational Mobility within the Firm', and P. B. Doeringer, 'The Determinants of the Structure of Industrial Type Internal Labor Markets', *Industrial and Labor Relations Review*, Vol. 20, 1966–7.

[3] Reynolds, *op. cit.*, pp. 149–51.

touched on internal mobility. There it was found that, 'while the great majority of workers with only one employer also had only one job, a tenth of the men, and less than a tenth of the women held two jobs in the period; a scattered few held three or more'.[1] In four of the cities over a fifth of the workers had experienced upward occupational mobility between 1940 and 1950, and 40 per cent of these had accomplished this through internal mobility.[2] The only comparable data for the U.K. are provided by Jefferys' study of Battersea and Dagenham. She found that 22 per cent of the occupational changes in Battersea and 31 per cent in Dagenham did not involve a change of employer.[3] Promotion to supervisory posts and upgrading from semi-skilled or labouring to skilled work were most commonly made within the plant.

These earlier studies clearly establish that the re-allocation of labour between occupations can take place without a change of employer, and that plant management may be able to increase its supply of a particular type of labour without recourse to the external market, but they provide little guidance as to the volume of internal mobility we might expect to find in the present enquiry. This is partly because of differences in institutional factors, in sampling procedures or in methodology, but mainly because it is difficult to relate these earlier findings on internal mobility to the industrial and technological context in which it took place, or even to the labour market environment which might have shaped its character and extent.

The chief aim of this chapter is therefore to measure the extent and the importance of internal mobility in the case-study plants, and to assess the relative importance of the factors thought to shape internal mobility. We begin by reviewing in Section 2 the main factors which appear likely to be influential in determining plant policies with respect to internal mobility. Section 3 outlines the nature of the data used in this study and begins the analysis of internal mobility by considering the extent to which individuals remained in the same occupational group throughout their employment with a plant. This section also takes up the question of mobility into the skilled occupational group, where recruitment difficulties have been most acute. In Section 4 we turn to certain other aspects of internal mobility, particularly that between unskilled and semi-skilled workers, while Section 5 contrasts the volume of occupational mobility which occurs as a result of employer-shifts and of internal mobility within the plants. Section 6 outlines the chief characteristics of internal market structures in the British engineering industry, and a brief summary of the discussion is provided by Section 7.

[1] Palmer, *op. cit.*, p. 35. [2] Parnes, *op. cit.*, p. 74. [3] Jefferys, *op. cit.*, p. 93.

2. INTERNAL LABOUR MARKETS

The analytical models which have been developed to explain the phenomenon of internal mobility were outlined briefly in Chapter 1.[1] Expanding ideas originally put forward by Kerr,[2] Doeringer has discussed three characteristics of internal labour markets. The first is concerned with the way in which the jobs in a plant are arranged into groups within which internal mobility takes place. At one extreme, in a loosely organized or unstructured internal market, workers may move between a wide variety of jobs, with no close linkages between particular pairs of jobs being displayed. In the opposite case where channels of promotion and transfer are tightly regulated, movement will conform to a clearly defined pattern. The less structured is the internal labour market, the greater the range of jobs within the plant between which labour can be interchanged, and hence the greater the freedom of management to meet changed needs through internal adjustments.

The rules or procedures governing promotion and transfers are a second distinguishing feature of internal labour markets, and here we find a major difference between British and American practice. Thus, none of the case-study plants operated a procedure for internal promotion in which seniority was the only guiding principle, and there was no evidence that managements were under pressure from unions or workers to regularize the very informal procedures used to determine such promotion. This contrasts with much American experience, where promotion has become an important issue in labour-management relations, and the criteria used in selecting candidates for promotion have become a matter for collective bargaining.[3]

A third characteristic of an internal labour market is its 'openness' to the external market. To the extent that jobs are reserved for present employees, they are closed to prospective applicants in the external labour market. Contact with the external market is represented by the points in the hierarchy of jobs at which new employees are engaged—what Kerr designated the 'ports of entry' to the internal market.[4] A 'closed' internal market is one in which all jobs above the least skilled are filled by internal promotion, so that the only port of entry is at the lowest job classification in the plant. The opposite extreme, an 'open' internal market, would arise if no internal move-

[1] See pp. 300–1 above.

[2] C. Kerr, 'The Balkanisation of Labor Markets', in E. W. Bakke, *op. cit.*

[3] Cf. S. H. Slichter, J. J. Healy and E. R. Livernash, *The Impact of Collective Bargaining on Management* (1960), esp. Chapter 7.

[4] Kerr, *op. cit.*

ment took place between jobs, so that all vacancies arising were filled by external recruitment and all jobs therefore constituted a port of entry linking the plant with the external market.

To some extent occupational mobility within a plant must be subject to the same influences as occupational mobility between plants. Thus internal as well as external occupational mobility will be obstructed if the prevailing technology produces a job structure with sharp discontinuities in skill, and will be facilitated if training, skill and experience can readily be transferred from one job to another. Again, just as labour shortages are likely to encourage occupational mobility as employers seek new recruits in the external market, so they may stimulate the employer to meet some part of his needs by upgrading labour within the plant.

The costs of external and internal sources of labour supply are unlikely to be equal and will vary with the employment conditions ruling in the market. For example, in a tight labour market a plant seeking to hire additional labour may have to adopt more active and expensive methods of recruitment (e.g. advertising), whereas in a slack market the supply of recruits seeking work on their own initiative may be quite adequate to meet the plant's needs. Probably more important than this is the likelihood that training costs will tend to be lower if a vacancy is filled by the external recruitment of a suitably qualified worker than if internal sources of supply are used. Suppose, for example, that the vacancy is for a skilled turner. If the management recruits a man with the necessary skill and experience in the external market, the only training required will be a short period of familiarization. If, however, it decides to transfer a semi-skilled employee to such work, then some period specifically devoted to training is likely to be necessary before the promoted employee can fill the new job satisfactorily. However, although we might expect employers to prefer to fill vacancies, and particularly skilled vacancies, through external recruitment, this option may not be open to them in a tight labour market situation, or at least might impose prohibitive costs through competitive wage bidding or higher recruitment expenditure. In a tight labour market one would therefore expect employers to encourage internal mobility, whereas external recruitment seems likely to be more common when the market is slack.

We have so far assumed that the employer is free to exercise his preferences between utilizing external and internal sources of labour supply. This need not, of course, be a valid assumption, particularly in the engineering industry with its highly skilled and unionized labour force. Unions may attempt to preserve certain types of work for their own members in a multi-union plant, or they may seek to restrict the work to those with particular qualifications. Thus it is customary in certain parts of the engineering industry to limit certain jobs to workers who have served a recognized engineering apprenticeship. Ostensibly, regulations or agreements of this sort limit the ability of management to choose the type of labour which it considers most suitable or economic for the job concerned. However, as suggested in the previous chapter, the practical effect of such conventions is more difficult to establish, as in many cases a plant may have no wish to move workers between the jobs in question. For example, there may be an understanding with the trade unions preventing the transfer of a craftsman from one craft job to another; but, given the retraining costs which would normally be involved, there may be few cases where it would be economic for the plant to do so anyway. In other cases, however, it appears more likely that trade union policies can seriously impede internal mobility, so that the impact of institutional restrictions cannot be ignored.

These considerations enable us to formulate certain predictions regarding the pattern of internal mobility likely to emerge in our labour market areas. For one thing, there were substantial differences in the technology, and therefore in the occupational structure of the plants, in the various markets. Most of the case-study plants in Birmingham were engaged in mass or large-batch production of standardized units, and although the products were rather different, the same was true of the newer engineering plants in North Lanarkshire, New Town and Small Town. In contrast, the Glasgow engineering industry was typified by plants specializing in custom-built or small-batch orders, often for the capital goods industry. Hence the technology of the Glasgow plants differed a good deal from that of the case-study plants in Birmingham and in the Other Scottish Areas.

These differences in technology will be reflected in differences in the occupational structure of the labour force. In the mass or batch-production industries, much of the work is repetitive and does not require a particularly high level of skill. This is likely to mean that one job is not markedly different from neighbouring ones in the job hierarchy, so that movement between jobs is relatively easy. The craft-based Glasgow engineering industry involved a different job structure. The varied nature of production required that a higher proportion of the labour force was able to understand and work from complex drawings, involving a high degree of skill. Progression from semi-skilled to skilled work was therefore less easily accomplished.

Secondly, employment conditions differed considerably between the markets, particularly between Birmingham and Glasgow. The tight labour market conditions in Birmingham were likely to have encouraged managements to make more effective use of their existing labour force by up-grading and training workers for more highly skilled jobs.[1] Although the labour markets of the Other Scottish Areas were not as tight as that in Birmingham, they too lacked any substantial unused supply of engineering labour. Consequently, the case-study plants may have depended heavily on recruiting labour into unskilled jobs, and then training such labour for semi-skilled and skilled engineering work. Again, then, one expects to find a considerable amount of upward movement within the plants in the Other Scottish Areas. In contrast, Glasgow experienced high unemployment and declining engineering employment throughout the study period, which ensured that most types of labour were relatively easy to recruit in the external market. In these circumstances one might expect upward internal mobility to have been less common.

Interviews with plant managers also indicate that the importance of institutional restrictions varied between the labour market areas. Manning questions were within the jurisdiction of the local district committees of the Amalgamated Union of Engineering and Foundry Workers (A.E.F.). In Glasgow the union had established in virtually all plants that certain types of work designated as skilled could only be undertaken by those who had served a formal engineering apprenticeship. This regulation usually covered, for example, toolroom work, horizontal and vertical boring, centre lathe turning and fitting. In contrast, relatively few plants in Birmingham were subject to strict requirements of this sort, and the competence of labour to undertake a particular task was to a greater extent a matter of managerial discretion. North Lanarkshire corresponded more closely to the Glasgow pattern, being a traditional engineering area, but in New Town and Small Town managements appear to have had greater freedom of action. This was partly but not wholly due to a low degree of unionization. In New Town, for example, managements were able to negotiate dilution agreements with the unions, enabling them to employ on skilled work men who had not served an apprenticeship.

On all counts, therefore—technology, employment conditions and institutional restrictions—we would expect relatively little internal

[1] In practice the matter may be more complicated than this, for it will be observed later that severe labour shortages such as experienced in Birmingham can affect the pattern of mobility either favourably or unfavourably. On the one hand, they can act as a stimulus to internal promotion and training; on the other, high turnover rates will both cause workers to leave before they have acquired the skills to move to other jobs, and will discourage plants from investing in training programmes.

mobility in Glasgow, particularly between semi-skilled and skilled jobs. Here plants would have met their needs for skilled labour either through their apprentice-training programmes or through the recruitment of trained workers in the external market. In the other areas, however, conditions seem to have been favourable to internal mobility. We must now see whether they produced substantially different results in practice.

3. INTERNAL MOBILITY IN THE CASE-STUDY PLANTS

Our discussion of internal mobility is based upon those male manual workers who joined the case-study plants over 1959–66. Details of the jobs held by these employees with the plants during this period were

TABLE 11.1

PERCENTAGE REMAINING IN SAME OCCUPATIONAL GROUP WITH PLANT, BY OCCUPATIONAL GROUP ON JOINING: MALES

Occupational group on joining	Percentage				
	Birmingham	Glasgow	North Lanarkshire	New Town	Small Town
Skilled	94·6	97·4	96·7	95·1	75·3
Semi-skilled	89·3	95·5	91·6	85·2	89·3
Unskilled[1]	71·8	69·7	43·6	39·5	35·7
Non-production	85·4	82·4	73·8	81·0	47·4
Total[2]	86·2	89·3	76·8	75·4	79·3

Note: 1. We have ignored all movements from unskilled work to apprenticeships in calculating the percentages shown. This is because boys taken on for apprentice-training in Glasgow and North Lanarkshire often began with a period of unskilled employment, while this practice was much less common in the other areas.

2. Calculated by excluding those who joined as apprentices.

obtained from plant personnel records. Although provision was made to record information for up to four jobs held with the plant, only the first and last known jobs are used in the present analysis.[1] The job titles used in the plant records were classified into one of the 86 job codes described in Chapter 10, which were arranged to yield the following occupational groups: skilled, semi-skilled production workers, unskilled production workers, semi-skilled and unskilled

[1] One exception to this is Table 11.4 below, which is derived from an analysis of *all* jobs.

non-production workers and apprentices.[1] For these groups we show in the Annexe to this chapter the total number of recruits over 1959–66 by their first and last jobs with the case-study plants. This material is drawn upon to construct the tables in the following text. We begin by looking at the overall amount of internal mobility with the aid of Table 11.1, which shows the percentage of all those employees who joined the case-study plant during the study period, and whose last job with the plant was in the same occupational group as their job on joining.

The five market areas fall into two distinct groups. In each of the Other Scottish Areas only a little over three-quarters of the labour force remained in the same occupational group during their employment with a plant, while in Birmingham and Glasgow those not changing their occupation with a plant amounted respectively to 86·2 and 89·3 per cent of all recruits. The greater tendency for employees in the Other Scottish Areas to move out of the occupational group into which they were recruited was, however, a product of the greater internal mobility displayed by unskilled and non-production workers, as the experience of skilled and semi-skilled workers varied relatively little between the areas. It can also be seen that, with the exception of Small Town, the percentage of workers remaining in the same occupational group with a plant declines as one moves through skilled, semi-skilled and unskilled employees, although internal mobility was less marked for non-production workers than for the unskilled. For example, in Birmingham 94·6 per cent of those starting in skilled jobs remained in that occupational group while with a plant, compared with 89·3 per cent of the semi-skilled, 71·8 per cent of the unskilled and 85·4 per cent of non-production workers. The evidence therefore confirms Palmer's view that 'skilled men are more strongly attached to their occupations than are semi-skilled men'.[2] In the internal market as in the external market skilled men are unlikely to move downwards in the occupational hierarchy, and their opportunities for upward movement to non-manual jobs are fairly limited.[3] Conversely, there are greater opportunities for promotion for manual workers with a lesser degree of skill, if only because they *can* move up the occupational ladder within manual work.

However, while the movement *out* of skilled jobs within the plant

[1] The classification system is explained in detail in the Appendix. For reasons already stated we have excluded those whose skill could not be specified and staff workers from the analysis.

[2] Palmer, *The Reluctant Job Changer* (1962), p. 14.

[3] Movements to more attractive jobs *within* the skilled group are not under consideration here.

was rare, movement *into* skilled jobs was a matter of some importance, because external recruitment difficulties were most acute for such workers. In view of this, plants will have had an incentive to attempt to increase their supply of skilled labour from internal sources, and their degree of success can be gauged from Table 11.2. The table shows the total number of workers whose last job with a plant was skilled and the percentage of this number who had joined the plant in the occupational groups shown.

TABLE 11.2

LAST JOB SKILLED: PERCENTAGE IN EACH OCCUPATIONAL GROUP ON JOINING THE PLANT: MALES

Occupational group on joining	Percentage				
	Birmingham	Glasgow	North Lanarkshire	New Town	Small Town
Apprentices	2·6	8·6	10·2	3·1	9·0
Skilled	74·9	86·2	75·6	74·7	65·2
Semi-skilled	20·3	2·7	7·5	19·2	18·0
Unskilled	1·8	2·2	6·4	3·1	7·9
Non-production	0·4	0·3	0·3	0·0	0·0
Total numbers	1,686	5,167	966	261	89

In each market area the principal source of skilled labour for the case-study plants was the recruitment of trained labour from the external market. This was most true of Glasgow, where 86·2 per cent of those whose last job with the plant was skilled had entered the plant in that occupational group. At the opposite extreme was Small Town where only 65·2 per cent of the skilled workers had joined the case-study plant at that skill level, while in the other markets about three-quarters of the skilled were recruited externally.

In some markets, apprentices completing their training were the main internal source of skilled labour, while in others this role was filled by the semi-skilled. In the former group are Glasgow and North Lanarkshire, where 8·6 per cent and 10·2 per cent of the skilled jobs were filled by those who had completed apprentice-training with the plant during the study period, with a smaller proportion of skilled jobs being filled by the upgrading of those recruited as semi-skilled. Apprentice-training programmes were as important in Small Town, but in this instance the biggest single *internal* source of skilled labour was those recruited as semi-skilled employees.

The reliance on semi-skilled employees to fill skilled vacancies through upgrading was particularly marked in Birmingham and New Town: for example, 20·3 per cent whose last job was skilled in Birmingham had joined the plant in a semi-skilled capacity, against only 2·6 per cent who had been recruited as apprentices during this period. This reflects greater upward mobility on the part of the semi-skilled, but it also reflects the fact that the plants' own apprentice-training programmes contributed relatively little to the supply of skilled labour coming forward over the study period. This is hardly surprising in New Town, where the plants' labour force expanded rapidly from a very small base, but in Birmingham the same situation did not apply. We shall see in Chapter 12 that relatively little apprentice-training was undertaken by the Birmingham units relative to their stock of skilled employees.

On the evidence of Table 11.2, one would conclude that there were certain quite clear differences between the markets in the volume and pattern of the internal mobility which occurred. Taking first New Town and Small Town, it is clear from a more detailed analysis of the data in Table 11.2, and from discussion with managements, that the movement of semi-skilled men to skilled jobs did represent genuine internal mobility between occupations, in the sense of a shift to a distinct and higher level of skill. The plants in New Town claimed that they were generally able to promote suitable semi-skilled workers to a wide range of skilled jobs and that in this process the negotiation of dilution agreements had been of considerable importance. Plants in Small Town could also promote suitable semi-skilled men to skilled jobs, such as turning, boring or toolroom work. Internal mobility from semi-skilled to skilled jobs was therefore an important means by which managements could react to a shortage of skilled engineering labour in these areas.

As Table 11.2 demonstrates, the larger markets of Birmingham, Glasgow and North Lanarkshire differed considerably in the extent to which skilled labour was drawn from those initially recruited to semi-skilled jobs. Thus 20·3 per cent of those whose last job was skilled had begun their employment with Birmingham plants as semi-skilled workers compared to only 2·7 per cent in Glasgow and 7·5 per cent in North Lanarkshire. Employment conditions and institutional factors could account for these contrasts, but we should first consider whether they also arise because of differences in production technology and consequently in occupational and job structure between the areas.

As we saw in the previous chapter, technological factors were important in causing a greater volume of occupational mobility in

Birmingham than in Glasgow as workers changed employers. We must expect that similar influences would be at work conditioning the volume of internal mobility in the two areas. The first possibility is that the amount of internal mobility into skilled work from semi-skilled differed because the two occupational groups were themselves of different significance in the various markets. This argument was discussed fully in Chapter 10, where we showed that some of the difference in the amount of external mobility from semi-skilled to skilled work could be explained in this fashion. We have followed the same procedure with regard to internal mobility, namely that of multiplying the percentage of skilled workers recruited from amongst the semi-skilled in Glasgow (2·7 per cent) by the ratio of semi-skilled to skilled workers in Birmingham *relative* to Glasgow.[1] The percentage of skilled workers we would 'expect' to be upgraded from the semi-skilled in Birmingham plants turns out at 5·7 per cent. Since the actual percentage of skilled drawn from semi-skilled was 20·3 per cent, it is evident that, although some part of the greater movement into skilled from semi-skilled work in Birmingham as against Glasgow was due to differences in the relative importance of these groups, the greater part of the difference remains to be explained.

The second way in which technology can affect the amount of mobility is through its effect on the job structure of the occupational groups themselves. It will be recalled from Chapter 10 that the problem centres on the classification of those designated as machine-men and machine-tool setter operators whom we have classified in the semi-skilled group, though there is obviously some affinity between their jobs and that of the skilled machine-tool setter. These groups were relatively much more important in Birmingham than in Glasgow, and an examination of the detailed matrix of job changes shows that a considerable proportion of the internal mobility from semi-skilled to skilled occupations in the former area did involve movement between these jobs. A precise estimate of the effect of such differences on the amount of internal movement from semi-skilled to skilled work is not possible, but it would seem safe to say that between a third and a half of the difference between Birmingham and Glasgow plants could be accounted for by technological factors, as distinct from market or institutional ones.

Our argument so far has been concerned with the factors which, in Birmingham, New Town and Small Town, appear to have assisted upward internal mobility from semi-skilled to skilled work. Yet even in these areas only a small fraction of the semi-skilled entrants to a plant actually moved to skilled work. This is demonstrated by Table

[1] I.e., we multiply by 2·1. See Table 10.3, p. 278 above.

11.3 which shows the total number recruited as semi-skilled workers and, of this number, the percentage who were employed in their last job in the various occupational groups.

In all areas the very great majority of those recruited to semi-skilled jobs remained at that skill level throughout their subsequent employment with a plant. For those who did move between occupational groups, upward mobility within the plant was more common

TABLE 11.3

SEMI-SKILLED RECRUITS, BY LAST OCCUPATIONAL GROUP WITH THE PLANT: MALES

Last occupational group	Percentage				
	Birmingham	Glasgow	North Lanarkshire	New Town	Small Town
Skilled	6·0	1·9	6·2	11·7	5·9
Semi-skilled	89·3	95·5	91·6	85·2	89·3
Unskilled	3·0	1·5	1·6	0·7	1·1
Non-production	1·6	1·0	0·6	2·3	3·3[1]
Total numbers	5,713	7,407	1,158	427	272

Note: 1. A further 0·4 per cent of the semi-skilled entrants became apprentices. This group has been excluded from the table as their numbers were either non-existent or trivial; see Annexe.

than downward mobility, but even the volume of upward mobility was fairly severely restricted, ranging from 1·9 per cent in Glasgow to some 6 per cent in Birmingham, North Lanarkshire and Small Town and to almost 12 per cent in New Town. Consequently, although Table 11.2 showed that a substantial percentage of the skilled men in Birmingham, Small Town and New Town were drawn from those originally recruited to semi-skilled jobs, this represented a very small part of all semi-skilled entrants, particularly in the first two areas. This apparent contradiction can again be explained in terms of the occupational structure of the labour force. As we have seen, an occupational structure with a high ratio of semi-skilled to skilled employees will usually mean that a high proportion of skilled jobs will be filled by upward internal mobility from semi-skilled jobs. This will be true even if the percentage of semi-skilled workers upgraded is no greater than that in other areas where semi-skilled workers are a smaller proportion of the total work force. We can see the effect of this in that, although there was a very considerable difference between Birmingham and Glasgow in the extent to which the skilled were upgraded from the semi-skilled (20·3 against 2·7 per cent), there was a much less marked difference in the percentage of semi-skilled entrants upgraded to skilled work (6·0 against 1·9 per cent). While the chances of promotion for a semi-skilled worker were undoubtedly worse in Glasgow, they were not so unfavourable as might be suggested by Table 11.2 at first sight.

This raises the question why the upgrading of semi-skilled workers was not more substantial, particularly in Birmingham where acute shortages of skilled labour persisted through most of the study period. One would expect managements to have been particularly vigorous in promoting suitable semi-skilled labour, both to solve immediate shortages and perhaps as a further inducement to semi-skilled labour to join the plant in the expectation of eventual promotion. Nor were they likely to be prevented from following this policy by institutional restrictions which did not, in Birmingham, pose insuperable obstacles to such transfers.

One possible explanation for the failure of managements to make good skilled labour shortages through internal mobility may lie in the wage structure of the engineering industry. We have already hypothesized[1] that certain peculiar features of occupational mobility involving a change of employers in Birmingham may have arisen because of the high earnings which could be obtained by semi-skilled employees, and internal mobility may have been similarly affected. As we saw in Chapter 5, it was not uncommon for less-skilled pieceworkers to have higher earnings than skilled timeworkers, and in these circumstances management may well have experienced difficulties in persuading competent semi-skilled workers to move to higher-skilled yet lower-paying jobs within the plant. Again, given the existence of substantial inter-plant wage differentials, it was perfectly possible that a semi-skilled worker in a low-wage plant could gain more by moving to another semi-skilled job at a high-wage plant, than by accepting internal promotion. That is, managements may have wished greater internal mobility, but may have been unable to secure it without prior action on the wages front.

Another possible explanation for limited internal mobility and continuing skill shortages stems from the analysis of the detailed job changes. This indicated that a high proportion of internal movement from semi-skilled to skilled jobs was accounted for by workers moving from machine-operating to machine-tool setting. These are closely related jobs between which it is relatively easy for men to move with a comparatively small amount of informal training. This,

[1] See pp. 284-5 above.

however, is relatively rare, for the experience acquired on most semi-skilled jobs is of only limited relevance to a wide range of skilled work, so that upgrading is not so easily accomplished. This naturally raises the question whether the plants concerned made a concentrated effort to overcome this difficulty by providing improved training facilities. All plants in Birmingham, as in the other market areas, undertook some training of semi-skilled labour, but in general the training period was short and the methods of instruction somewhat informal. Most training was 'on-the-job', but this often amounted to no more than 'learning-by-doing', and any instruction given was not sufficient to provide the expertise necessary for many skilled jobs. A number of managements seemed to be aware of their deficiencies in this field, but by their own admission this realization usually dated from the Industrial Training Act, which seems to have been responsible for a fairly widespread review of training provisions, particularly for semi-skilled work. If this is so, the quality of such training seems likely to improve, and this could result in a greater volume of internal mobility than in our study period.

While inadequate training facilities may have impeded internal mobility in all areas, there was a further factor at work which was peculiar to Birmingham. Some time must elapse before a semi-skilled (or unskilled) worker can gain sufficient experience to justify upgrading. Hence, to the extent that employees leave a plant before such experience has been acquired, internal mobility will be limited. High labour turnover is a feature of tight labour markets such as Birmingham, and the very conditions of scarcity which should encourage plants to develop their internal sources of labour supply also make the achievement of this objective more difficult by causing a high turnover amongst potential candidates for promotion.

To ascertain whether high turnover is likely to be a serious impediment to internal mobility we have to establish when internal mobility is likely to occur in an employee's service with a plant. This question was examined by means of an analysis of the jobs held by those persons recruited as semi-skilled workers by the case-study plants during 1959 and 1960, and who remained with a plant for at least six years.[1] The occupational group of the recruits on joining the plant was compared with their occupational group after each successive year of service, and from this the percentage promoted to skilled work was calculated. The results are shown for Birmingham and the Other Scottish Areas in Table 11.4. No results are shown for Glasgow, where very few semi-skilled workers were upgraded.

[1] All jobs held by an individual were taken into account. See p. 302, n. 1 above.

A comparison of Tables 11.3 and 11.4 will show that, of those recruited to semi-skilled jobs with at least six years' service, a higher percentage had been upgraded to skilled jobs at the end of that period of service than was the case with all semi-skilled recruits. The reasons for this contrast can be found in Table 11.4 below. As length of service rose, and presumably acquired experience increased, a higher proportion of semi-skilled recruits were upgraded to skilled work. Thus, while only 2 per cent of semi-skilled recruits were upgraded to skilled work by the end of their first year of service, no less than 10

TABLE 11.4

PERCENTAGE OF SEMI-SKILLED RECRUITS WITH SIX YEARS' SERVICE PROMOTED TO SKILLED WORK, BY YEARLY PERIODS: MALES

Length of service (years)	Birmingham	Other Scottish Areas
	Percentage promoted	Percentage promoted
1	2·3	2·6
2	4·1	6·7
3	5·9	8·2
4	8·3	8·2
5	9·6	10·4
6	10·2	10·4
Total numbers	1,278	269

per cent had been upgraded after six years. The average length of service of all semi-skilled recruits on which Table 11.3 is based was, however, much shorter than six years, especially in Birmingham, and many semi-skilled employees simply did not stay long enough with a plant to acquire the skill and experience which might have justified upgrading to skilled work. By itself, this must have limited internal mobility, quite apart from any secondary effects which high turnover rates might have had on the willingness of plants to invest resources in training such highly mobile assets.

In explaining differences in the amount of internal mobility within plants account must also be taken of differences in trade union policy. In Glasgow and to a lesser extent North Lanarkshire, the A.E.F. had established the principle that certain skilled jobs could usually only be filled by those who have completed an apprenticeship. This was seldom the case in the three other labour market areas. This may appear to give considerable significance to institutional restrictions but must be qualified on two grounds. First, because the type of work was different in the markets, managements in Glasgow and North Lanarkshire may well have *preferred* to employ apprentice-trained labour on skilled work rather than upgrade semi-skilled workers. A second, related point is that, even in Birmingham with its relative absence of institutional restrictions, upward mobility within the plant was insufficient to overcome shortages of skilled labour. In fact, detailed examination showed that semi-skilled men were upgraded into a very narrow range of skilled jobs. There was little or no mobility into skilled jobs such as toolmaking or boring where labour shortages were particularly acute. This simply reflected the view of management that such work could only be undertaken by apprentice-trained labour, and so once again we come back to the importance of technological factors because, in Glasgow particularly, the jobs which were normally filled by apprentice-trained labour were especially important. Hence, while the influence of institutional restrictions cannot be ignored, they were not the major factor bringing about the differences between areas in the volume of internal mobility.

The influence of employment conditions on the labour market is still problematical. The relatively slack conditions in Glasgow will presumably have served to reduce the amount of internal mobility to skilled jobs, because of the availability of skilled labour in the external market. Yet the effect of labour shortages on the amount of internal mobility in Birmingham was not as marked as we initially expected. Much of the greater volume of internal mobility from semi-skilled to skilled jobs was accounted for by differences in occupational job structures and was largely confined to a narrow range of skilled jobs. Such behaviour seems to be largely attributable to the absence of adequate training facilities, at least prior to the effective operation of the Industrial Training Act, but in turn this only raises the question why, in conditions of persistent scarcity of skilled labour which raises the cost of recruitment, plants had not invested in additional training facilities to make greater use of internal sources. A major reason seems to be that a tight labour market also leads to high turnover. This will mean that relatively few workers stay long enough to acquire the skill and knowledge necessary to qualify for promotion which, as we have seen, seldom occurs within the first year of service. Furthermore, the increased prospect of a man leaving either during or soon after completing a training course, thus denying the plant a return on the training costs, may be sufficient to deter it from undertaking such programmes.[1] It was, of course, this type of

[1] See Becker, *op. cit.*, p. 21.

argument which led to the Industrial Training Act, the effects of which are discussed more fully in the following chapter.

4. OTHER ASPECTS OF INTERNAL MOBILITY

Thus far we have been concerned with internal mobility into skilled jobs. Yet the Annexe tables show that considerable internal mobility has also taken place between other occupational groups. The present section will consider such mobility between semi-skilled, unskilled

TABLE 11.5

LAST JOB SEMI-SKILLED: PERCENTAGE IN EACH OCCUPATIONAL GROUP ON JOINING THE PLANT: MALES

Occupational group on joining	Percentage[1]				
	Birmingham	Glasgow	North Lanarkshire	New Town	Small Town
Skilled	1·0	1·4	1·4	1·6	5·5
Semi-skilled	90·8	92·0	74·8	72·5	84·1
Unskilled	6·9	5·0	20·0	23·9	6·9
Non-production	1·3	1·6	3·8	2·0	3·5
Total numbers	5,625	7,692	1,419	502	290

Note: 1. The handful of employees who began as apprentices and whose last job was semi-skilled have been excluded.

and non-production workers. Table 11.5 shows the number of workers whose last job with the plant was semi-skilled and, of this number, the percentage who had joined the plant in the occupational groups shown.

In Birmingham and Glasgow more than 90 per cent of the semi-skilled workers entered the case-study plants in that occupational group. This is the result one would have expected in Glasgow, since the decline in engineering employment enabled the plants to fill semi-skilled vacancies through external recruitment. In Birmingham, however, given the tightness of the market, one might have expected a rather different outcome. As it is, the volume and the character of internal mobility into semi-skilled jobs were remarkably similar in both conurbations and the contrast is provided by the Other Scottish Areas. In North Lanarkshire and New Town less than three in four of those whose last job was semi-skilled were recruited at that skill level by the case-study plants. A relatively greater number of semi-skilled workers were obtained through external recruitment in Small

Town, but even here 16 per cent of those who finished in semi-skilled jobs had undertaken some internal mobility within the plant.

The different behaviour of the plants in the Other Scottish Areas is explained by two factors. First, they were building up their labour forces rapidly over the study period, and the process of growth created a large number of vacancies for semi-skilled workers. Such an environment was favourable for internal mobility, but the second, crucial influence was the fact that external recruitment could not meet all the plants' labour needs because of the relatively small supply of

TABLE 11.6

UNSKILLED RECRUITS BY LAST OCCUPATIONAL GROUP WITH THE PLANT: MALES

Last occupational group	Birmingham	Glasgow	North Lanarkshire	New Town	Small Town
	Percentage[1]				
Skilled	1·6	4·5	7·7	3·4	16·7
Semi-skilled	19·8	15·1	35·2	50·8	47·6
Unskilled	71·8	69·7	43·6	39·5	35·7
Non-production	6·7	6·3	5·9	5·5	0·0
Total numbers	1,968	2,531	807	236	42

Note: 1. Percentages may not add to 100 per cent because those moving to apprenticeships have been excluded for the reasons stated in Table 11.1, n. 1.

labour with experience of the engineering industry. In consequence, managements turned to their internal market, so that a large proportion of semi-skilled vacancies were filled by upgrading, especially of unskilled employees.

The counterpart to this process was, of course, the promotion opportunities offered to the unskilled. These can be illustrated in Table 11.6 which shows the total number recruited as unskilled workers and, of this number, the percentage who were employed in their last job in the various occupational groups.

As we would expect from our previous analysis, the five areas again divide into two groups, the two conurbations on the one hand, and the Other Scottish Areas on the other. In Birmingham and Glasgow, some 70 per cent of those recruited to unskilled jobs remained unskilled throughout the remainder of their employment with a plant, with 19·8 and 15·1 per cent respectively moving to semi-skilled jobs. In the Other Scottish Areas, however, the majority of unskilled recruits did undertake some form of occupational mobility within the plant, and of these the bulk were upgraded to semi-skilled work. Indeed, in Small Town and New Town more had been upgraded to semi-skilled jobs than had remained in the unskilled group.

Our explanation for the greater internal mobility into semi-skilled jobs in the Other Scottish Areas plants has run in terms of the rapid growth of employment and the lack of trained engineering labour in the markets. While both of these considerations were extremely important there was a further factor which appears to have been influential. Six of the twelve plants in the Other Scottish Areas were under American control and/or management, and these plants were particularly active in adopting policies aimed at promoting internal mobility. As far as was possible the plants followed a deliberate policy of engaging labour at the bottom of the occupational ladder, with the intention of promoting such labour to semi-skilled jobs when the opportunity arose. Upgrading from unskilled to semi-skilled work was indeed usually carried through fairly quickly in the Other Scottish Areas, and more than one-half of those upgraded from unskilled to semi-skilled jobs had made the transition within their first year of employment with a plant. Active measures to promote internal mobility were not confined to American units, for it was not uncommon to find British plants adopting such policies both within and without the Other Scottish Areas. Nonetheless a large proportion of British plants had given little thought to developing measures which might facilitate internal mobility, unlike their American counterparts who regarded such mobility as an important method of meeting their labour requirements.

5. INTERNAL AND EXTERNAL MOBILITY

In Chapter 10 we discussed the extent to which managements filled vacancies by external recruitment of persons who had already acquired the necessary skill and experience and the extent to which such external recruitment was accompanied by occupational mobility. In this chapter we have, of course, been examining how far employers met their needs through encouraging occupational mobility within the plant, thus avoiding the need to recruit externally, or reducing the skill level required of the new recruit. The discussion of external and internal occupational mobility has been carried through quite separately, but it is obviously useful to attempt to put the analysis into perspective by considering which form of occupational mobility was most common. Did employees change occupations more readily on changing employers, or was occupational mobility within the plant more easily accomplished?

We consider the question by means of an examination of those persons recruited over 1959–66 for whom occupational group with the previous employer was known as well as first and last occupational group with the case-study plants. This information was analysed to produce two matrices each for Birmingham, Glasgow and North Lanarkshire. The first shows the recruits' occupational group with the previous employer against their occupational group on *joining* the case-study plants, while the second shows occupational group with the previous employer against the *last* occupational group with the case-study plants. By comparing these matrices we can establish how much occupational mobility resulted from employer-shifts, and how much from internal mobility within the case-study plants.

This is best seen by reference to Tables 11.7 and 11.8. Table 11.7 deals with mobility into skilled, semi-skilled and unskilled jobs on *joining* the plants by those previously employed in one of these three occupational groups. Thus we take in Birmingham the total number previously employed as skilled workers and calculate the percentage who became semi-skilled or unskilled on joining the case-study plants. We can see that 43·5 per cent took up semi-skilled jobs on recruitment, and 4·0 per cent unskilled jobs. While Table 11.7 shows the amount and the direction of occupational mobility which accompanied employer-shifts, Table 11.8 shows the total amount and the direction of occupational mobility which occurred as a consequence of *both* external *and* internal mobility. This is done by showing, for those employed in each occupational group with their previous employer, the percentage who had changed their occupational group in their *last* employment with the case-study plants. By comparing the results of Tables 11.7 and 11.8 we can see whether external or internal occupational mobility was more frequent.

Let us begin by examining downward occupational mobility by those engaged as skilled workers by their previous employers. Such downward mobility on changing employer was considerable in Birmingham, especially into semi-skilled jobs, and 43·5 per cent of those who were skilled took up semi-skilled jobs with the case-study plants. Table 11.8 shows that 42·7 per cent of this group were still in semi-skilled jobs in their last job with the case-study plants. This indicates that some of those who moved down the occupational ladder on joining the case-study plants subsequently moved back up the occupational ladder *within* the plants. However, while such downward mobility accompanying employer-shifts was sometimes followed by some upward mobility, relatively few of those who moved from skilled

TABLE 11.7

PERCENTAGE[1] OF THOSE WHO CHANGED OCCUPATIONAL GROUP: OCCUPATIONAL GROUP WITH PREVIOUS EMPLOYER AGAINST OCCUPATIONAL GROUP ON JOINING THE CASE-STUDY PLANT: MALES

Occupational group with previous employer[2]	Occupational group on joining								
	Birmingham			Glasgow			North Lanarkshire		
	Skilled	Semi-skilled	Un-skilled	Skilled	Semi-skilled	Un-skilled	Skilled	Semi-skilled	Un-skilled
Skilled	—	43·5	4·0	—	9·1	0·7	—	20·0	5·9
Semi-skilled	6·6	—	11·1	8·2	—	11·1	5·9	—	18·7
Unskilled	1·3	33·5	—	2·9	29·7	—	0·5	35·5	—

TABLE 11.8

PERCENTAGE[1] OF THOSE WHO CHANGED OCCUPATIONAL GROUP: OCCUPATIONAL GROUP WITH PREVIOUS EMPLOYER AGAINST LAST OCCUPATIONAL GROUP WITH THE CASE-STUDY PLANT: MALES

Occupational group with previous employer[2]	Last occupational group								
	Birmingham			Glasgow			North Lanarkshire		
	Skilled	Semi-skilled	Un-skilled	Skilled	Semi-skilled	Un-skilled	Skilled	Semi-skilled	Un-skilled
Skilled	—	42·7	3·6	—	9·9	0·6	—	19·3	2·1
Semi-skilled	16·8	—	21·6	9·0	—	4·6	9·0	—	11·0
Unskilled	1·6	35·1	—	3·9	35·5	—	1·8	50·8	—

Note: 1. In calculating these percentages we have ignored those whose job on joining or last job was as apprentices or in the 'skill not specified' category.

2. In Birmingham 1,946 were employed as skilled employees by their previous employer, 4,029 were semi-skilled and 2,061 were unskilled. The corresponding figures for Glasgow were 2,127; 1,105; and 439 and for North Lanarkshire 1,121; 756; and 448.

to semi-skilled jobs on changing their employer subsequently returned to skilled jobs as a consequence of internal mobility.

Now let us turn to those who moved from semi-skilled to skilled jobs. It is convenient in this instance to begin with Table 11.8, which shows the end result of both external and internal occupational mobility. Of those engaged by their previous employer as semi-skilled workers, 16·8 per cent were employed as skilled men in their *last* job with the Birmingham plants compared to 9·0 per cent in both Glasgow and North Lanarkshire. To establish whether such upward mobility was largely a consequence of external or internal mobility we compare these figures to those of Table 11.7. This shows that 6·6 per cent of those engaged by their previous employer as semi-skilled workers were *recruited* to skilled jobs in Birmingham compared to 8·2 and 5·9 per cent in Glasgow and North Lanarkshire. In other words, movement from semi-skilled to skilled work was relatively infrequent in Glasgow, and any such mobility which did occur was largely a consequence of employer-shifts. In North Lanarkshire upward mobility was again restricted but occurred more frequently within the plant than in Glasgow, while in Birmingham the greater degree of mobility from semi-skilled to skilled jobs was largely a consequence of internal mobility rather than of employer-shifts. The latter conclusion also applies to New Town and Small Town.

With the unskilled the chief interest is in movement into semi-skilled work, because only a handful make the transition to skilled jobs. In each area the *total* amount of movement from unskilled to semi-skilled jobs as a result of external and internal mobility was considerable, but in all cases the former type of mobility was much more common. In Birmingham, 35·1 per cent of those employed as unskilled workers by their previous employer had moved to semi-skilled work in their last job with the case-study plants, but most of this had been accomplished as a consequence of employer-shifts, since 33·5 per cent of this group had been *recruited* to semi-skilled work with upward internal mobility amounting to only 1·6 per cent. The situation in Glasgow was not much different. The total amount of mobility from unskilled to semi-skilled work was almost identical to Birmingham's and this was again mostly due to external mobility, although upgrading within the plant was somewhat more important in this case. The main contrast, however, lies in North Lanarkshire, where the volume of external mobility from unskilled to semi-skilled work was not much greater than in the two conurbations but where upgrading between these skill levels was much more considerable. Much the same conclusion applies to New Town and Small Town.

The distinction between the Other Scottish Areas and the two conurbations is the result of three factors—a rapid expansion of employment, a small reserve of labour with engineering skills, and the policies adopted, particularly by American managements, to increase internal mobility. Yet even in the Other Scottish Areas external occupational mobility was more important than internal mobility. This is not to say that we can safely neglect internal mobility, for it occurred even within the established engineering plants in the conurbations. Yet when all relevant considerations are taken into account, surprisingly little occupational mobility took place within the case-study plants. Even in Birmingham where conditions seemed to have been exceptionally favourable to internal mobility, a comparatively small proportion of the semi-skilled actually moved to skilled jobs, and the volume of upgrading was not sufficient to alleviate significantly the shortages of many types of skilled labour.

6. THE STRUCTURE OF INTERNAL LABOUR MARKETS

It is clear from our earlier discussion that there were considerable differences between our market areas in the volume and the characteristics of occupational mobility within the plant. To take upgrading to skilled jobs as an example, we find that in Glasgow and North Lanarkshire entry to skilled jobs for those currently engaged by the plant was largely restricted to those completing apprentice-training. In Birmingham, New Town and Small Town, however, the volume of movement from semi-skilled to skilled work was greater than the supply of skilled workers coming forward through the plants' apprentice-training programmes. The evidence of the present study indicates that the main cause of these differences in internal mobility lay in the occupational and job structure of the respective market areas. Thus the mix of skilled and semi-skilled labour will itself influence the volume of internal mobility, but a more important factor is the nature of work which the prevailing production technology demands. Movement between certain jobs is much easier than between others, depending on the extent to which the skill and experience acquired in the one job fits a man for others. The more common elements there are in different jobs, the easier it will be for workers to move between them.

To a considerable extent the jobs to be performed and the relationships between them in terms of skill content are effectively established by the capital equipment of the plant. When new equipment is being installed, therefore, management may have considerable discretion in the way in which the jobs are designed. Jobs can be designed as distinct entities, so that little mobility between them is possible, or they can be arranged so that some or all of the jobs form part of a hierarchy of skills, with linkages between them. The scope for action in this field is fairly large,[1] and although attention to job design is often seen only as a means of increasing 'job satisfaction', the approach can also be applicable to assisting internal mobility. By a careful design of the content of jobs, particularly when new plants are installed, managements may find that they are able to make much greater use of internal sources of labour supply. This was certainly the case in the Other Scottish Areas, especially in the American-controlled units.

There are, however, limits to the extent to which it is possible to 'link' jobs together, so that shortages for certain types of labour may be difficult to make good through internal mobility. In Birmingham, for example, toolmakers were the type of skilled labour most difficult to obtain and certainly more difficult to recruit than machine-tool setters, yet a much higher proportion of the semi-skilled moved to the latter job. Indeed the upgrading of semi-skilled workers to toolmakers was extremely rare and contributed little to a solution of the recruitment difficulties plants experienced for this job. The reason for this was simply that the majority of plants in Birmingham themselves preferred to use apprentice-trained labour for toolmaking jobs, so that the solution of this shortage could not be effectively resolved by upgrading.

Perhaps the single most important finding of the study of internal mobility is that in the British engineering industry internal labour markets are relatively unstructured. In other words, in a large number of plants little attention has been paid to measures which might promote internal mobility, nor have plants been subject to pressure from unions or employees to formalize the procedures used to govern internal mobility.[2] We might illustrate this conclusion by reference to the three characteristics of internal labour markets outlined in Section 2. These concerned the way in which the jobs in a plant are arranged into groups within which, and between which, internal mobility takes place; the rules governing the selection of workers when opportunities for promotion arise; and the extent to which jobs are filled by internal promotion as against external recruitment.

As regards the first, there were patterns to the internal mobility which took place, and these patterns differed between our market areas. Where movement into skilled jobs was concerned, this in part

[1] L. Davis, 'The Design of Jobs', *Industrial Relations*, Vol. 6, 1966.

[2] This was certainly so as regards upgrading. We shall see in Chapter 14 that action in redundancy situations was more closely ruled by convention, but even here formal agreements were rare to the point of being non-existent.

reflected trade union policies, but union restrictions were confined to this sphere, and even here appeared to be a less important factor in shaping internal mobility than the prevailing technology. In none of the areas was there any evidence of rules having been established by collective bargaining or any other means, to restrict managements' freedom of action in the selection of men for promotion. Thus while management might take seniority into account when determining promotion, they were not obliged to do so. Seniority might, of course, have been influential because of notions of 'equity' or 'justice', but more important appears to have been the consideration that an employee must complete some minimum time with the plant before acquiring the skill and experience which would justify promotion to more demanding work. This, however, remained a matter of management discretion, and there was no indication that management prerogatives were under attack in this area.

The third characteristic of an internal labour market is the extent to which jobs are filled by internal promotion or external recruitment. Again, we found no evidence that any market studied had developed formalized practices in this respect so that, for example, managements were compelled to attempt to fill a vacancy internally before undertaking recruitment in the external market. Certain managements, particularly those in the Other Scottish Areas, preferred to fill vacancies through internal recruitment, but such a policy was usually qualified by the phrase 'whenever possible' and reflected a management decision arrived at independently of union pressure. The overall result was that occupational mobility within the plant was less frequent than the occupational mobility which accompanied employer-shifts. Internal labour markets in the plants studied, and in the British engineering industry generally, do not then appear to have reached a very advanced and formal stage of development.

7. CONCLUSIONS

(i) The overall volume of internal mobility was greater in the Other Scottish Areas than in either of the conurbations, but in each area, save Small Town, there was a distinct pattern determined by occupational attachment. Internal mobility became more frequent as one moved through skilled, semi-skilled and unskilled employees. Non-production workers were less likely to change their skill level within the plant than the unskilled, but were more occupationally mobile than the skilled or the semi-skilled.

(ii) In each market area, the plants obtained most of their skilled labour by external recruitment, although the proportion obtained in this way varied a good deal between the markets. The volume and the pattern of internal mobility into skilled jobs also differed between the areas. In Glasgow and North Lanarkshire, those completing apprenticeships provided the main internal source of skilled labour, while in Birmingham, New Town and Small Town the chief internal source of supply was the semi-skilled. This contrast is partly due to the fact that plants in Birmingham and New Town had not developed substantial apprentice-training programmes, but in the latter area and in Small Town the plants had put considerable emphasis on developing policies to promote internal mobility.

(iii) The higher proportion of skilled jobs filled through the upgrading of semi-skilled workers in Birmingham relative to Glasgow was in substantial measure due to differences in their occupational and job structures, reflecting the contrasting technologies used by the plants in these areas. In other words, a relatively high proportion of skilled jobs were filled by the semi-skilled in Birmingham because the former comprised only a small proportion of the labour force, and because the job structure in Birmingham contained a large number of those jobs between which internal mobility was relatively easily accomplished. The counterpart of this was that in Birmingham movement into the skilled occupational group was largely restricted to a narrow range of skilled jobs, and there was very little mobility into those skilled jobs where recruitment difficulties were particularly acute. It should also be noted that only a small fraction of semi-skilled recruits were upgraded to skilled work in all market areas. In Birmingham, and in New Town and Small Town particularly, the limitation on upgrading did not seem primarily due to institutional restrictions. It was more probably due to wage-structure problems and the inability of on-the-job training systems to train semi-skilled workers for most skilled jobs.

(iv) In Birmingham and Glasgow only a small percentage of semi-skilled jobs were filled by the upgrading of unskilled employees within the plant. Upward mobility into semi-skilled jobs was considerable in North Lanarkshire, New Town and Small Town, and in the two latter areas a greater proportion of those recruited to unskilled work were promoted to semi-skilled work than remained as unskilled. The greater volume of internal mobility in the Other Scottish Areas appears to be due to three factors—a rapid expansion of employment in the case-study units, a lack of trained engineering labour, and the deliberate policy, especially of American units, of promoting internal mobility.

(v) Comparison of the volume of occupational mobility resulting from an employer-shift and from movement within the plant demonstrates that in each labour market area most occupational mobility was a consequence of an employer-shift and that internal mobility was, with certain exceptions, severely restricted in volume. In sum, internal labour markets in the British engineering industry are generally unstructured, i.e. the volume of internal mobility is limited, there is a port of entry at every level of skill, and such internal mobility as does occur is subject to few formal rules. Nor is there any indication that managements' prerogatives in this field are being seriously challenged by unions.

ANNEXE TABLE 11A

OCCUPATIONAL GROUP ON JOINING AGAINST LAST OCCUPATIONAL GROUP WITH THE CASE-STUDY PLANTS (NUMBERS)

Occupational group on joining plant	Last occupational group at plant					
	Apprentices	Skilled	Semi-skilled	Un-skilled	Non-production	Total
Birmingham						
Apprentices	615	43	4	0	0	662
Skilled	0	1,262	54	14	4	1,334
Semi-skilled	1	343	5,104	172	93	5,713
Unskilled	4	31	390	1,410	133	1,968
Non-production	0	7	73	32	656	768
Total	620	1,686	5,625	1,628	886	10,445
Glasgow						
Apprentices	1,059	444	3	0	0	1,506
Skilled	7	4,452	105	4	5	4,573
Semi-skilled	1	141	7,077	111	77	7,407
Unskilled	370	113	381	1,507	160	2,531
Non-production	19	17	126	21	855	1,038
Total	1,456	5,167	7,692	1,643	1,097	17,055
North Lanarkshire						
Apprentices	262	99	0	0	0	361
Skilled	0	730	20	3	2	755
Semi-skilled	0	72	1,061	18	7	1,158
Unskilled	108	62	284	305	48	807
Non-production	0	3	54	13	197	267
Total	370	966	1,419	339	254	3,348
New Town						
Apprentices	36	8	0	0	0	44
Skilled	0	195	8	0	2	205
Semi-skilled	0	50	364	3	10	427
Unskilled	3	8	120	92	13	236
Non-production	0	0	10	1	47	58
Total	39	261	502	96	72	970
Small Town						
Apprentices	25	8	1	0	0	34
Skilled	0	58	16	2	1	77
Semi-skilled	1	16	243	3	9	272
Unskilled	0	7	20	15	0	42
Non-production	0	0	10	0	9	19
Total	26	89	290	20	19	444

PART V: PERSONNEL AND MANPOWER POLICIES

CHAPTER 12

MANPOWER FORECASTING AND APPRENTICE-TRAINING

1. INTRODUCTION

The ability to anticipate forthcoming events so as to shape current action to future needs is a skill crucial to the art and science of management. As an art, it depends on hunch and guesswork; as a science, on using more sophisticated, although not necessarily more reliable, techniques to forecast the future. The art and the science are not mutually exclusive, and in recent years there has been some tendency to supplement hunch by more 'scientific' methods. In labour management, this has involved a more extensive use of manpower forecasting, which may be described as a process of estimating the plant's future manpower requirements on a systematic basis and over specified periods. Accurate manpower forecasts are useful because they provide advance warning of impending changes in labour requirements—changes which may involve recruitment and selection problems, provision for possible redundancies, new training and re-training requirements and so forth. Failure to anticipate such changes does not avoid the need for such adaptation, but it is likely to imply that adjustments are carried through less efficiently.[1] The first question we must then consider is how plants attempt to forecast future labour needs. Do they, indeed, attempt to anticipate the future, or do they adjust to change on a day-to-day basis as and when the need for change is thrust upon them?

The derivation of manpower forecasts is not an end in itself; they are useful in practice only if they form the basis of specific decision-taking. This has two implications. First, the forecasts must be sufficiently accurate to allow judgments to be made between possible

[1] The advantages which may arise from effective manpower forecasting are described in Department of Employment and Productivity, *Company Manpower Planning*, *op. cit.*

alternative courses of action, and, second, they must provide sufficient advance warning to allow necessary adjustments in the size or structure of the labour force to be made smoothly and in time. In Britain, manpower forecasting is a fairly recent innovation, and one which does not so far appear to be widely applied. Even in larger, more technologically advanced firms, according to a survey of the metal industries carried out by the Ministry of Labour, 'less than one in four of all firms approached said they had made some sort of forecast for more than two years ahead, and only about one in forty said that this covered all categories of workers'.[1] This chapter will show that even where manpower forecasts are made they are often based on methods which inspire little confidence in the results obtained.

The range of employment problems with which plants are faced varies to some extent from market to market, and for this reason there are likely to be differences in the emphasis placed on particular aspects such as the occupational coverage and the period covered by manpower forecasting. If a plant employing semi-skilled men or women knows that new recruits will require in-plant training, often lasting from three or four weeks to three months, then forward employment requirements must be estimated over at least this period to enable the necessary recruitment and training programme to be carried through. Where the plant is attempting to gear apprentice-training to future skilled labour needs, however, the forecast should relate to a much longer period, generally extending to four or five years. Such considerations are discussed in Section 2 which examines the extent and the methods of manpower forecasting employed by the case-study plants.

In Section 3, we come to the heart of the matter, which is the response of managements to their own forward estimates. For example, where a forecast has identified a sector where labour is likely to be in short supply, this should enable management to take the necessary action to resolve any anticipated recruitment difficulties. This is particularly relevant where labour may have to be recruited from other areas or where an extensive period of training is required. It might be expected that in areas of high or expanding employment managements would have a greater incentive to act along these lines, and we shall see whether the response of the Birmingham and Other Scottish Areas case-study plants fulfils this expectation. Equally, advance notice of a need to reduce the size of the labour force provides management with an opportunity of avoiding or reducing the scale of a redundancy by the redeployment of workers, by reducing overtime or by stopping recruitment to allow

[1] Ministry of Labour, *The Metal Industries*, *op. cit.*

normal wastage to help towards the desired adjustment. We shall examine in Section 3 how far plants with manpower forecasts made effective use of these methods when faced with an impending redundancy.

Relating apprentice-training programmes to future needs for skilled labour poses particular difficulties for manpower forecasting; for, given the period of training required, it is necessary to anticipate skilled labour needs over a period as long as five years ahead. Yet the benefits of such forecasts could be considerable, primarily because the training period, and hence the adaptation period to changed labour needs, is so long. In the event, the formidable problems of relating long-term forecasting to current training needs appear to outweigh any putative benefits. As Section 4 will show, manpower forecasts were seldom acted on as a basis for determining the recruitment of apprentices. It is in fact difficult to ascertain the factors which were crucial in this field, but it would appear that the number of apprentices recruited was usually decided on the basis of some conventional rule-of-thumb or in response to conditions *currently* ruling in the labour market or in the plant itself.

In this examination, it is necessary to bear in mind the extensive shortages of skilled labour which have prevailed for many years in the survey areas, particularly Birmingham, and nationally. Such shortages are not attributable simply to deficiencies in the training field. They may result, for example, from rapid technological change, and may be aggravated by institutional pressures which require that particular jobs should be carried out by certain grades of skilled workers. Nevertheless, the supply of skilled workers is to a large extent a function of the input of trained apprentices, and the arrangements made by plants for determining the scale of apprentice-training are thus a matter of considerable importance in any study of the operation of labour markets.

The problem of securing adequate apprentice-training is by no means a new one. The Carr Committee's report in 1958[1] found serious qualitative and quantitative deficiencies in the provision of apprentice-training, but its strictures brought no response until the Industrial Training Act of 1964. The Act was based on the premise that an insufficient supply of skilled labour was forthcoming, because those plants which undertook the costs of apprentice-training did not earn a sufficient return on these costs when training was completed.[2] Instead, other plants which had not borne any of the costs of

[1] Ministry of Labour, *Training for Skill* (H.M.S.O., 1958).

[2] See B. J. McCormick and P. S. Manley, 'The Industrial Training Act', *Westminster Bank Review*, February 1967.

training 'poached' the trained labour and reaped the benefits. To rectify this supposed inequity, the Act set up Industrial Training Boards which were empowered to impose a levy on firms on the basis of the numbers employed, and from these levies to make grants to those who undertook training approved by the Boards. The levy was the stick intended to spread the costs of training more 'equitably' over an industry as a whole, while the grants were the carrot designed to encourage plants to improve the quality and the quantity of training. The period between the setting up of the Engineering Industry Training Board and the end of our survey, December 1966, was a comparatively brief one, so that no final conclusion can be reached on the effects of the Act. Nonetheless, we can measure the extent of apprentice-training prior to 1964, and indicate some short-term changes in apprentice-provision which appear to have resulted from the Act. A summary of the conclusions reached in Section 4, together with those from the previous sections, is provided in Section 5.

2. EXTENT AND METHODS OF MANPOWER FORECASTING

Manpower forecasting is concerned with identifying and measuring probable future changes in the level and the structure of employment. Such changes may arise in three ways: through fluctuations in product demand, through technological change, and through rationalization. In the short term, changes in product demand generally assume most importance, and it is this factor which tended to be predominant in manpower forecasting by the case-study plants. In other words, where manpower forecasts were made they generally assumed the existing organization and level of technique as given, so that the forecast was concerned with estimating future output which could then be converted directly into manpower equivalents.[1] Technical and organizational changes, although they may to some extent have been the result of stimuli emanating from the demand side, were distinct in the sense that they were the result of conscious management decisions which emerged only rarely, or else did not have major consequences for manpower utilization in the short term. Again, while rationalization could produce a substantial change in labour demand, the

[1] No example was found of a plant modifying manpower forecasts in view of decisions regarding rationalization and only one of technical change which affected a manpower forecast. The latter arose from the introduction of tape-controlled machines. In general, however, technical change was continuous rather than discrete, so that it was absorbed without any marked short-term impact on employment.

period of advance warning was usually substantial, so that appropriate measures could be adopted to ease the process of adaptation.

In our sample of plants the period over which manpower forecasts were made was usually dependent on the length of the order book. Relatively few plants attempted to predict product demand beyond this point by reference to past trends and/or by explicit consideration of developments in the economy as a whole. The 'typical' manpower forecast simply consisted of converting forward orders into labour equivalents on the assumption of fixed technical coefficients. The period over which the forecast applied was, therefore, largely dependent on the nature of the work undertaken and tended to be longer

TABLE 12.1

PERIODS COVERED BY MANPOWER FORECASTS

	Number of plants				
Area	No forecasts	Less than one year	One year	Three or more years	Total
Glasgow	3	18	6	0	27
Birmingham	9	7	4	7	27
Other Scottish Areas	2	5	4	1	12

for those engaged in producer-durable production, for plants working on long-term government and local authority contracts, and for establishments which were branch factories of larger organizations with a centrally determined production programme. A tabulation of the results of our enquiries may give a somewhat spurious precision to the various points which emerged, but Table 12.1 gives some indication of the situation in the respective labour market areas.

Of the plants in all the areas which made some forecast, thirty, or more than half, did so for periods of less than a year, and in 18 of these cases the interval was three months or less. These facts reflected the length of forward order books and the reluctance of managements to put much weight on forecasts beyond this period, even where order books were longer. Partly for this reason, one-year forecasts were normally subject to review at monthly or quarterly intervals. Perhaps the most interesting point which is brought out by Table 12.1 was the apparently different attitude to forecasting among Birmingham plants from that in Glasgow and the Other Scottish

Areas. In Birmingham more than one-third of the case-study plants stated that they did no formal forecasting, but operated on a day-to-day basis. Yet this does not appear to have put them at any substantial disadvantage. For example, such plants, from a general feel of the situation based on past experience, were able to start recruitment procedures in advance of the time when additional labour was required on the shop floor. Their rule-of-thumb methods of proceeding may, therefore, not have been substantially different from those obtaining in plants where more formal methods of short-term forecasting were practised.

The difficulties of forecasting for one year ahead and even for shorter periods may be illustrated by the experience of one plant which was able to furnish detailed records of forecasts and out-turns over a yearly period. This plant made a detailed manpower forecast for 12 months ahead, based on the estimates contained in an annual financial budget. A monthly forecast was also prepared which attempted to estimate employment in *each* of the six subsequent months. Not only were the trends in the annual budget not realized, but in the monthly forecasts the plant failed to anticipate correctly the direction of change in employment in all cases over 1966. It is noteworthy that these difficulties could arise even where considerable care was being exercised in making detailed forecasts (individual skilled occupations such as turners, fitters and toolmakers were analysed separately) and where there was consultation with the plant's sales, design and production departments. In large part, the errors arose because of over-optimism in a period when employment levels were tending to fall. This tendency to anticipate the best rather than the worse was apparent in other instances and was not confined to managers. In one plant where forecasts were prepared by specialist sales economists, product and labour demand were consistently overestimated. The predilection to view the future through rose-tinted glasses is not apparently unique to Britain. An American study of manpower forecasting suggests that plants tend to be over-optimistic in estimating future employment, that forecasts of the trend in employment seldom improve on the results expected by chance, and that plants are particularly poor at anticipating a fall in employment.[1]

The procedures followed by the case-study plants in making forecasts were, with a few honourable exceptions, rudimentary. In most plants, personnel officers normally consulted sales and production control departments in formulating forecasts, but the translation of

[1] R. Ferber, *Employers' Forecasts of Manpower Requirements: A Case Study*, University of Illinois (1958).

the sales and production data into manpower requirements was seldom carried through in a systematic fashion. It was common for the works manager to 'form an impression' of labour needs from consultation with first-line management or to judge 'from experience' the manpower requirements of the plant in accordance with the work schedules. The personnel officer would have to rely on such guidance in attempting to make a forecast. The uncertainty of these procedures appeared to be particularly marked in the case of plants in or associated with the motor-car industry, where work schedules could not be planned beyond a forward period of three months, and even then were tentative and liable to cancellation at short notice. In contrast, one American-controlled engineering firm with two case-study plants had developed forecasting techniques which were more advanced than those normally adopted in British units. In this case standard times had been established for production tasks, which were used to convert sales forecasts into individual machine operations and thence into labour targets.

A further weakness in the preparation of forecasts was that little account was taken of labour wastage. Usually the forecasting took the form of estimating the future level of employment rather than calculating the recruitment necessary to achieve that level. In a few cases, wastage was not specifically allowed for because it was usually low, and in some recently established plants it had been taken into account only for the period when the scale of initial recruitment and losses had been considerable. More generally, plants either did not know their wastage rates or, somewhat illogically, they maintained turnover statistics separately from their manpower forecasts, and claimed that these were 'borne in mind' by personnel officers when considering whether advertising of vacancies or other advance action was necessary.

One or two plants confined their forecasts to total labour requirements. Usually, however, forecasts were more detailed. Some plants made estimates for all or most occupational groups, but others did so only for skilled occupations, especially certain key trades whose volume of output determined the pace of work for the factory as a whole. In the Other Scottish Areas some plants detailed their semi-skilled workers by jobs, whereas in Glasgow some half a dozen plants did not consider that semi-skilled or unskilled categories merited consideration. In Birmingham, forecasts tended to cover all types of worker, usually analysed by department. A distinction was drawn by some Glasgow plants between direct and indirect categories rather than between occupations as such. In these cases, a detailed forecast was made of direct labour needs to which some percentage figure for indirect labour would be added to give the total manpower requirements of the establishment.[1]

3. RESPONSE TO MANPOWER FORECASTING

The question we must now discuss is whether manpower forecasting as generally practised was adequate to the needs of the plants concerned. This is best considered in the light of the reaction, or lack of reaction, by management to the forecasts obtained. Our criterion is whether or not subsequent management policy was geared to the forecasts made, and whether these forecasts assisted management to overcome or minimize potential labour problems. This is examined in relation to the three spheres in which action by personnel departments might be expected to be related in some measure to forecasting—recruitment, redundancy, and the establishment of a new plant.

The first general point to be made is that, except in the case of newly established plants, there was a less direct relationship between forecasting and recruitment than might have been expected. The basis of recruitment action was usually not a formal forecast, even where this was made, but rather the personnel officer's general 'feel' of the market. This was often supported by a substantial amount of documentation, but this usually consisted of other personnel records such as a detailed statement of the numbers employed currently and in the recent past; the number recruited and leaving over previous periods; and current 'requisitions' for additional labour from departmental managers or foremen. This was the kind of information to which the personnel officers attached most importance. It was studied in the light of any particular recruitment difficulties which might have arisen, and any unusual volume of separations in the previous month would also be taken into account. The formal forecasts at most only supplemented these other statements, in the sense that a projected increase in employment would perhaps encourage the personnel officer to intensify his efforts to deal with the current situation or to advertise vacancies in advance of production needs. Normally, however, the personnel officer did not appear to initiate advance recruitment action primarily on the basis of manpower forecasts, as distinct from these other personnel records and the current notifications of labour requirements from production

[1] For some details of this method, see D. H. Walker, 'Labour Budgeting—An Approach to Manpower Planning', *Personnel Management*, Vol. 48, 1966, and Vol. 49, 1967.

departments. This indeed is hardly surprising, as to base recruitment on forecasts would require a degree of detail and accuracy which few forecasts are likely to approach.

In Birmingham, half of the plants which prepared forecasts for a year or more ahead normally took action to recruit additional skilled or other labour only as the need for additional labour arose on the shop floor or a week or so in advance of that need. Only four plants initiated action for certain categories of labour three months ahead of need. As we have seen, the actions of those plants which prepared short-term manpower forecasts were indistinguishable from those of units which did no formal forecasting. Amongst Glasgow plants, too, recruitment policy was seldom closely related to forecasts of labour need. This was partly because the plants, like those in Birmingham, had insufficient faith in the accuracy of forecasts to base recruitment policy on them, but an additional factor was the easier labour market situation which allowed demands for many types of labour to be filled as and when the need arose. In the Other Scottish Areas, the position was similar, although in those areas with little previous experience of engineering employment where in-plant training for semi-skilled labour was necessary, forecasts were sometimes used to determine recruitment policy.

Managements appeared more likely to react to forecasts of future labour needs when a run-down in labour requirements was foreseen than when an increase in employment was anticipated. It was quite common, therefore, for management to take early action to reduce or eliminate the need for redundancies through a cessation of recruitment, a reduction of overtime or the introduction of short-time working. Manpower forecasting was, in the relatively easy labour market conditions which obtained in Glasgow and the Other Scottish Areas, often regarded by managements as having greater significance and value in the context of redundancy than in that of helping to deal with recruitment problems. There were comparatively few redundancies among the newer plants in the Other Scottish Areas, but one of these plants first introduced a detailed scheme of forecasting in 1966, because two earlier redundancies had arisen which the management considered in retrospect could have been avoided. Another plant, which attached considerable importance to forecasting and to personnel management generally, claimed that by budgeting a year ahead and carrying out monthly reviews they had been able to avoid redundancy except on one occasion. In Birmingham also, managements tended to emphasize the importance of forecasting in minimizing redundancies, but there was no objective evidence which indicated that those plants which employed the most sophisticated forecasting techniques did, indeed, experience a relatively low rate of redundancies.

It is interesting to categorize the plants according to the type of manpower forecasting undertaken (as in Table 12.1) and compare this to redundancy experience. This has been done for Glasgow where redundancies were more important and more widespread than in Birmingham. The result is that those plants with no manpower forecasts of any description were the units which least frequently experienced redundancies in the sense that, for the period 1959–66, they had the highest proportion of quarters with no redundancies at all. The frequency of periods in which redundancy occurred was greatest for those plants with the most sophisticated forecasting techniques and the longest forecasting periods, i.e. one year or three years or more. The picture is not as black as painted, for when redundancies did occur they did tend to be more important, *pro rata*, in plants with no forecasting. Hence, it may be that plants with forecasts were able to reduce the impact of a redundancy through early warning and advance action. The whole exercise is, of course, somewhat hazardous, as a host of other factors may be producing such results quite independently of the type of forecasting or the lack of forecasting. Yet it must be said that, although managers who made manpower forecasts felt the process assisted them to foresee, and hence reduce, redundancies, there was no clear evidence that the forecasts made any noticeable difference.

The experience of the case-study plants underlines the difficulties of deriving accurate manpower forecasts, particularly over extended periods. It must be said, however, that the degree of accuracy attained could be increased by improvements in method. In many cases, plants appear to have constructed manpower forecasts not because they are considered useful but because management textbooks imply that no 'progressive' or 'good' management should be without one. The forecast could, then, be proudly displayed to the visiting academic as proof that the time spent at management conferences was not wasted. To be fair, there appears to have been over the period 1959–66 a move towards forecasts attempting a more detailed breakdown of labour requirements by skills and, in particular, an attempt to identify groups where labour shortages would have adverse effects on the over-all production level of the plant. Hence the greatest attention was paid to skilled workers where the potential labour supply available to the plant was least plentiful. Labour market conditions were instrumental in explaining certain other aspects of behaviour. Thus, in Birmingham, all types of workers tended to be included whereas in Glasgow, where semi-skilled and unskilled workers were easy to obtain, fewer plants attempted to forecast their future needs for these groups in any detail. In other respects the techniques adopted were unsatisfactory, particularly the fairly general failure to take cognizance of wastage rates as an essential element in the calculation of recruitment needs over the forecasting period.

The previous discussion has been mainly concerned with the methods of manpower forecasting employed by established plants. It is useful to set against this the experience of plants in the Other Scottish Areas which were newly established or substantially expanded during the survey period and which, before embarking on such policies, made fairly extensive use of manpower forecasting to evaluate labour demand and supply prospects. Manpower forecasting for this purpose was undertaken by eight case-study plants. The normal practice was to work out a target for the maximum employment likely and to consider this target in relation to the likely reserves of labour available in the local labour market. As regards the availability of skilled labour to match their forecasts, managements in the Other Scottish Areas recognized that there was virtually no pool of skilled men unemployed. So far as North Lanarkshire was concerned, however, it was anticipated that there would be a sufficient flow of skilled men becoming available on turnover from that area or Glasgow. In New Town, the convenience of the situation, its rapid growth and the ready availability of new housing encouraged the plants to expect that their forecast requirements of skilled men could be attracted to the area. For semi-skilled and unskilled male and female labour, plants were able to take into consideration the experience of other plants, the size of the population within daily travelling distance, and the existence of other industries where employment was declining.

The estimates of labour requirements made by firms prior to establishing or extending plants generally covered, as was to be expected, a longer time-span, usually about two to five years, than was found to be the normal practice with established plants. Another distinctive element of the planning process in the new or rapidly expanding establishments was that top management was actively involved. However, once the plant was operating normally, manpower forecasting was carried out at the level of the personnel manager and was not of great interest to the higher echelons of management.

4. EXTENT OF APPRENTICE-TRAINING

The main point which has emerged from the foregoing section has been the very limited extent to which the case-study plants endeavoured, by systematic forward planning, to anticipate and make provision for the labour market situations which they were likely to encounter at a later date. This general assessment is applicable to the attitudes of the managements both in relation to recruitment difficulties and, possibly to a lesser extent, to redundancy. We now examine the policies of the plants in regard to apprentice-training, another field where the needs of the future can be met only if adequate provision is made in advance. In this case, current action has to be related to some estimate of labour needs in the future where the future must be seen in years rather than months. Any failure on the part of management to anticipate future events may have serious consequences not only for the individual plant concerned but also for the industry as a whole.

As has been shown, shortages of skilled labour arose in all areas from time to time, but were most acute and persistent in Birmingham. This being so, one might expect that Birmingham plants would have been particularly active in training apprentices especially as shortages of skilled labour had been evident for a long period prior to 1959. In the event, apprentice-training was much more widespread in Glasgow and North Lanarkshire. This difference, and especially that between Glasgow and Birmingham, can partly be explained by differences in production methods which required in Glasgow a higher proportion of skilled men. Again, plants in Birmingham were more able to meet labour shortages through internal promotion, thus avoiding the need, in some cases at least, for apprentice-training. Yet internal promotion to skilled jobs was not very common and was not sufficient to surmount skilled labour shortages. Nor was there any indication that the position would change in the future. Despite this, there was no movement by Birmingham plants to increase apprentice-training until the advent of the Industrial Training Act of 1964. Until that year, the case-study plants in both Glasgow and Lanarkshire employed almost three times as many apprentices in proportion to their skilled labour force as was the case in Birmingham. That this situation should have continued despite the greater difficulty of obtaining skilled labour in the latter area reflects the force of tradition. In Glasgow and North Lanarkshire, the great bulk of engineering plants had always trained apprentices, and they continued to do so. In Birmingham, many plants had never trained apprentices or had trained very few relative to the size of their skilled labour force, and they too continued in the established manner despite acute shortages of skilled labour.

The position in Glasgow and Birmingham can be seen from Table

12.2. The first column takes as a measure of apprenticeship provision the number of apprentices employed by all plants at October 1st of any year as a percentage of the number of skilled workers employed by these plants at July 1st in that year. The stock of apprentices is measured at October 1st because plants generally regard the 'apprentice year' as running from October to October and because the largest intake of apprentices covers the months immediately preceding October. In view of the latter factor, it appears reasonable to suppose that, if plants determine apprentice recruitment in the light of their current employment of skilled labour, then recruitment policy over the crucial summer months may be most affected by the stock of employees some months earlier. Hence the choice of July 1st.[1] This procedure was also applied to each individual plant to give for each

TABLE 12.2

APPRENTICE-PROVISION AND EXTERNAL MARKET CONDITIONS: GLASGOW AND BIRMINGHAM, 1959–66

Year	Apprentice provision, all plants	Coefficient of variation of apprentice provision by plants	Apprentice intake, all plants	Ratio of unemployed to vacancies, skilled (June)	Percentage male unemployment, engineering (June)
Glasgow					
1959	28·0	69·5	4·9	12·1	2·5
1960	28·0	68·9	6·3	3·7	2·1
1961	25·1	71·6	5·9	1·8	1·8
1962	23·3	68·2	4·4	10·1	2·8
1963	23·1	67·8	4·9	30·4	4·5
1964	22·4	70·7	5·0	4·4	2·8
1965	23·9	76·8	8·1	0·5	1·9
1966	23·8	78·4	6·9	1·1	1·9
Birmingham					
1959	8·1	175·5	2·2	0·5	0·7
1960	7·0	147·8	2·5	0·1	0·3
1961	7·4	137·0	2·6	0·2	0·7
1962	9·3	162·6	3·7	1·0	1·3
1963	9·5	165·2	2·9	1·9	1·5
1964	11·9	165·1	4·9	0·2	0·7
1965	16·2	158·7	7·5	0·1	0·5
1966	21·0	123·2	7·6	0·1	0·4

[1] Alternative periods were tried but none had any significant effect on the results obtained.

unit the number of apprentices at October as a percentage of the stock of skilled employees at July. The coefficient of variation for the resultant figures was then calculated showing the extent to which provision for apprentice-training varied between plants in the same labour market area. Changes in policy towards apprentice-training will only make their full impact on the number of apprentices over a period of years. In Glasgow at least, few boys failed to complete an apprenticeship once they began their training, so that the number of apprentices at any time reflected past decisions as well as current policy. This being so, the third column of the table shows, for all plants, the apprentice 'intake' in each year. This is the total number beginning apprenticeships each year as a percentage of the average stock of skilled men in that year.[1] Finally, the table includes two indicators of employment conditions in the labour market—the ratio of unemployed to unfilled vacancies for skilled men in June of each year and the percentage of male engineering workers wholly unemployed in that month.

The contrast between the extent of apprentice-training in Glasgow and Birmingham is brought out clearly by Table 12.2. The ratio of apprentices to skilled men never fell as low as one to five in Glasgow prior to the Industrial Training Act of 1964 and never reached one to ten in Birmingham. Examination of the yearly intake of apprentices shows no tendency for the position to change prior to 1964. In Glasgow, the intake of apprentices seldom fell below 5 per cent of the stock of skilled males, while in Birmingham the percentage was always in the 2–3 per cent range. The second column of the table provides part of the explanation for this difference. We find that the coefficient of variation of the percentage of apprentices to skilled men was always lower in Glasgow than in Birmingham. In simple language, this means that in Glasgow there was less variation between plants in the proportion of apprentices employed. Thus, apprentice-training was more evenly spread between plants in Glasgow, while a considerable number of plants in Birmingham made little or no attempt to meet their skilled labour needs through their own apprenticeship programmes. Of the 25 Birmingham plants for whom data were available, six did not recruit any apprentices over the eight-year period and ten plants recruited only 48 apprentices in total. The remaining nine establishments, which employed less than 80 per cent of the skilled labour force, hired 895, or 95 per cent of all apprentices taken on by Birmingham plants over 1959–66.[2]

[1] Obtained by averaging the four quarterly stock observations of the number of skilled men employed.

[2] These figures provide support for the view that prior to the Industrial

The Industrial Training Act of 1964 marks a clear break with past practice in Birmingham. The percentage of apprentices to skilled men rose from 9·5 per cent in 1963 to 11·9 per cent in 1964 and to 21·0 per cent in 1966. The change in the apprentice intake was, as would be expected, even more substantial. It rose from 2·9 per cent in 1963 to 4·9 per cent in 1964 and to 7·6 per cent in 1966, exceeding the rate of intake in Glasgow for the first time. This increase in apprenticeship provision was accompanied by a narrowing in the spread of the apprentice/skilled-worker ratios in Birmingham plants so that the coefficient of variation of apprenticeship provision fell from 165 per cent in 1964 to 123 per cent in 1966. Hence, while apprentice-training was still more widespread in Glasgow, the 1964 Act did, in Birmingham, fulfil its aim of increasing the number of apprentices by inducing more plants to take up such training. In Glasgow, the Act did not have much effect, but this was only to be expected as the Act presumably was attempting to move apprentice-training *towards* a position such as that which had already been achieved in that city.

Although events in Birmingham were consistent with the stated objectives of the 1964 Act, it remains an open question whether those objectives are justified in economic terms. After all, if a plant which has undertaken little apprentice-training is compelled to pay a levy to the Engineering Industry Training Board, it has an incentive to expand its training programme to recover some part of that levy. This does not necessarily involve an increase in efficiency. It may merely imply that the plant prefers one sub-optimal position, a lesser difference between the levy and the grant, to another sub-optimal position, a greater difference between the levy and the grant. The plant may prefer, on rational economic grounds, the initial position where it did little training and where no system of levies and grants was imposed upon it.

However, although the objectives of the Act may be criticized from first principles[1] it is as well to bear in mind the fact that first principles in economics assume that managements act rationally; that they weigh against each other the present and future costs and benefits arising from apprentice-training, and act accordingly. On the evidence available, it is difficult to believe that the decisions of Birmingham plants were based on economic rationality rather than simply on tradition. The number of apprentices taken on by a plant

Training Act small plants were likely to contribute relatively little *pro rata* to the supply of trained apprentices. There was, however, no systematic relationship between plant size and apprentice-provision in Glasgow.

[1] See D. Lees and B. Chiplin, 'The Economics of Industrial Training', *Lloyds Bank Review*, April 1970; and M. Oatey, 'The Economics of Training', *British Journal of Industrial Relations*, Vol. 8, 1970.

was often determined by a rule-of-thumb—on the number taken on in the past, on some 'accepted' relationship between the number of apprentices and the number of skilled men, on the number which could be accommodated given present training facilities, etc.[1] In certain circumstances, such conventions might produce a supply of apprentice-trained labour sufficient to meet future skilled labour needs, but these circumstances were not met in Birmingham.

With few exceptions, throughout 1959–66 the reserves of unemployed skilled labour in Birmingham were meagre. Most plants on their own admission found such labour extremely difficult to recruit and yet made no attempt until the 1964 Act to expand apprentice-training. Nor can the plants be credited with an expectation that the supply of skilled labour would be adequate in the future. Nothing in their own actions or in labour market trends justified such a view and, indeed, the plants seldom had any coherent view of the future—or, if they had, it was incorrectly formed. We have seen that few plants made manpower forecasts over a long enough period to relate current apprentice-training to expected future needs. Moreover, those plants which did formulate long-term manpower forecasts did not use these to decide current intakes of apprentices.

Although convention was important in determining the amount of apprentice-training undertaken, there was some variation over time in the ratio of apprentices to skilled men and in the rate of intake of apprentices. As these variations were not based on forecasts, however crude, of future skilled labour needs, it seems reasonable to suppose that plants may have acted on the assumption that current conditions would continue to rule in the future. The view taken of current conditions could be based on two separate sets of factors—on the general employment conditions prevailing in the labour market as a whole, and on the specific employment situation in the plant itself. In the first case, we would expect that apprentice provision and the rate of inflow of apprentices would be inversely related to the unemployment/vacancy ratio for skilled men and to the male unemployment rate in engineering. The logic behind this argument is that, if plants judge the future in terms of the present, they would increase apprentice-training when skilled labour was difficult to recruit in the expectation that such recruitment difficulties would persist in future. Hence, the figures for apprentice provision and apprentice intake would be high when unemployment and the unemployment/vacancy ratio was low, and vice versa.

[1] It is interesting to note that these conventions were seldom based on a maximum ratio of apprentices to journeymen laid down by trade unions. This was true of Glasgow as well as of Birmingham.

In Glasgow there does appear to be some relationship between employment conditions in the market and the inflow rate. As Table 12.2 shows, the rate of inflow increased after 1959 as the unemployment/vacancy ratio and the unemployment rate fell. Conversely, the rate of inflow tended to be lower when employment conditions worsened over 1962–3, and rose again as unemployment and the unemployment/vacancy ratio fell thereafter. There is, then, some evidence that in Glasgow the rate of recruitment of apprentices may have been affected by general employment conditions in the market, but the evidence is not very strong and finds no echo in Birmingham, especially if we ignore the rather special circumstances after 1963.

It may be, however, that decisions as to apprentice-training are influenced not by employment conditions currently ruling in the market but rather by the specific employment conditions ruling in the plant. In other words, if the plant is finding it necessary to expand its employment of skilled men, and probably meeting recruitment difficulties, it may decide to increase apprentice-training to meet expected skill shortages in the future. The obverse argument would also apply, so that in this case a positive relationship would be expected between the number of apprentices and the number of skilled men employed by a plant. This hypothesis was tested for Glasgow, where 26 of the 27 plants undertook apprentice-training,[1] by correlating for each plant the number of apprentices employed at October 1st of each year with the number of skilled males employed at July 1st. In 20 of the 26 cases, the resulting coefficient was positive and, of these, seven were significant at the 5 per cent level. There is some evidence, therefore, that the volume of apprentice-training was adjusted to employment conditions in the plant or, if such a policy was not consciously applied, the same end result was achieved through the application of an inflexible apprentice-to-journeymen ratio. The future supply of skilled men therefore depended to a large extent on tradition and on current events. That these did not guarantee an adequate supply of skilled labour is well illustrated by the behaviour of the Birmingham plants.

The low proportion of apprentices in Birmingham may, in part, have been due to a high 'drop-out' rate amongst those who undertook such training. In circumstances where the training is not completed, the plant will incur costs which are not offset by the benefits expected, and this will naturally discourage the adoption of more extensive training programmes. If we take those who began apprentice-training over 1959–61, and who would, had they continued their

[1] The analysis was not repeated for Birmingham because only a handful of plants undertook apprentice-training on any significant scale. See p. 338 above.

training, have become skilled men by 1966, we find that less than 15 per cent failed to complete their training in Glasgow and the Other Scottish Areas compared to no less than 53 per cent in Birmingham. Some of these apprentices had been transferred to other jobs within the plant, but the great majority had left the plant voluntarily without completing their training. This difference in experience no doubt reflected the more buoyant labour market conditions in Birmingham, and particularly the availability of semi-skilled jobs with earnings levels substantially above those for apprentices. In Glasgow and the

TABLE 12.3

PERCENTAGE OF APPRENTICESHIP TERMINATIONS BY STAGE OF TRAINING REACHED

	Percentage of terminations		
Stage reached	Glasgow	Birmingham	Other Scottish Areas
Less than 1 year	42·9	52·5	56·4
1 year up to 2 years	29·9	27·7	17·9
2 years up to 3 years	14·8	14·8	15·4
3 years plus	12·4	5·0	10·3

Other Scottish Areas, this alternative was less attractive and less frequently available. In addition, the completion of apprentice-training offered greater security of employment in the future and higher status in communities in which the craftsman was still highly regarded.

An analysis of all those apprentices over 1959–66 who left their plant *without* completing their training will indicate whether there were any critical periods in which the wastage rate was particularly high. The relevant data are assembled in Table 12.3 which shows the percentage of all those leaving before completing their training by the stage of training reached.

In all areas, approximately half of those who failed to complete their apprenticeships had given up that training within one year, and about three-quarters within two years. There are two possible explanations of the high proportion leaving in the first year. Either the first-year training was inadequate and boring, or the boys may have found the effort to develop manual skills and to assimilate the theoretical instruction and associated education too much for them. Whichever of these may have been the case, the figures spotlight the need for a well-devised course of first-year training under competent instructors and teachers in order to reduce wastage. The emphasis placed by the Engineering Industry Training Board on off-the-job first-year courses would thus appear to be justified, and it would be interesting to review the position after the measures taken by the E.I.T.B. have had time to take effect.

It is also apparent that although the Industrial Training Act may have defects and may be capable of improvement, it was not replacing a system which guaranteed an adequate future supply of skilled labour. Plants had little or no idea of future requirements and determined current apprentice-training by convention and with regard to current events rather than in response to some assessment of future needs. In this field, the failure of the plants to make greater use of manpower forecasting is perhaps most disappointing, but in view of the methods of forecasting currently adopted, it is perhaps as well that they are not faithfully accepted as an infallible guide to action.

5. CONCLUSIONS

(i) The forward period over which manpower forecasts ran was usually determined by the length of the order book. Very few plants attempted to forecast orders, and hence manpower requirements, beyond this point in time. It follows from this that very few plants made forecasts of labour requirements beyond a period extending one year ahead. The forecasts usually assumed that current levels of labour productivity would hold in the future and did not make explicit adjustments for technical change and rationalization.

(ii) Managements in established undertakings attached little importance to manpower forecasts. In Glasgow and the Other Scottish Areas, this might have reflected the easier labour market conditions, which allowed most recruitment needs to be met without substantial advance warning. The same argument did not, however, apply in Birmingham. It seems likely, therefore, that managements had insufficient confidence in forecasting to use it as a basis for determining recruitment policy. To some degree, this might reflect the rudimentary techniques adopted—for example, the failure to take account of wastage and the tendency to be over-optimistic about future trends—but it may also reflect the difficulty, even with more refined methods, of obtaining a sufficient degree of accuracy for day-to-day recruitment needs. In any event, plants with manpower forecasts tended to postpone recruitment action until near the time when additional labour needs arose, and it is difficult to find any major difference in behaviour on the part of those units which did not attempt any manpower forecasting.

(iii) Managements regarded manpower forecasts as most useful in avoiding or reducing the need for redundancies. In a number of cases, it was claimed that advance action on the basis of forecasts had been useful in assisting adjustment in this field. While it is not possible on the evidence to reach any conclusive judgment, it must be said that there is little objective evidence which suggests that plants with forecasts were particularly successful in reducing redundancies.

(iv) Before the advent of the Industrial Training Act in 1964, apprentice-training was much more widespread in Glasgow than in Birmingham, and there was no indication that the relative position was changing despite persistent shortages of skilled labour in the latter area. Convention appeared to be particularly important in determining the amount of training undertaken by a plant. When fluctuations in the number of apprentices did occur, they seemed to be a response to short-term factors, and particularly to changes in the number of skilled men employed by the plant. Current training programmes were not linked to estimates of future manpower requirements. This applied even in plants where manpower forecasts extended over a long forward period.

(v) The passage of the 1964 Act had a substantial impact on the amount of apprentice-training in Birmingham. The rate of recruitment of apprentices increased substantially and in 1966 even exceeded the rate of recruitment in Glasgow. There were also signs that the Act had reduced disparities between Birmingham plants in the degree of provision made for apprentice-training.

(vi) The drop-out rate amongst apprentices in Birmingham was extremely high relative to that in other areas. This must have had some effect in discouraging plants from investing more heavily in such training and reflected the greater availability and attractiveness of alternative employment in that area. Failure to complete apprentice-training was concentrated amongst those in their first year of apprenticeship, a fact which underlines the decision of the E.I.T.B. to pay particular attention to this aspect of training.

CHAPTER 13

METHODS OF RECRUITMENT AND SELECTION

1. INTRODUCTION

A market can only work effectively if there are adequate channels of communication between the buyers and the sellers, and a labour market is no exception to this rule. The efficient communication of information on job vacancies and conditions of employment will minimize frictional unemployment and the costs of job-search incurred by the employee, and reduce recruitment, selection and production costs for the employer. It is therefore important to ascertain how information on job vacancies is communicated in the market and whether the mechanisms used do, indeed, meet these requirements.

Previous studies have suggested that the job information available to persons seeking employment is severely limited, so that imperfect knowledge prevents the market from operating smoothly. One writer has concluded that 'the typical worker has . . . no idea of the full range of jobs, wage rates and working conditions prevailing in the area'.[1] The general impression conveyed by empirical research is that the job-search is often casual and unsystematic, restricted in scope and highly dependent on methods, such as casual calling at factory gates and the advice of friends and relatives, which severely limit the job horizons of those seeking employment.[2]

In existing circumstances, such behaviour may be perfectly understandable. The employee is likely to make some implicit valuation of the possible costs and benefits of alternative methods of job-search and can be expected to continue to seek additional information until the point at which the likely value of additional information is less than the cost of obtaining it.[3] Hence, if employees do rely on informal channels, this may simply indicate that these channels are the most

[1] Reynolds, *op. cit.*, p. 85.

[2] See, for example, Parnes, *op. cit.*, pp. 162–74; H. L. Sheppard and H. Bilitsky, *The Job Hunt* (1966); and G. R. Horne, W. J. Gillen and R. A. Helling, *A Survey of Labour Market Conditions, Windsor, Ontario, 1964: A Case Study* (1965).

[3] G. J. Stigler, 'The Economics of Information', *Journal of Political Economy*, Vol. 69, 1961.

efficient methods available despite their apparently haphazard nature. Yet this is hardly the end of the matter. If employers have 'gentlemen's agreements' to avoid 'poaching' labour from each other, if the information they provide about earnings prospects and working conditions is meagre, if job offers must be accepted or rejected 'on the spot', or if they prefer to rely heavily for recruitment on casual callers or on a network of friends and relatives, then the employee may be forced to adopt a certain pattern of behaviour. This may be understandable given the circumstances in which the employee finds himself, but it is hardly the best of all possible worlds, nor is it support for the view that informal channels are necessarily effective means of communicating information simply because they are often used by employees and are many and varied.[1]

There is plenty of evidence which suggests that some employers act in a manner that reduces the impact of competitive forces,[2] and that serious imperfections in knowledge exist on the part of employees which prevent the labour market from operating with maximum efficiency. The existence of the public employment service is, indeed, predicated on the assumption that the market by itself will not necessarily secure an efficient matching of labour demand and supply. The attitudes of employers to and their relationships with employment exchanges form part of the subject matter of Section 2 below. The section also reviews the qualitative evidence supplied by plants on the manner in which they communicate information on job vacancies to the market, in an attempt to establish whether there is a marked preference for particular methods of recruitment. This discussion of the relationship between plants and the employment exchanges is taken a stage further in Section 3, which considers the extent to which job vacancies are notified to the exchanges and the degree of success achieved by the exchanges in filling them. Section 4 reviews the empirical information available on the methods through which successful applicants obtained jobs, while hiring standards and selection procedures are discussed in Section 5. The conclusions from these prior sections are summarized in Section 6.

2. MANAGEMENT ATTITUDES TOWARDS RECRUITMENT

Interviews with plant managers suggested that plants relied heavily

[1] This is the essence of the argument put forward by Rees, who seems content with the present system simply because it is there, and does not consider how it may be improved. See 'Information Networks in Labor Markets', *American Economic Review*, Vol. 56, 1966.

[2] Myers and Maclaurin, *op. cit.*, pp. 40–5; Sheppard and Bilitsky, *op. cit.*, pp. 192–3 and Lester, *Adjustments to Labor Shortages*, pp. 46–9 and 75–7.

on informal methods of recruitment, particularly on casual callers seeking jobs at the factory gate and on applicants with relatives or friends in the establishment. Most plants had a stream of persons seeking jobs in these ways either on the 'off chance' of obtaining employment, or in response to notification of job openings on bill-boards, or through the 'bush telegraph' of friendship and kinship networks. Because of this, personnel managers tended to make use of the public employment exchanges only when informal methods did not produce applicants, and there was clear evidence that employers preferred to meet recruitment needs in the latter fashion.

Casual enquiries produced many applications for work. Sometimes these would result from a prominent display of current vacancies on a strategically placed notice-board at the gate, but just as often stemmed from persons who sought jobs through calling at factories in their neighbourhood on the 'off chance' of obtaining employment. The advantage in seeking a job in this way is not limited to the possibility of finding a vacancy before its existence is generally known. Many plants invited applicants to file an application even when no jobs were immediately available. In this manner, plants acquired a 'bank' of potential recruits who could be called upon when an appropriate vacancy occurred. All except six Glasgow plants maintained records of previous applicants who had been considered suitable for employment, as did three-quarters of the Birmingham establishments and all the units in the Other Scottish Areas.

Kinship and friendship networks also played an important role in the labour market for manual workers. The personnel manager, or much more frequently the foreman, could often obtain applicants simply by letting it be known that there were job vacancies available. Such information was then communicated by word of mouth amongst the acquaintances of those workers employed by the plant. The process often worked even where the management had not given any indication of outstanding recruitment needs. Individuals, on the recommendation of friends or relatives, simply applied in the hope of obtaining employment and, if they did not obtain work immediately, were usually added to the queue of potential recruits.

Informal methods have the advantage of cheapness. It may be possible to reach a fairly large number of recruits at low cost, and other benefits *may* also arise, especially where individuals obtain jobs through friends or relatives. Employees may be able to gain more detailed information from this source than from advertisements or employment exchanges, and may prefer to work in a unit where they already have acquaintances. The recommendation of an employee may 'provide good screening for employers who are satisfied with their present work force' and is likely to produce recruits with short travel-to-work journeys and, possibly as a result of this, lower wastage, absenteeism, etc.[1]

Because of such factors, but mainly because of low cost, the employer is likely to rely heavily on informal methods when the labour market is slack. American evidence[2] does, however, suggest that, as employment conditions improve and labour becomes more difficult to obtain, some employers will turn increasingly to more formal and more expensive methods of recruitment, particularly advertising. Interviews with managers did seem to confirm this view, as Birmingham plants put particular stress on the use made of advertising. Eight plants reported that they normally advertised vacancies in the local press, especially for skilled labour, and a further seven claimed to have used such advertisements from time to time. Advertising for labour through notices in retail shops and through subscription to advertising pamphlets distributed from door to door was also fairly common. In Glasgow, a large proportion of plants reported that they made no use, or only infrequent use, of advertising, although nine units had relied heavily on newspaper advertisements to recruit skilled labour.

A number of authorities have stressed the possibility that the efficient operation of the labour market may be impeded by the existence of 'anti-pirating' or 'no poaching' agreements between employers. For example, an unwritten understanding not to hire applicants employed by other plants will reduce mobility on the part of those already employed, for they will have to relinquish their current job, and accept the possibility of unemployment, if they wish to find alternative work. That collusion of this nature may be fairly widespread has been demonstrated by many American studies,[3] but few British data are available although they might be expected to display rather different features given the generally low levels of unemployment which have prevailed in this country.

It is useful to distinguish between management policy towards the hiring of persons currently employed by other units, where the management is essentially passive until approached by a recruit, and that towards more aggressive recruitment tactics, which we might term 'labour enticement', where management actively seeks to attract labour employed in rival units. While there appears to be, in the United States, a fairly widely observed agreement amongst plants not to hire labour already in employment and whose services were re-

[1] Rees, *op. cit.*, p. 562. [2] See, for example, Malm, *op. cit.*

[3] 'An employed worker who wants to change jobs . . . must usually win the acquiescence of his present employer': Reynolds, *op. cit.*, p. 52.

quired by another establishment, such considerations were rare in all our market areas and almost non-existent in Birmingham and Glasgow. The general view was that, given the labour market conditions prevailing, such a policy would in the words of one Birmingham manager, 'mean that it was impossible to hire any labour at all or, at least, any labour which was employable'. In the smaller, more isolated labour markets of Small Town and New Town where personal and business contacts between managers were closer, it was accepted practice to contact the current employer of an applicant before engaging him. Such a procedure was not universal in these areas, however, and as far as could be ascertained it was rare to refuse an applicant a job on this count.

Aggressive recruitment tactics, approaching former employees now working for other plants, asking employees to seek 'experienced' workers, and delivering job advertisements from door to door were also quite common in Birmingham and, more surprisingly, in Glasgow. The most blatant examples, indeed, occurred in the latter area. In one plant a foreman with 'a nose for this work' was paid an honorarium for his recruitment activities, which apparently included approaching employees at other factories during the lunch-hour, on their way to and from work, or even by door-to-door canvassing. A number of other plants also encouraged foremen or employees to approach former employees or persons who might be interested in a job, even if it were known that these individuals were employed in other neighbouring factories. The attitude of one Glasgow establishment which had experienced difficulty in meeting its specialized labour requirements was to do 'anything underhand to recruit labour so long as we are not caught'. The available evidence does not suggest that collusion between employers reduced labour mobility or introduced imperfections on other counts. On the contrary, a number of establishments employed very aggressive recruitment policies that would increase the knowledge of job opportunities available, and the practice of insisting that applicants should have already relinquished any existing employment was not important.

We turn now to a consideration of the attitudes of plants towards the employment exchanges whose *raison d'être* is to improve the efficiency with which the labour market operates. As the exchanges provide a focus to bring buyers and sellers of labour together, the attitudes of both groups towards the services provided will be important. Only nine plants in Glasgow claimed to notify the exchanges of all vacancies for manual workers, and closer examination suggested that even these claims were unfounded.[1] Nonetheless, it is possible to

[1] See p. 354, n. 1 below.

identify plants which made frequent use of the exchanges and those which relied almost exclusively on other channels. The statement by one personnel manager that the exchanges 'were incapable of supplying skilled men' was typical of a widely held view. In consequence, managements reported a greater tendency to notify semi-skilled and unskilled vacancies to the exchanges, a pattern which also seemed typical of Birmingham. Disillusionment with the service provided by the exchanges was even more widespread in the latter area, and very few plants claimed to notify all vacancies to the exchanges: only one obtained the majority of its recruits from this source. Managers in both Birmingham and Glasgow often regarded applicants from the exchanges as 'loafers and layabouts', persons who 'had no initiative', who 'could not get a job on their own', and who did not want a job but simply 'wanted their card signed to continue on the dole'. The exchanges were also criticized for sending for interview 'just anyone who is on their books'.

These views allow us to consider how the number of unfilled vacancies might compare with the actual or real demand for labour at a given point in time. The data on unfilled vacancies published by the D.E.P. are often taken as a proxy for the outstanding, unsatisfied demand for labour, and we used vacancy data in this manner in Chapter 7. We observed there that this procedure must be viewed with some reservation, for the true demand for labour may be greater than or less than the number of unfilled vacancies. If employers are finding difficulty in securing labour, then they might notify more vacancies than they actually have in the hope that this will ensure for them a greater part of the available supply of labour. Equally, they might simply fail to notify vacancies to the exchanges in the belief that the exchanges cannot meet their needs.

On the evidence of the interviews, the former situation seldom arises. Only one instance was found, in Birmingham, of a plant notifying more vacancies to the exchanges than the job openings actually available, and even this was restricted to a particular class of female labour. On the other hand, there was plenty of evidence that plants did not notify all vacancies to the exchanges. Only a minority of personnel managers in Birmingham and Glasgow claimed to notify all vacancies, and in Glasgow, where it was possible to check such statements against objective evidence, they were not substantiated. There seems little doubt that the number of unfilled vacancies is considerably less than the actual demand for labour in all market areas irrespective of the employment conditions prevailing.

However, while we may accept the proposition that unfilled vacancies understate the true demand for labour, we may yet find unfilled vacancies useful as a proxy for true demand if the relationship between unfilled vacancies and actual job openings is stable over the cycle and between market areas. As we have seen, this relationship may be modified by employment conditions, the exact nature of the change depending on whether employers react to labour shortages by notifying a higher or a lower proportion of actual vacancies to the exchanges. To investigate this problem we have only the comments made by managers backed up by some indirect evidence such as the 'penetration rates'[1] achieved by the exchanges. The former source would indicate that plants in Birmingham tended to notify to the exchanges a relatively small proportion of their true demands for labour. Plants in that area were most dissatisfied with the quality of the service the exchanges provided and seldom claimed to notify a high percentage of actual job openings. We shall also see, although this is by no means conclusive, that the proportion of jobs filled through the exchanges was low in Birmingham relative to Glasgow.[2] It seems possible, therefore, that the proportion of job vacancies notified in the tight Birmingham labour market was lower than the proportion in Glasgow, despite the fact that even in the latter area the plants appeared to notify only some one-fifth of actual job vacancies.[3]

Vacancy information is not, of course, collected mainly to provide an indicator of aggregate labour demand. This is simply a by-product of the use of vacancy notifications for the day-to-day placement operations of the employment exchanges. If these placement operations are to be carried through efficiently it is essential that plants should notify their labour needs accurately to the exchanges, and that the information provided to the exchanges should be up-dated on a regular basis. That plants notify only a small proportion of actual vacancies must reduce the effectiveness of the exchanges in their placement operations. This is not, unfortunately, the end of the matter, for it appears that even those vacancies actually registered at the exchanges often fail to match the real requirements of the plants. From each case-study plant recruiting labour at the time of interview a detailed list was obtained of outstanding manual labour requirements on that day. On the conclusion of the interview, contact was made with the relevant employment exchange through pre-arranged channels to secure a list of the notified vacancies currently outstanding for the plant. Comparison of these two lists showed, as expected,

[1] The penetration rate is simply the number of job placements made through the exchanges as a percentage of total placements through all channels of recruitment.

[2] Table 13.3, p. 357 below.

[3] See Table 13.1, p. 354 below.

a consistent tendency for actual demands to exceed notified vacancies, but, more important than this, it was extremely common to find that the occupational characteristics of the lists did not match in important respects. In plain language, the vacancies notified at the exchanges were often for types of labour which the plant was not seeking to recruit.

It is difficult to reconcile the above finding with the previous observation that, over the period 1959–66, the occupational distribution of vacancies and unemployment suggested a pattern of labour shortages consistent with observations made by the plants.[1] This suggests that, for the market as a whole, vacancy/unemployment ratios may be a good indicator of demand/supply relationships, while the above paragraph shows that at the level of the individual unit vacancy information may be a poor guide to labour requirements. We do not wish to amend either argument. There is strong evidence that vacancies notified to the exchanges may have serious deficiencies for placement purposes,[2] but provided the errors are not biased in any particular direction by occupations, vacancy data used in conjunction with unemployment statistics may provide some guidance to conditions in the market. Nonetheless, since vacancy data are primarily gathered for placement purposes, deficiencies in notification are particularly serious at the level of the individual plant. We found no evidence that the exchanges had recorded information received from plants incorrectly or that the exchanges were at fault in some other direction. On the contrary, where vacancies notified to the exchanges by the plants failed to match in number or characteristics the plant's actual requirements, the fault appeared to be with the plant itself. It was common for plants to complain of the poor quality of the services provided by the exchanges, but until they themselves provide a more adequate flow of information the placement activities of the exchanges must be seriously impaired.

3. VACANCY NOTIFICATION AND THE EMPLOYMENT EXCHANGES

The above discussion of the role played by the employment exchanges in meeting plants' labour requirements was largely based on qualitative information provided in interviews by plant managers. While this

[1] See pp. 61–3 above.

[2] This failing is not confined to Britain. See National Bureau of Economic Research, *The Measurement and Interpretation of Job Vacancies* (1966).

method allows us to reach some tentative conclusions, it has certain dangers unless we can check the results obtained. This can be accomplished for a restricted period by matching vacancy notifications against recruitment for case-study plants in Glasgow, North Lanarkshire and New Town. The results confirm the argument of the previous section, that plants notified only a small proportion of their actual demands for labour.

Employment exchanges were able to provide information of all vacancies notified by, and all placements made with, 23 Glasgow case-study plants during 1966. By comparing notifications and placements to the actual number of new starts with these plants over the period, it was possible to examine the relationship between vacancy notifications and actual labour requirements, and also the penetration rates achieved by the exchanges. An identical analysis for three plants in North Lanarkshire and five plants in New Town had to be restricted to the last six months of 1966. Unfortunately, it is not possible to provide comparable data for Birmingham as the employment exchanges could not furnish the necessary information on vacancy notifications and placements.[1]

Table 13.1 has been constructed by expressing the number of vacancies notified by plants as a percentage of all recruitment over the relevant period. Three groups of workers are distinguished: skilled, semi-skilled and 'others', the latter category comprising largely unskilled and non-production workers but excluding staff. There is one point of procedure which should be referred to before considering the results of Table 13.1. The table does not take into account any standing orders or 'informal understandings' that plants may have had with their local exchange. To the extent that these arrangements existed, the results shown below will underestimate the degree of vacancy notification. However, standing orders, where the plant in effect asked the exchanges to forward job applicants of a particular type, were of minor importance, and only one 'informal understanding' of any importance was discovered. This occurred with one plant to which, by longstanding practice, the exchange submitted job-seekers without prior notification of available vacancies. Where these persons were hired, the transaction was recorded both as a placement and as a vacancy notification. The results for this plant have been excluded in what follows.

The proportion of job vacancies actually notified to the exchanges was only one in five in Glasgow and one in eight in North Lanarkshire. This is ample confirmation of our previous suggestion that vacancies notified substantially understated the true demand for

[1] The numbers for Small Town were too small to allow meaningful analysis.

labour.[1] There was also a clear pattern by skills in these areas, with a relatively low proportion of actual vacancies notified for skilled workers. Managements were aware that most persons available for work and registered at the exchanges were semi-skilled and unskilled and, believing that the exchanges could not meet their skilled labour requirements, they acted accordingly. The exchanges, then, were caught in the classic vicious circle. Few skilled vacancies were notified to them, so that skilled men, unless unemployed for a period long

TABLE 13.1

VACANCIES NOTIFIED TO EMPLOYMENT EXCHANGES AS A PERCENTAGE OF ALL RECRUITMENT

Area	Occupational Group			Total
	Skilled	Semi-skilled	Others	
Glasgow (1966)	16	25	22	20
North Lanarkshire (1.7.66 to 31.12.66)	9	20	12	12
New Town (1.7.66 to 31.12.66)	43	32	51	41

enough to claim benefit, sought work through other channels. This perpetuated the tendency of managements to seek skilled employees by other means.

In New Town, the links between the plants and the exchanges were much stronger. No less than 41 per cent of all vacancies were reported to the exchanges. The situation was, however, atypical. It reflected the rapid increase in employment and population fed by a high rate of immigration from other areas often separated from New Town by substantial distances. In these circumstances the plants could not rely to the same extent on informal methods of recruitment, as these are normally only effective in securing labour from within the immediate area of the factory. The plants therefore made substantial use of the employment exchange network to widen their potential area of recruitment. It is a moot point whether the high rate of vacancy notification would continue as the employment and population structure became more established.

The penetration rates achieved by the exchanges will depend to a considerable degree on the extent of vacancy notification, but other

[1] Of the 23 plants included in the Glasgow analysis, only six notified more than 50 per cent of all vacancies.

factors may also be important as can be seen from a comparison of the results in Table 13.2 with those of Table 13.1. Penetration rates have been calculated by expressing the number of placements made by the exchanges as a percentage of the total number of persons recruited by the plants over the relevant period.

As might be expected given the results in Table 13.1, the penetration rates achieved by the exchanges were highest in New Town and lowest in North Lanarkshire. Nor is it surprising to find that in each

TABLE 13.2
PENETRATION RATES OF THE EMPLOYMENT EXCHANGES

	Occupational group			
Area	Skilled	Semi-skilled	Others	Total
Glasgow (1966)	8	19	17	13
North Lanarkshire (1.7.66 to 31.12.66)	3	7	11	6
New Town (1.7.66 to 31.12.66)	8	28	31	21

area the penetration rate was particularly low for skilled employees. Both these findings are perfectly understandable given the pattern of vacancy notifications. The penetration rate of the exchanges tended to be high when the exchanges were well informed as to job vacancies, and vice versa. This was not, however, the only influence at work, for the penetration rate for skilled men was low even when due allowance is made for the low proportion of vacancies notified for this group. The most striking example of this is in New Town, where only 8 per cent of skilled job vacancies were filled through the exchanges although no less than 43 per cent of these vacancies were notified to the exchanges. In contrast, the exchanges achieved a penetration rate for the semi-skilled of 28 per cent although only 32 per cent of actual semi-skilled vacancies were notified.[1] The low proportion of notified vacancies for skilled employees filled by the exchanges emerges in each labour market area. It appears likely, therefore, that skilled employees prefer to seek work through other channels, and that increased notification of skilled job vacancies to the exchanges would not necessarily have a substantial impact on their penetration rate for this group.

[1] Indeed in all areas the exchanges filled a respectable proportion of those semi-skilled and 'other' vacancies which were notified to them. This can be seen by comparing Tables 13.1 and 13.2.

One last point should be made about vacancy notification. In Glasgow, plants with less than 500 employees tended to notify a higher proportion of all vacancies to the exchanges than larger units. The penetration rate achieved by the exchanges was also greater in the case of the smaller units, and this still holds true if allowance is made for the greater degree of vacancy notification in this instance. The contrast between small and large units probably reflects a preference of well-staffed personnel departments in the larger units to seek labour through other channels, particularly relatives and friends and casual callers. Smaller units with small personnel departments, or, indeed, no personnel officers at all, find the services of the exchanges more valuable. They may also apply lower hiring standards and, as the exchanges deal primarily with labour of rather poor quality, their penetration rates tend to be higher for smaller units than in the case of large units with stricter hiring standards.

4. CHANNELS OF RECRUITMENT

It was not possible, from plant personnel records, to ascertain how new recruits had obtained jobs over 1959–66, and few of the plants were able to provide detailed information on their channels of recruitment. Each plant on being interviewed was therefore asked to complete over a forward period a schedule showing how those recruited had first heard about the job opening they obtained. Completed schedules were eventually received from 12 units in Birmingham, eight in Glasgow, four in North Lanarkshire, three in New Town and two in Small Town.[1] The results relate to the period April–October 1968 in Birmingham and to June 1967–June 1968 in the other areas. As these periods fall outside our main study period, the results have to be interpreted rather cautiously, particularly in Birmingham where the rate of unemployment, although still below that in the other areas, had risen above 1959–66 levels.[2] Despite this reservation, it is possible to highlight a number of important conclusions as to the manner in which job information is disseminated through the market. Table 13.3 shows, for skilled, semi-skilled, unskilled and total recruits, the percentage of engagements by methods of job-search.

The most striking feature of Table 13.3 is the large proportion of workers who found jobs through the agency of relatives and friends,

[1] The data for the two Small Town plants are not reproduced in Table 13.3 but are referred to in the text where it is considered relevant.

[2] Because of the difference in period and the smaller number of plants covered by Table 13.3, the results are not directly comparable with Tables 13.1 and 13.2.

or by casual enquiry at the factory gate. In each labour market area, more than half of the recruits heard of jobs in these informal ways, although the importance of kinship and friendship networks was especially pronounced in the smaller, more compact areas of North

TABLE 13.3
PERCENTAGE OF ENGAGEMENTS BY METHODS OF JOB-SEARCH: 1967–8

Method of job-search	Area			
	Birmingham	Glasgow	North Lanarkshire	New Town
1. *Relatives and friends*				
Skilled	20·6	14·1	37·4	13·3
Semi-skilled	25·5	22·2	45·7	31·3
Unskilled	24·9	25·9	53·9	57·7
Total	24·8	19·7	45·8	34·6
2. *Casual callers*				
Skilled	29·4	25·4	30·8	17·8
Semi-skilled	39·0	39·2	48·6	32·0
Unskilled	48·8	32·7	30·4	26·6
Total	41·3	33·5	43·2	28·5
3. *Employment exchanges*				
Skilled	13·2	29·6	7·7	20·0
Semi-skilled	12·6	18·4	3·5	16·0
Unskilled	20·2	39·5	13·7	9·4
Total	15·2	28·3	5·7	14·0
4. *Advertisements*				
Skilled	20·6	25·4	23·1	44·4
Semi-skilled	7·4	17·8	2·2	15·3
Unskilled	1·4	0·9	2·0	3·1
Total	6·8	15·0	5·1	17·3
5. *Other methods*				
Skilled	16·2	5·4	1·1	4·4
Semi-skilled	15·5	2·5	—	5·3
Unskilled	4·7	0·9	—	3·1
Total	11·9	3·5	0·2	4·6
Total numbers	1,073	630	644	259

Lanarkshire and New Town. The smaller the area the more likely is a job-seeker to have a contact in a plant seeking labour, and the more quickly will the 'grapevine' operate. Even in the larger urban areas, this source of job information was important, with about a fifth to a quarter of workers hearing of jobs in this manner. Not only was this a relatively inexpensive way for establishments to notify the market of job openings, but also, as was stressed in interviews, management often had a preference for those applicants who were personally recommended by one of the current work force, in the belief that such applicants would find it easier to establish themselves in the environment and develop into reliable workers.

The extent to which plants depended on informal contacts through relatives and friends to fill vacancies varied with the skill of the job-seeker. This source of job information was most important in the case of unskilled workers and least important in the case of skilled employees. A distinction between skilled groups was also apparent in the case of casual callers where, once again, skilled employees made relatively less use of this channel of recruitment. Despite this, enquiries at the factory gate were for most groups the most important single source of job information in the conurbations of Birmingham and Glasgow, and remained extremely important even in close-knit communities such as North Lanarkshire and New Town.

The proportion of workers obtaining work by means of casual calling and friends and relatives strongly influenced the placement activities of the employment exchanges. This was most noticeable in North Lanarkshire where virtually 90 per cent of all employees were recruited through informal channels. The flow of applications from these sources was sufficiently large to relegate to a minor role the placement activities of the employment exchanges, to which vacancies were notified only occasionally. The greater importance of the exchanges in New Town reflected the problems of engineering plants in recruiting a labour force in an area with little previous experience of engineering work. Whereas the flow of unskilled applicants through informal channels was generally sufficient to meet the plants' needs, greater use was made of the employment exchanges to recruit skilled workers. This arose because of the lack of skilled engineering labour in the vicinity of the new establishments, and because of the advantages of using the network of employment exchange services to contact skilled labour in other areas.

The exchanges were also a fairly important source of new recruits in the conurbations because relatively few recruits in these areas obtained jobs through referrals by friends and relatives. The penetration rate achieved by the exchanges was higher in total and by occupational groups in Glasgow than in Birmingham, but this may be due to the operation of rather special circumstances.[1] Of more interest is the difference in the exchange penetration rates in Birmingham and Glasgow, on the one hand, and North Lanarkshire, on the other. The exchanges were responsible for a much lower proportion of placements in the latter area, reflecting the nature of the community

[1] See below.

and the importance of informal methods of recruitment. It appears likely that in small communities plants rely relatively little on the exchanges. The data for New Town are somewhat misleading in this respect because of factors already mentioned, but it is interesting to note that in Small Town the exchanges also had a low penetration rate, accounting for only 10 per cent of all new recruits.[1] While these results require further confirmation through research in other areas before being accepted at face value, they do imply that both absolutely and relatively the exchanges are most important as a source of job information in the large urban markets. In smaller, localized communities informal channels of recruitment are normally adequate and the exchanges satisfy only a small part of a plant's labour requirements.

In view of the evidence received from the plants,[2] perhaps the most surprising result of Table 13.3 is the low proportion of recruits obtained through advertising in Birmingham. Only 6·8 per cent of all employees were recruited through this method in Birmingham compared to 15·0 per cent in Glasgow and 17·3 per cent in New Town. The explanation for this phenomenon is that the Birmingham data from Table 13.3 relate to a period in which plants were re-hiring workers who had been declared redundant or 'laid off' during late 1966 and 1967. This being so, plants were able to meet a substantial part of their labour needs through re-hiring without extensive recourse to advertising. This explains the relatively high proportion of employees recruited through 'other methods' (largely re-hirings) in Birmingham and, compared to Glasgow, the greater importance of relatives and friends and casual callers. In more normal circumstances, Birmingham plants probably made more use of advertising, but it seems unlikely that even then this channel of communication would replace casual callers and relatives and friends as the most important method of contacting likely recruits. In New Town, plants made frequent use of advertising to obtain labour, especially skilled labour, from other areas.

5. SELECTION PROCEDURES AND HIRING STANDARDS

Once applicants for a given vacancy have been obtained, the plant faces the difficulty of selecting the 'right' person for the job. There were normally two stages in the selection procedure followed for

[1] Informal channels, casual callers and relatives and friends, accounted for 81·6 per cent of all recruits in Small Town.

[2] See p. 348 above.

manual workers. In the first instance, applicants were referred to a personnel officer for an initial screening interview. At this stage it was unusual for a representative from the relevant production department to be present. The function of this first stage was to reject 'undesirables' who were deemed unsuitable for the job and to prepare a short list of suitable candidates. Any applicant who survived the initial screening interview was then referred to the relevant foreman, whose task was to assess the applicant's skill and competence to perform the tasks associated with the vacant job before deciding whether or not to hire.

The selection procedures employed were not very refined, at least in Birmingham and Glasgow. Few plants in these two conurbations applied any formal test at the first stage of selection, and decisions as to the short list were made by the personnel officer on the basis of his ability to gauge suitability through a short, often informal, interview. It seldom fell to the personnel officer to attempt any judgment of skill and technical competence. Instead, factors such as age, personal appearance, colour, and previous employment experience were the hiring standards applied in this initial screening process. Formal tests of technical competence and proficiency were also rare at the second stage. The foreman usually decided whether the individual possessed the necessary degree of experience or ability by ascertaining his previous job experience, and, at this stage also, subjective assessments were extremely important.

Hiring standards were numerous but may broadly be classified into those that are economic in the sense of being related to the job, those that are subjective, and those that are institutional in nature. Those standards that can be regarded as economic may be applied in either of the two interviews. The personnel officer may reject an applicant during the screening interview if it becomes apparent that the person does not meet some specific need that is essential to the job and of which the personnel officer is aware. In order for the personnel officer to be aware of any such requirements, it is usually necessary to have some procedure for specifying the particular skills and abilities required by the job. Detailed specifications of job requirements were quite common in the Other Scottish Areas, particularly in plants with American management, but were comparatively rare in Birmingham and Glasgow. In most cases, the personnel officer's knowledge of the job was often limited, and it is therefore not surprising that at this stage little stress was placed on technical competence. The personnel officer's function was primarily to eliminate manifestly unsuitable candidates, which in effect meant those individuals whose attitudes and values were not felt suitable given the requirements of the plant. Particular stress was placed on the need to identify and reject those groups who were likely to display a high wastage rate. Hence the importance of subjective judgments.

The hiring standards applied at the first stage of selection are difficult to define because only a handful of plants were prepared to lay down rigid rules to be applied in all circumstances. Such an attitude is only to be expected, given changing labour market conditions. Flexibility in such matters increases the power of the plant to adjust to different circumstances, but the attendant disadvantage is that it is difficult to identify hiring standards and to establish whether these standards have a rational, economic basis. In general, most plants denied the existence of explicit hiring standards, and while these were, indeed, relatively rare, it was common to find certain standards implicitly applied by personnel officers. Of these, the most important related to age, previous industrial experience, and colour.

The age of an applicant was important in a number of respects. As has been indicated in previous studies,[1] there was a widespread reluctance to hire older workers when other applicants were available. Here a number of factors were mentioned, such as lack of adaptability, a slow pace of work, an inability to withstand strenuous physical conditions and the difficulties imposed by superannuation schemes. However, in a few cases, older workers were at an advantage over younger applicants. This occurred particularly in unskilled jobs, where the plants occasionally preferred to recruit older men, who were held to be less ambitious and hence less likely to leave for alternative employment. There was a more widespread tendency to discriminate against applicants in their teens and early twenties, particularly where semi-skilled jobs were concerned. The reason for this is obvious enough. This group has a high wastage rate, and the plants are naturally reluctant to recruit such workers and undertake the costs of on-the-job training where the individual has a high propensity to seek other employment. Particularly for semi-skilled work, therefore, there is a tendency to discriminate against young and old applicants in favour of those, say, aged 25–40, especially if they are married.

The previous employment of the applicant was also considered an important factor in judging suitability by many plants—especially so in Birmingham and Glasgow. For example, plants in Birmingham appeared to be reluctant to hire the ex-employees of motor-car plants, while engineering establishments in Glasgow tended to discriminate against employees from shipyards. This hiring standard

[1] See, for example, O.E.C.D., *Promoting the Placement of Older Workers* (1967).

was by no means absolute but it was nevertheless real. Discrimination in both cases arose because individuals from these industries were considered to be less skilled than their equivalents in engineering, particularly as regards their ability to work to fine tolerances. They were also regarded as particularly militant 'troublemakers' and persons who generally sought out temporary employment outside their industry only when it was going through a period of lower activity. It was held that such employees could not be relied upon to stay with the plant; they would return to their previous employers as soon as new work was available.

The extent to which colour discrimination was practised was difficult to judge, as plants, particularly in Birmingham with its large immigrant community, were very sensitive to any questioning on this topic. Only three plants would privately admit to colour prejudice, but it certainly did exist, in greater or lesser degree, in a larger number of units. Such discrimination was often 'justified' on the grounds that immigrants, especially first-generation, Asian immigrants, were difficult to employ because of bad health, absenteeism, or inability to speak English. Whatever the truth of these allegations, casual inspection suggested that low-wage units in Birmingham tended to have an exceptionally high proportion of coloured workers.[1]

Some of these hiring standards may have had a rational economic basis. For example, car workers in Birmingham were indeed extremely likely to return to that industry when employment conditions there improved,[2] and it is perfectly sensible for plants to discriminate against young applicants who have a high quit rate, especially where the job requires a substantial investment by the plant in training. The adoption of these hiring standards simply reflects an attempt by the plant to recruit employees who will form part of a stable and contented work force.

Yet while the dividing line is difficult to draw, it seems fairly certain that the degree of discrimination applied to certain classes of workers —particularly older workers and coloured workers—was greater than that which would be justified by an objective assessment of costs and benefits. Moreover, the selection procedures adopted, although not as irrational as they appeared at first sight, were hardly very satisfactory. Apart from plants in the Other Scottish Areas, the outstanding impression of selection procedures is that they were extremely casual and informal. Judgments were made 'on the spot', often after a rather perfunctory interview, and no attempt was made to

[1] It is worth noting that our observations antedate the Race Relations Act of 1968.

[2] See Kahn, *op. cit.*, pp. 58–61 and 143.

subject hiring standards to objective tests. Interviewers, once again excepting plants in the Other Scottish Areas, seldom made any attempt to assess whether the recruit had potential for 'upgrading', possibly because the process of internal promotion was relatively little developed. Hiring was undertaken for a particular job with little thought to long-term considerations, and the requirements of the initial job had seldom been thought through systematically.

An experienced personnel officer backed by a foreman to judge technical competence may be able to 'weed out' undesirable applicants without recourse to elaborate formal procedures, but there is little evidence that such a heavy reliance on 'experience' and 'instinct' produces satisfactory results. As we have seen, a fairly high proportion of recruits were discharged for unsuitability and misconduct, and the vast bulk of such dismissals arose when applicants were found to be unsuitable after a short probationary period, usually some 1–3 months in duration. This is an indication of the inability of the selection system to discriminate accurately between applicants, and the failure to do so was likely to be costly in terms of lost production and outlays on training. Also, quit rates were extremely high during the induction crisis despite attempts by plants to discriminate against those groups with a high potential rate of mobility. The quit rate is, of course, partly or mainly a function of employment conditions, but the success of plants in the Other Scottish Areas in reducing wastage amongst short-service employees may not simply be attributable to the higher levels of unemployment in that area. They may also reflect the considerable care which these plants, especially the American units, took to evolve more systematic methods of recruitment. In this respect, British management appeared to have a lot to learn.

The last subject which we might consider is whether hiring standards varied significantly between plants. It is difficult to be precise on this point since hiring standards were implicit rather than explicit, and it was not possible to make direct comparisons between the plants. Nor can one reach any definite conclusions by merely 'observing' the work force on visits to plants without having any recourse to detailed output, capital investment and productivity data. The views expressed by managers, however, suggest that the hiring standards applied by plants did vary, and with them the 'quality' of the labour force. Moreover, as economic theory would suggest, high-wage plants appeared to be able to enforce stricter hiring standards. Thus, the manager of a high-wage unit could remark that 'we can set high standards and still get plenty of people' while his opposite number in the low-wage unit 'has to scrape the barrel and take who we can to make up our labour force'. Such extreme cases can be identified, but they were extremes. While hiring standards did vary, and with them the quality of the labour force, they were not thought through strictly enough or applied rigorously enough to cause substantial variations in the 'quality' of the labour force between units except those at the top and the bottom of the wage hierarchy. In most other establishments the individual's chance of obtaining a vacancy depended as much on prejudice and luck as on his ability to fill the particular job in hand. From the plant's point of view, more careful consideration of selection procedures might eliminate unwanted biases and, more important to its operating efficiency, might provide greater certainty that hiring standards which had a rational basis were properly enforced.

6. CONCLUSIONS

(i) Managements relied heavily on informal methods of recruitment—casual callers and employee referrals by friends and relatives—in all labour market areas. This reflected the cheapness of such methods of recruitment, but preferences for such methods were also based on other grounds. Managers preferred casual callers who had shown 'initiative', and those referred through friends and relatives whose friendships might inhibit subsequent mobility, to those job applicants referred by the employment exchanges. There was widespread criticism of the quality of labour seeking work through the exchanges.

(ii) There was no evidence to support the view that mobility of labour, and hence the efficient operation of the labour market, was seriously impeded by collusion between employers in enforcing anti-pirating agreements. On the contrary, a number of plants, especially in Birmingham and Glasgow, adopted active recruitment policies to entice labour from neighbouring establishments.

(iii) Only one instance was found of a plant deliberately exaggerating its labour requirements to the employment exchanges in an attempt to obtain a larger share of the available labour supply. Few plants claimed to notify all vacancies, and the objective evidence shows even these claims to be exaggerated. It appears likely that the inability of the exchanges to meet the plants' labour requirements in a tight labour market resulted in a situation where an exceptionally low proportion of vacancies were notified. While this contention cannot be demonstrated by statistical evidence, it is clear that the vacancies registered by a plant at the exchanges often did not match its true requirements; that only a small proportion of vacancies were notified in Glasgow and North Lanarkshire, and that this underreporting was most substantial for skilled labour. The penetration rates achieved by the exchanges were, to a considerable degree, determined by the extent of vacancy notification, but the exchanges were particularly poor at filling those skilled vacancies registered with them. Small plants made greater use of the exchanges, and the public employment service proved particularly valuable to units in New Town where labour was recruited over a wide geographical area.

(iv) Informal channels accounted for the bulk of all recruitment in each labour market area. Casual callers were the most important single source of recruitment in Birmingham and Glasgow and were an extremely important source of labour even in the tightly knit communities of North Lanarkshire and New Town, where employee referrals by friends and relatives accounted for a higher proportion of all recruits. These channels were least important for skilled labour. The degree of dependence on informal channels was greatest in the smaller communities where the employment exchanges played a minor role in recruitment.

(v) Most plants adopted a two-stage selection procedure. At the first stage the personnel manager was responsible for selecting a short list from the applicants while the final selection of the successful applicant was left to the foreman. Few formal tests were applied and subjective judgments were extremely important. Hiring standards were seldom inflexible and were almost always implicit. Nonetheless, considerable weight was given to factors such as personal appearance, age, colour and previous employment experience. Some discrimination on each of these grounds was probably justified by strictly economic considerations, but most hiring standards had not been subjected to critical evaluation. Selection procedures were often extremely casual, with adverse consequences subsequently through high quit rates during the induction crisis and a substantial proportion of dismissals through unsuitability. Most plants hired for a particular job without regard to longer-term considerations, and personnel officers were seldom instructed as to the skills and aptitudes required to complete the necessary tasks successfully. Hiring standards differed between high-wage and low-wage units, but in most plants differences in hiring standards did not appear to be very important.

CHAPTER 14

REDUNDANCY

1. INTRODUCTION

Much of this volume has been concerned with the ways in which plants sought to overcome the recruitment problems associated with different types of labour market, particularly those in which labour with the required skills has been difficult to obtain. In this penultimate chapter, however, we look in greater detail than hitherto at the opposite situation; that is, the situation in which a plant's labour force is found to be in excess of present or foreseen production requirements. There are several courses of action which management can adopt to deal with this problem, one of which, the subject of the present chapter, is to discharge labour considered to be surplus to production needs.

Since the Acton Society Trust's survey of the problem published in 1958[1] there have been several major redundancy studies in Britain:[2] A feature of most of these studies is that they have each been concerned with single instances where large-scale redundancies have occurred, involving either the complete closure of an establishment or a substantial reduction in its labour force. Thus while valuable insights have been gained into the procedures adopted in redundancy situations and the job-search conducted by workers, it is unlikely that these 'one-off' situations are representative of the redundancies which normally occur in industry. In this instance, however, we have the opportunity of observing the redundancy situations which developed over 1959–66 in a substantial number of plants. This should assist us to widen our understanding of the problems which arise. We are also able to examine the incidence of redundancy, i.e. the frequency with which redundancy occurred, the numbers involved,

[1] *Redundancy, A Survey of Problems and Practices.*

[2] The main ones are D. Wedderburn's two studies, *White-Collar Redundancy*, and *Redundancy and the Railwaymen*, University of Cambridge, Department of Applied Economics, Occasional Papers Nos. 1 and 4 (1964 and 1965); A. Fox, *The Milton Plan* (Institute of Personnel Management, 1965); Kahn, *op. cit.*; R. Thomas, *An Exercise in Redeployment* (1969).

its relationship to plant size and so forth. This forms our first major area of enquiry.

The second main area of interest concerns the manner in which the necessary decisions which are required in a redundancy situation are made. They can either be decisions arrived at unilaterally by management, or they can be the result of collective negotiations with the trade unions.[1] In contrast to American practice where redundancy is frequently the subject of detailed collective agreements,[2] the procedures adopted in Britain are seldom established in detail, or even in general principle, by formal arrangements. Instead redundancy situations are usually approached in an *ad hoc* fashion, often with a backward glance to previous practice and convention. This has often been regarded as inadequate. Hence the Donovan Report[3] recommended that companies should conclude formal agreements with trade unions laying down a recognized procedure for handling redundancies. In view of this it is worth while examining with some care the mechanism through which decisions are arrived at, to determine, for example, the locus of power in the management and the unions, the degree of consultation between the parties, provision for notice and severance payments, and the choice of those to be declared redundant.

In the next section, we discuss some of the earlier studies of redundancy procedures and indicate the questions which can profitably be pursued in greater detail. Section 3 looks at the incidence of redundancy in the Glasgow plants covered by the present enquiry in an attempt to establish whether the typical redundancy situation involved a 'large' or a 'small' number of employees in absolute or relative terms. The section also takes up the vexed question of 'labour hoarding' and looks at the relationship between redundancy and plant size. The redundancy practice of the plants is investigated in Section 4 and the length of service and age of redundant workers are discussed in Section 5. Following our normal convention a summary of the conclusions is contained in Section 6.

[1] There is, of course, now a third source of such decisions, namely that of statutory regulation. Both the Contracts of Employment Act (1963) and the Redundancy Payments Act (1965) lay down certain minimum conditions which must be observed in redundancies.

[2] For example, a study by the U.S. Bureau of Labor Statistics has shown that about three-quarters of collective agreements contain provisions describing the procedure to be followed in the event of redundancy or lay-off: cf. Acton Society Trust, *op. cit.*, pp. 43–50. The severance pay and lay-off benefit provisions in U.S. agreements are summarized in the *Monthly Labor Review*, Vol. 88, 1965. See also Slichter, Healy and Livernash, *op. cit.*, pp. 142–77.

[3] *Royal Commission on Trade Unions and Employers Associations, Report* (Cmnd. 3623, 1968), para. 182.

2. EARLIER STUDIES OF REDUNDANCY PROCEDURES

Investigations of redundancy procedures indicate that, in its relations with unions, British 'management has in most cases reserved to itself the right to make the final decision: the process is one of consultation, and not negotiation'.[1] The emphasis on consultation, as distinct from negotiation, is also apparent from the Ministry of Labour's study of redundancy arrangements in 371 firms. It was found that 'about 45 per cent of the policies were adopted after consultation with employees' representatives but only a few—18 in all—are embodied in formal signed agreements. All the other policies were adopted by managements acting on their own.'[2]

One reason why so few firms have negotiated formal redundancy procedures seems to be that some unions are reluctant to conclude such agreements or consider them relatively unimportant.[3] The former attitude springs from a view that redundancies are never necessary (or at least that it is unwise to anticipate them) and a consequent unwillingness to enter into negotiations upon the matter. Hence the lack of formal redundancy agreement at the Milton works[4] and the inability of the Confederation of Shipbuilding & Engineering Unions to arrive at a policy acceptable to all its constituent unions.[5] Although the matter has been raised periodically within the C.S.E.U., often in the form of a demand for the negotiation of a national agreement on redundancy, internal differences, quite apart from any reluctance on the employers' side, have prevented the subject becoming a serious matter for negotiation. Consequently, it has been easy for the Engineering Employers Federation to maintain their long-established view that redundancy questions are essentially ones to be resolved by unilateral management decision; they are regarded as falling within the area of managerial prerogative, rather than of collective bargaining.

Two of the main substantive provisions in redundancy situations, length of notice and severance payments, vary from case to case, although statutory regulation has established certain minimum standards which must be observed by all plants. It does appear that most plants attempt to provide advance warning or notice of dismissal in excess of the minimum legal requirement; but, while

[1] Acton Society Trust, *op. cit.*, p. 30.
[2] *Ministry of Labour Gazette*, Vol. 71, 1963, p. 50.
[3] A. D. Smith, *Redundancy Practices in Four Industries* (O.E.C.D., 1966), p. 90.
[4] Fox, *op. cit.*, p. 11.
[5] See Marsh, *op. cit.*, pp. 330–7.

supplementation does occur in certain instances,[1] it is not yet clear whether a substantial number of plants make redundancy payments in excess of those required by the Redundancy Payments Act. The third major substantive issue concerns the criteria used to select redundant workers, and here too there is some uncertainty as to the principles normally applied. Thus the Ministry of Labour's enquiry found that 40 per cent of redundancy policies specified that certain categories of workers should be the first to be declared redundant (e.g. employees over retiring age, married women, part-time workers); a further 20 per cent were based on the seniority principle of 'last in, first out', and the remaining 30 per cent on a combination of seniority and efficiency. In the latter instance there is obviously scope for differences in interpretation, but even when seniority is applied on the basis of length of service the principle can be applied in many different ways, e.g. on a factory, department or job basis.[2]

In considering the substantive aspects of redundancy procedure we shall pay particular attention to the latter subject, the criteria by which those to be declared redundant were selected, for two reasons. First, because existing knowledge in this area is rather limited and, second, because the criteria adopted will to a considerable degree influence the characteristics of redundant employees, and hence the ease with which they can obtain new employment. Quite apart from the procedural and substantive issues, however, we also wish to explore certain other areas of interest such as the incidence of redundancies in industry. This forms the subject matter of the next section; but before proceeding to the analysis we can consider what we might expect to find in our labour market areas.

The first thing to be said is that our expectations have to be formulated in a very broad and impressionistic manner. Public attention and academic research have tended to concentrate on large-scale redundancies often arising from plant closures. This is understandable given the number of workers involved and the special circumstances of many of these situations, but it does mean that we have relatively little knowledge of the range of redundancies in terms of absolute and relative size. We cannot indicate what a 'typical' redundancy involves in terms of the number and type of employees affected, so that we must simply see what emerges in our market areas. All we can say is that, given the more buoyant state of the Birmingham labour market and the rapid expansion of employment in the

[1] For example, in the closure of A.E.I.'s Woolwich factory. See *The Times*, January 30, 1969.

[2] For an interesting case study which illustrates this point see J. E. T. Eldridge, *Industrial Disputes* (1968), pp. 206–28.

Other Scottish Areas, we would expect few redundancies in those markets relative to Glasgow, where engineering employment was declining. Indeed redundancies were so infrequent in New Town and Small Town that we have excluded these areas entirely from our analysis of the incidence of redundancy and have concentrated our attention on Glasgow and Birmingham.

Related to the question of the size and frequency of redundancies is the much-debated issue of 'labour hoarding'. It seems to be widely accepted that managements, especially in tight labour markets, have been reluctant to release labour when faced with situations where the labour force is greater than that demanded by current production needs. This is thought to be due to a belief that if the labour force is cut back to the minimum level necessary, and 'surplus' labour released, it might be difficult to build up the labour force again when conditions subsequently improve. In other words, the redundant employees will find new jobs and the plant may in the future experience severe recruitment problems. This proposition is difficult to test, for if we find, let us say, that redundancies have been less frequent in Birmingham than in Glasgow, this could be a consequence of labour hoarding in the tighter market or could simply reflect the more buoyant Birmingham labour market. However, if labour hoarding does take place it seems likely to apply especially to skilled labour which has been in shortest supply in the market. We might, then, hypothesize that in the tight Birmingham market, and in North Lanarkshire where the supply of engineering labour has been restricted, plants will have been especially reluctant to dismiss skilled labour.[1] Redundant workers may then have different occupational characteristics in the separate markets which indicate the presence or relative absence of labour hoarding. We take up this possibility in the next section.

3. THE INCIDENCE OF REDUNDANCY

From our analysis of the personnel records of the case-study plants we can show, by plant and by quarter, the number of redundancies and the rate of redundancies relative to the stock of employees. We have treated the number *known* to be redundant in Birmingham as the total number of redundancies and have not adjusted this number (or the rate of redundancies) as in Chapters 6 and 7 to allow for those leavers where reason for leaving was not known. Such an adjustment would have had no appreciable effect on the results obtained, but the

[1] This is best explained by W. Oi, 'Labor as a Quasi-Fixed Factor', *Journal of Political Economy*, Vol. 70, 1962.

reader might bear in mind that the rate of redundancy is likely to be understated by some one-tenth for Birmingham males and females.

Following previous practice, we have broken down our data to four quarterly periods in each year. This means that if a plant declared redundancies in, say, February and March of 1959 this was regarded as 'one' redundancy, as both months fell in the first quarter. On the other hand, if the redundancies had occurred in March and April, this would have been taken as 'two' redundancies because the months fell in different quarters. There is evidently an arbitrary element in our method of treatment. No doubt we have on occasion treated a redundancy phased over a number of weeks as two (or more) separate events. Equally, we must in certain cases have lumped together redundancies occurring independently of each other within a three-month period. Such decisions do not, of course, affect the measurement of the total number of redundancies, but they do influence our measurements of the incidence and frequency of redundancy. Nonetheless, some dividing line has to be taken, and one which deals with quarterly periods seems as reasonable as any other.

Let us begin with the simplest measure of the frequency with which redundancy occurred. All quarters in which a plant discharged one or more persons as redundant were summed for all plants in a market and expressed as a percentage of the total number of quarters 'at risk' (the number of case-study plants times 32). Such an exercise for males shows that in Birmingham redundancies occurred in 16 per cent of the total number of quarters, while the corresponding figure for Glasgow was 37 per cent. Expressing this rather differently, the 'typical' plant in Glasgow experienced a redundancy of some size about every third quarter, and the 'typical' Birmingham plant only every sixth quarter.

It is also of interest to repeat an analysis similar to that above for individual units, to establish whether some plants successfully avoided redundancies while others were 'hire and fire' establishments, taking on workers and discharging them after short periods of employment. Of the 25 plants in Birmingham for whom data are available, 21 had male redundancies in 10 or less of the 32 quarters over 1959–66. Four plants had some redundancies in 11–21 quarters and none exceeded this frequency. In Glasgow only 12 of the 27 case-study plants had 10 or less redundancy quarters, ten had between 11 and 21 redundancy quarters, and five had a redundancy in 22 or more of the 32 quarters. It is apparent that in Glasgow redundancies were an almost ever-present feature of certain units. It is true that many of the redundancies involved only small numbers of employees, so that in 66 per cent of redundancy quarters in Birmingham and in 58 per cent of such quarters in Glasgow, no more than four male employees were dismissed. Nonetheless, in a handful of Glasgow units redundancies involving larger numbers were fairly frequent throughout 1959–66.

We can pursue this question of the frequency and the magnitude of male redundancies in Glasgow plants with the aid of Table 14.1 below which allows us to investigate the influence, if any, of plant

TABLE 14.1

FREQUENCY AND QUARTERLY RATE OF REDUNDANCY, BY SIZE OF PLANT: GLASGOW MALES

Number of males employed	Frequency of redundancy	Rate of redundancy
0–199	21·9	0·76
200–499	52·8	1·29
500–999	46·1	1·13
1000+	28·6	0·21
Total	36·7	0·54

size. The procedure adopted for each quarter was to allocate plants to the size ranges shown in the table according to the number of males employed at the beginning of the quarter. The total number of quarters in which there was a redundancy amongst the plants in each size range was obtained and expressed as a percentage of the total number of quarters 'at risk', giving the frequency of redundancy amongst plants of varying sizes. The numbers made redundant in each size range were also totalled and compared to total employment in the quarters in which the redundancies occurred, so as to give the quarterly *rate* of redundancy by size range. The results are shown in Table 14.1.

Among the plants in the size ranges 200–499 and 500–999 male employees there was a greater frequency and a higher rate of redundancies than in the smallest and largest plants. Taking first frequency, which we have defined as being the percentage of quarters in which *any* redundancy occurred, the result for the small plants with 0–199 employees is not altogether surprising. After all a quarter with even one redundant employee has been taken as a redundancy quarter and, other things being equal, such a definition would yield a greater frequency of redundancies as the size of the unit increases. Yet the contrast between the smallest plants and those in the two medium-size bands is not altogether a matter of definition, as the *rate* of redundancies was lower in the smallest units, and this is unaffected by our definition of frequency. At the other end of the scale we find that the frequency of redundancies was less in units with 1,000-plus employees than in those of the two medium-size bands, whereas in view of the above argument we would expect the greatest frequency of redundancies in the largest units. Plants with 1,000-plus employees had a

TABLE 14.2
DISTRIBUTION OF QUARTERLY REDUNDANCY RATES, BY SIZE OF PLANT: GLASGOW MALES

Number of males employed	Percentage of redundancies in each rate range		
	Under 2%	2%–4·9%	5%+
0–199	67·6	16·9	15·5
200–499	66·9	16·9	16·1
500–999	81·3	9·3	9·3
1000+	89·7	3·4	6·8
Total	72·8	13·8	13·5

much lower rate of redundancies than every other size range, and this leads to the conclusion that employees in such units were least likely to experience redundancy. However, in view of the fact that the rate of redundancies was lower in the smallest plants than in the two medium-size ranges we cannot assume that there is any simple relationship between plant size and the redundancy rate.

Further light on this can be gained from Table 14.2 which also allows us to investigate in greater detail the magnitude of the redundancies which occurred over 1959–66. Table 14.2 is constructed by the same method as Table 14.1, but in this instance the number of redundancies by plants in each quarter is expressed as a percentage of the stock of employees and the resultant quarterly rates are allocated to the ranges shown. Thus we can see that, of all the redundancies occurring in plants with 0–199 employees, 67·6 per cent involved less than 2 per cent of the stock of employees, 16·9 per cent involved 2–4·9 per cent and 15·5 per cent involved 5 per cent or more.

While the smallest plants, as Table 14.1 shows, had the lowest frequency of redundancy, the redundancies which did occur were similar in their relative size to those experienced by plants with 200–499 employees. Moreover, when units in the two smallest size groups did have redundancies, they normally affected a larger proportion of their labour force than did redundancies in plants with 500–999 or 1,000-plus employees, the contrast being particularly marked in the latter case.

We can conclude then that plants in the 200–499 size range were most vulnerable to redundancies on all counts. They had the greatest frequency of redundancies, and a high proportion of these redundancies involved 2 per cent or more of their labour force, so that the rate of redundancies averaged 1·29 per cent per quarter compared to 0·54 for all units (see Table 14.1). On the other hand, the largest units with 1,000-plus employees were least affected by redundancies. It is true that the frequency of redundancies was somewhat greater than in plants with 0–199 employees, but this was only to be expected given the manner in which frequency has been defined. When due allowance is made for this, the largest plants had no redundancies at all in a surprisingly large number of periods, and when redundancies did occur they seldom involved a large fraction of the labour force. Hence almost 90 per cent of all redundancy quarters in the largest units were quarters in which less than 2 per cent of the stock of employees lost their jobs. In fact the small relative size of redundancies, rather than their relationship to plant size, is perhaps the most striking feature of Table 14.2. Almost three in every four redundancy situations in all plants involved less than 2 per cent of male employees, and only 13·5 per cent resulted in 5 per cent or more of employees losing their jobs.

We can now widen our analysis from a consideration of the redundancy experience of individual Glasgow plants to a broader analysis of the average or market rate of redundancies. This is derived by expressing the number of redundant employees in all plants as a percentage of the total stock of employees in the relevant quarter. If this is done for Birmingham and Glasgow, we find that the quarterly market rate of redundancies for males was less than 1 per cent of the stock in all 32 quarters over 1959–66 in Birmingham and exceeded 1 per cent in only three quarters in Glasgow. This confirms our previous suggestion that the typical redundancy usually involved only a small number of employees. However, there was an undercurrent of redundancies in both markets over almost the entire period, and also some fluctuation in the market rate, which was positively related, as one would expect, to the level of unemployment.[1]

Of greater interest is the possibility that the market rate of redun-

[1] Correlating the quarterly market redundancy rate for males to the unemployment rate for wholly unemployed males, the coefficients obtained are +0·52 for Birmingham and +0·62 for Glasgow. Both are significant at the 1 per cent level.

dancies might vary between markets and by occupational groups. This can be investigated through Table 14.3 which shows the average of market quarterly redundancy rates over 1959–66 for the areas and groups shown.[1]

The average market redundancy rate for males was clearly highest in Glasgow, where it ran at about twice the level of Birmingham and North Lanarkshire. This is as expected given the reduction in engineering employment in the former area, but it should be noted that the employment security of employees in Birmingham was not quite as favourable as a comparison of male redundancy rates would suggest. In Birmingham there was a widespread tendency, especially amongst motor-car plants, to 'lay off' men on a temporary basis rather than to declare them redundant. This seldom happened in Glasgow. Plants lay off employees because of what is considered to be a temporary fall in demand, hoping to re-engage those laid off within a fairly short period. The promise of re-employment should conditions improve acts to prevent the worker from seeking another job and in a sense can be regarded as a mild form of labour hoarding. Nonetheless, a temporary loss of employment is involved, and although this is less serious than the permanent loss of employment through redundancy, it must be taken into account when comparing redundancy data in the two conurbations.

TABLE 14.3

AVERAGE QUARTERLY MARKET REDUNDANCY RATES

Area	Males				Females
	Skilled	Semi-skilled	Un-skilled	All	
Birmingham	0·1	0·3	0·3	0·2	0·9
Glasgow	0·6	0·5	0·7	0·5	0·6
North Lanarkshire	0·1	0·3	0·8	0·3	0·5

The low rate of male redundancies in Birmingham and North Lanarkshire relative to Glasgow was also to some extent due to the fact that in the two former areas females made up a relatively high proportion of the labour force in engineering. It can be seen from Table 14.3 that the rate of female redundancies was higher than that for males in each market, reflecting a greater propensity on the part of plants to dismiss females in preference to male employees in a

[1] As in Tables 7·1 and 7·2, p. 169 above, an unweighted average is used, i.e. the market redundancy rate for the 32 quarters is summed and divided by 32.

redundancy situation. Thus, as we shall see shortly, plants often operated redundancy procedures in which married women and part-time workers (who are usually married women) were first to be discharged. It was widely held that such a policy was 'equitable', because many married women were secondary workers who accounted for a relatively small proportion of family income and whose redundancy would therefore cause less financial hardship than would have arisen through male redundancies. Practical considerations may, however, also have been important. The plant may find its own interests best served by discharging females because their services can more easily be dispensed with and/or because they can more easily be re-engaged at a later date should this prove necessary. Thus the plant may wish to retain skilled males rather than semi-skilled females, and may be able to re-engage females previously declared redundant if such females have left the labour force temporarily on losing their jobs.[1]

It can be seen from Table 14.3 that there was a marked contrast between Glasgow on the one hand and Birmingham and North Lanarkshire on the other, in that in Glasgow the rate of male redundancies was much the same for all occupational groups, whereas in Birmingham and North Lanarkshire the rate of redundancies for skilled employees was low relative to that for the semi-skilled and the unskilled. Even if we take the data for males on their own, therefore, there is on our previous argument[2] *prima facie* evidence of labour hoarding, in that plants in Birmingham and North Lanarkshire were reluctant to declare skilled men redundant, presumably in case this might aggravate recruitment difficulties in the future. However, this argument becomes much stronger if we bear in mind our previous remarks about female redundancies. As we have seen, the rate of female redundancies was relatively high in all markets and especially in Birmingham. Moreover, females were largely semi-skilled and accounted for a large proportion of the engineering labour force in Birmingham and North Lanarkshire relative to Glasgow. If these factors are taken into account, the contrast between Glasgow and the two other areas, in terms of the impact of redundancies on occupational groups, was even more substantial than the male data suggest. Plants in Birmingham and North Lanarkshire when faced with the need to reduce their labour force cut back

[1] A number of plants in Birmingham employed females on a 'twilight' shift and such workers were usually the first to be declared redundant. The plants claimed that when production picked up again it was relatively easy to re-hire their former employees.

[2] See p. 370 above.

severely on male and female semi-skilled and unskilled labour and were extremely reluctant to make skilled males redundant. In Glasgow, the hoarding of skilled labour was not so apparent, perhaps because it could be more readily re-engaged if subsequently required.

4. REDUNDANCY PRACTICES

We saw in Section 2 that in the engineering industry there is no formal negotiated agreement at national level covering the procedures to be adopted in a redundancy situation. The only guidance given to employers is that emanating from the Engineering Employers Federation whose 'Recommended Policies and Procedure for Dealing with Redundancy' was drawn up in 1963.[1] Apart from confirming the Federation's view that, while managements should consult and consider suggestions put forward by representatives of the work force, the fundamental decisions are the responsibility of management, this document only lays down very general guidelines on procedural aspects and makes no attempt to grapple with substantive issues. In consequence management at the plant level has, within the bounds set by legislation, considerable discretion subject only to local pressures from employees and their representatives.

Formal redundancy agreements arrived at through collective bargaining were extremely rare throughout the case-study plants. Indeed only three units (one each in Birmingham, Glasgow and North Lanarkshire) had negotiated such an agreement. The other plants had a variety of other arrangements, such as a written document outlining the procedures to be followed which had been unilaterally drawn up by the management, an unwritten but strictly observed agreement between management and the unions, a 'general understanding' on early consultation and, in three cases, an 'understanding' to undertake *ad hoc* negotiations on the procedure to be followed when each redundancy came up. The practice is evidently very widely followed that the arrangements made in the event of a redundancy are determined unilaterally by management. The freedom of action of management may, of course, be circumscribed by past convention and practice and it may, and almost invariably does, provide some mechanism for consultations with union officials, shop stewards, etc., but this does not produce a set procedure which is followed in all circumstances. Certain generalizations are possible, but they are generalizations which usually require some qualification.

The first generalization is that, in the procedures adopted by the case-study plants, union officials and shop stewards were usually

[1] Reproduced in Marsh, *op. cit.*, pp. 338–42.

brought into the discussion only after certain key decisions had been made by management. These usually related to the number to be made redundant, the incidence of redundancies by department or jobs, and the phasing and timing of the redundancies. It was not uncommon, however, for management to go further and to compile an interim list of the individuals to be made redundant *before* entering into any consultations. This is not to say that these decisions could not be modified thereafter as a consequence of consultations, but the broad framework once determined by management was seldom subject to fundamental alteration, and consultation usually centred around certain substantive issues, particularly length of notice, severance pay arrangements and the criteria used to determine dismissals. The plants reflected the philosophy of the E.E.F.'s recommended procedure, which is that broad strategy is a matter for management decisions. Thus: 'The workpeople's representatives should be *informed* of the measures already taken to avoid redundancy, the reasons why it has become inevitable, the number and classes involved and the method to be used in determining those redundant.'[1]

The substantive elements in redundancy arrangements varied a great deal between the case-study plants. As regards the period of advance warning and notice given to redundant workers, managements' views had often been influenced by the consideration that too early a warning could lead to a loss of 'key' workers and possibly also excessive wastage of other workers. Thus, where a plant knew that its labour requirements must be reduced some 3–6 months ahead, this information was not always communicated to the labour force immediately, although even in Birmingham, where wastage could be a particularly serious problem, some of the plants were willing to do so. In other cases, the managements themselves appeared to have had little forewarning of a sudden fall in product demand which necessitated discharges, and thus had no opportunity to consider how long a warning should be given. In practice, 1–2 months' advance warning (where practicable) generally seemed to be regarded as 'fair' to the work force and also in the best interests of management. In addition to this, however, there were a number of cases both in Glasgow and Birmingham where a longer effective, if indirect, warning had been given by short-time working being introduced some time before the redundancy became imminent. It may be noted that the period of 1–2 months' warning usually related to the notification of the impending redundancy to the workers' representatives. In addition, individual notice of termination had to be given to the

[1] Marsh, *op. cit.*, p. 341 (our italics).

persons to be discharged, and these notices, which were usually of shorter duration, generally conformed to the national Termination of Employment Agreement, 1964 and were not less than the minimum legal requirements laid down by the Contracts of Employment Act, 1963.

As regards compensation payable to redundant workers, practice again varied a good deal between the case-study plants. Many of the redundancies referred to in this study had occurred prior to the Redundancy Payments Act (1965), which laid down certain minimum terms of compensation for redundant workers. Consequently, at the time of most of these redundancies there was no legal requirement upon employers to make any severance payments, and those who did make such payments tended to emphasize that they were *ex gratia* and did not imply any obligation on the plant in any future redundancies. Nonetheless, while the amount of severance pay varied substantially from unit to unit, the great majority of plants did provide some compensation to redundant employees prior to the Act of 1965 and a number of these units continued to supplement the minimum provisions of the Act thereafter.

By 1966, which was the end of the survey period, the Redundancy Payments Act had not had a marked impact on procedural aspects of redundancy policies, and only one case was noted where the Act had encouraged a plant to enter into a formal redundancy agreement with the trade unions. There was, however, some evidence that managements might, in future, pay less attention to the seniority principle based on length of service as a method of determining those to be declared redundant.[1] This development could well be strengthened by the experience of a number of plants, that there was a noticeable tendency after the Act for employees, and especially long-service employees, to volunteer for redundancy in order to obtain the lump sum redundancy payment. This was accompanied by an increased unwillingness of employees to accept transfers within the plant as an alternative to redundancy. Sufficient time had not elapsed since the Act became operational for us to check this view against our data, but if it is well grounded we might expect that in the future redundancies will be increasingly composed of long-service and older employees. As such workers are difficult to place in alternative work, and as even before the Act they accounted for a relatively large proportion of all redundancies,[2] this might have serious consequences for the operation of the labour market.

[1] Because, it was argued, the lump-sum benefits tied to length of service absolved the plant from any duty to protect long-service employees.

[2] See pp. 210–1 above and pp. 381–4 below.

Finally there is the question of the choice of the individuals to be made redundant. We have already noted that in almost all units initial management decisions on the scale and on the broad incidence of redundancies were arrived at before consultations took place, and that in many units a provisional list of the individuals affected had been compiled before consultations. The responsibility for compiling such a list usually fell to the immediate supervisor of the work group, operating under general instructions as to the number to be declared redundant and the criteria to be implemented. Where other management representatives were involved in an executive role at this stage they were normally drawn from those with direct responsibility for production (e.g. works managers). It was noticeable that in this process the personnel manager was usually restricted to an administrative function. He had little say in deciding who should be included in the provisional list and was generally confined to providing the necessary information—the pay-roll and details of length of service, age, and personal circumstances—or proffering some broad guidance on the procedure and criteria to be adopted.

Once a provisional list had been drawn up detailed negotiations began, usually with shop stewards. In these negotiations the stewards were extremely reluctant to specify which individuals should be declared redundant, and usually confined themselves to the negative argument that certain individuals should be removed from the list without nominating a replacement. The only exception to this rule was where it was felt that a point of principle was involved, e.g. seniority, the dismissal of dilutees before skilled men, or married women and part-time workers before male, full-time workers. In Birmingham, there were a few examples where more conclusive discussions took place regarding the workers to be selected and also on internal transfers, and in one case after the Redundancy Payments Act, the stewards furnished a list of volunteers to replace the management's choice of individuals to be declared redundant. In general, however, the common practice seems to have been that major decisions had already been taken before the trade unions were brought into the discussion, namely the size of the redundancy, the principle of selection, and the names of those who were to go. Pressure from shop stewards and union representatives did in a number of instances bring about modifications, particularly in the two latter areas, but this seldom amounted to substantial revision.

The principles on which selection of redundant workers was based varied from unit to unit and also, to some extent, between areas. The seniority principle of 'last in, first out' was applied much less rigidly than is often supposed, and was subject to modification in most cases.[1] In Glasgow, only two plants had applied the principle strictly and the majority of plants adopted some test of efficiency, modified by 'L.I.F.O.', although it was rarely clear upon what basis, or by whom, 'efficiency' was measured. Furthermore, in almost all cases, length of service, while calculated from the date of recruitment, was applied on a departmental or job basis. This accounts for the fact that 'bumping', the process by which a long-service employee in one part of the works is moved to another part, thus replacing a shorter-service employee, was extremely rare. Thus, although employees might be transferred within a plant, this almost invariably occurred either when the transferee was taking up a vacancy, or when the transferee took the place of someone who had volunteered for redundancy. Equally it was extremely rare for employees to move down the occupational ladder, 'bumping' out less skilled employees with short service.[2] As a consequence of the narrow manner in which L.I.F.O. was applied it was normally the case that some of those declared redundant had longer periods of service than certain of those who remained with the plant. The latter possibility could of course be compounded where emphasis was placed on efficiency, and by the fact that in many plants certain groups of workers would be made redundant before seniority or efficiency criteria were applied. It was usual practice for a plant in a redundancy situation to discharge those over normal retiring age first, and this was often accompanied by redundancies amongst those close to normal retiring age (e.g. males aged 60–64 and females of 55–59).[3] We shall see the effect of this in the following section.

5. SERVICE AND AGE OF REDUNDANT EMPLOYEES

In Chapter 8 we examined some of the more important characteristics of redundant employees,[4] but it is worth while reiterating, and

[1] It is of interest to note that, when asked about selection criteria, the initial reaction of management was almost invariably to specify 'L.I.F.O.' On detailed discussion of a specific situation, however, it became apparent that the application of this principle was often qualified. As a general rule for researchers we could suggest that the number of exceptions to 'L.I.F.O.' is directly proportional to the length of the discussion!

[2] Thus the hoarding of skilled labour which we have noted in Birmingham was accomplished by retaining surplus skilled labour on skilled jobs. It did not involve 'bumping' to any major degree.

[3] It should be pointed out that when this policy was pursued the great majority of plants made compensating lump sum payments and/or made special provisions for the payment of pensions out of a company scheme.

[4] See pp. 210–1 above.

where necessary extending, our previous findings in order to relate them to the above discussion. As we have seen, the length-of-service distribution of redundant leavers is rather surprising at first sight, for despite the emphasis which observers have often placed on the seniority principle by which short-service employees are the first to be discharged, a high proportion were in the long-service categories and relatively few in the short-service categories. For example, of those Birmingham males declared redundant over 1959–66 only 9·2 per cent had less than 12 weeks' tenure with a plant while 34·1 per cent had a length of service of two years or more. The comparable figures for voluntary quits were 42·0 per cent and 16·2 per cent. In other words, relative to the voluntary quits redundant male workers in Birmingham were predominantly *long*-service employees. The same was true for males in Glasgow and for females in both conurbations.

How do we explain these results? First, we must bear in mind the fact that although the typical redundancy affected only a small proportion of a plant's labour force, a high proportion of all the employees discharged were involved in a handful of large redundancies where a substantial slice of a plant's labour force lost their jobs. Thus only 13·5 per cent of all redundancies in Glasgow involved 5 per cent or more of a plant's male employees, but these accounted for just over half (51·7 per cent) of all male redundancies over 1959–66. In this case we might expect some long-service employees to be discharged even when redundancies were strictly determined by length of service.[1] This, then, is part of the explanation for the long periods of service of redundant employees.

The second factor at work was the manner in which the incidence of redundancies was determined. Our earlier discussion has shown that it was usual practice to deal with a redundancy by initially discharging those over retiring age, and where this occurred it was fairly common to declare redundancies amongst those close to retiring age. A large number of such workers would have completed long periods of service with their employers. Again, length of service was not the deciding principle where married females or part-time employees were the first to be declared redundant, and although these groups

[1] Especially if the plants had some forewarning of impending redundancies (which was likely where a high proportion of the labour force was involved) and accordingly stopped or reduced recruitment so that the labour force ran down through wastage. The cessation or reduction of recruitment, together with the fact that wastage was highest amongst short-service employees, would mean that when redundancies became necessary a high proportion of the work force would be long-service employees.

contained a lower proportion of long-service employees than older workers, some would no doubt have completed a long period of service with a plant.

On balance it seems likely that those groups particularly liable to redundancies would have contained a high proportion of long-service employees, but even where length of service is the determining influence the manner in which it is applied will be important. The great bulk of the case-study plants applied the seniority principle on a departmental or even on a job basis rather than across the plant. This means that, where the incidence of redundancy was uneven across the range of jobs in a plant, the redundant employees were likely to contain a higher proportion of long-service employees than would have been the case if seniority had been applied on a plant-wide basis with attendant 'bumping'. With the data at our disposal we cannot quantify the effect of this practice, but it does seem likely to contribute significantly to the relatively long period of service of redundant workers. We are not, of course, suggesting that long service offered no protection to an employee against redundancy. Almost all the plants took some account of seniority when determining who should be discharged, so that a long-service employee was less likely to become redundant than a short-service employee other things being equal. However, as is often the case, other things were not so obliging, so that the difference was not as substantial as might have been expected.

Before leaving the discussion of seniority it is worth while noting that the length-of-service distribution of redundant employees by skill did show some variation between Glasgow and Birmingham. In Glasgow, those with at least two years' service amounted to 49 per cent of the skilled males declared redundant and to 54 and 43 per cent of the semi-skilled and unskilled respectively. The comparable figures for Birmingham were 45, 35 and 26 per cent. The percentage of long-service redundant workers was always higher in Glasgow, but more interesting is the fact that in Glasgow skill had little influence on the percentage, while it had in Birmingham where the percentage of long-service employees declined sharply with skill level. This is of a piece with our earlier finding that the quit rate (and hence the separation rate) was inversely related to skill in Birmingham while no strong relationship emerged in Glasgow. In consequence, the length-of-service distribution of the stock of employees was much the same for all skill groups in Glasgow, and this was reflected in the length-of-service distribution of redundant employees by skill. On the other hand, in Birmingham the stock of skilled men had a much higher average length of service than the stock of unskilled men, so that when redundancies did occur the skilled males discharged were those with a relatively long period of service.

The age characteristics of redundant employees can be dealt with more briefly as they were strongly influenced by length of service. Hence, in the case of Glasgow males 15·1 per cent of redundant employees were aged 24 years or less and 34·3 per cent 45 years or more. This compares with 28·8 per cent and 14·7 per cent for voluntary quits. Redundant male employees in Glasgow tended, then, to be concentrated amongst the higher age groups relative to other categories, of leavers, and the same conclusion applies to males in Birmingham and to females in both conurbations. We can conclude, therefore, that the redundant employee has generally completed a fairly long period of service with a plant and has a high average age. On two counts therefore he (or she) is particularly vulnerable to unemployment. His personal characteristics are 'unfavourable' because the hiring standards adopted by other employers often discriminate against older applicants, and he comes on to the labour market at a time when unemployment is usually high. Of course, in Birmingham, or in another tight market, employment conditions may be favourable relative to other areas, but then it would seem that in a tight market the redundant employee is more likely to be unskilled. In short the redundant employee is a born loser.

6. CONCLUSIONS

(i) Representing the frequency of redundancy by total number of quarters in which any male redundancy occurred as a proportion of all quarters at risk, we find that on the average a Glasgow plant had a redundancy every third quarter and a Birmingham plant every sixth quarter. Of the 25 case-study plants in Birmingham, only four had 11–21 redundant quarters in the 32 quarters over 1959–66 and no plant exceeded this frequency. Amongst the 27 plants in Glasgow, ten units had redundancies in 11–21 quarters and five had redundancies in 22 or more quarters.

(ii) The bulk of all redundancies in both Birmingham and Glasgow was small in absolute terms, involving no more than 4 male employees. For Glasgow plants the frequency and the rate of male redundancy was greatest for plants with 200–499 and 500–999 employees. The *rate* of redundancy was lowest for the largest units with at least 1,000 employees while plants with 200–499 employees were most vulnerable to redundancies on all counts. Almost three-quarters of all redundancies involved less than 2 per cent of a plant's stock of employees, but although for most units the typical redundancy was small in absolute and relative terms, redundancies involving 5 per cent or more of a plant's labour force accounted for more than 50 per cent of all male redundancies in Glasgow.

(iii) The market redundancy rate for males was higher in Glasgow than in Birmingham and North Lanarkshire, and while this rate showed little variation by skill groups in Glasgow, there was a definite tendency for the market rate of redundancies to be lowest for skilled men in both Birmingham and North Lanarkshire. This suggests that in the two latter areas plants 'hoarded' skilled men rather than declare them redundant and risk subsequent recruiting problems. This impression is strengthened if one considers the redundancy experience and skill characteristics of female employees. Within each market, females had a higher rate of redundancies than males, reflecting the tendency of plants to discharge married females, and especially part-time employees before males. The rate of female redundancies was especially high in Birmingham, and here, and in North Lanarkshire, females formed a higher proportion of the labour force than in Glasgow. As females were largely semi-skilled it is apparent that in the two former areas a redundancy situation was usually met by a cut-back in semi-skilled and unskilled employees while skilled workers were relatively little affected.

(iv) As in British industry as a whole, formal redundancy agreements were extremely rare amongst the case-study plants. The key decisions (size of redundancy, timing, etc.) affecting the strategy adopted in a redundancy situation were made by management before consultations with unions or shop stewards began, and these decisions were only infrequently altered in fundamentals as a result of such consultations. The Redundancy Payments Act may have the effect of increasing the proportion of redundant employees with long periods of service and hence the proportion in the older age groups, but in its first year of operation the Act did not produce any noticeable movement towards the conclusion of negotiated agreements on redundancy procedures.

(v) It was usual practice to discharge workers over normal retiring age before other categories of employees, and certain other groups were also particularly liable to redundancy, e.g. persons near retiring age, married women and part-time employees. The seniority principle was seldom applied without some modification by considerations of efficiency, and even where seniority was decisive it was commonly applied on a department or job, rather than on a plant, basis. 'Bumping' was therefore extremely rare. Partly as a consequence of this, and of the fact that a large proportion of all redundant employees were discharged in redundancies involving at least 5 per cent of a plant's work force, redundancies were concentrated amongst older employees with long periods of service. Such individuals therefore have characteristics which reduce their prospects of alternative employment, and the effect must be compounded by the fact that they are thrown on to the labour market when unemployment is especially high.

PART VI: CONCLUSIONS

CHAPTER 15

IMPLICATIONS FOR THEORY AND POLICY

1. INTRODUCTION

In this, the concluding chapter, we shall attempt to draw together the threads of our previous argument so as to consider the relevance of our findings for economic theory and labour market policy. We do not intend to reiterate our detailed conclusions, for a synopsis of these has been provided in the final sections of Chapters 4–14. Instead, our intention is to pick up those points which are felt to be most significant and to use these as a launching-pad for a broader and more generalized discussion. We are interested in evaluating whether we can satisfactorily explain and predict the labour market behaviour of employers and employees from the usual assumptions made by economists, in assessing the impact on such behaviour of the employment conditions ruling in the labour market, and in considering any policy implications which arise from our analysis both in the public and private fields.

Throughout this book we have adopted a model of labour market behaviour that assumes labour markets to be populated by employers and employees who act rationally in the strictly economic sense, maximizing profits and net advantages. To a greater or lesser degree, these assumptions are known to be unrealistic, as behaviour, especially where wage and employment issues are concerned, is never solely determined by a strict evaluation of economic costs and benefits. However, when one advances beyond the description of a particular set of events to generalization and prediction, then some simplification and retreat from reality is inevitable. It is not suggested that these assumptions capture all the nuances of labour market behaviour but simply that they help us to explain and predict some aspects of such behaviour.

In this light our general impression would be that the economist's model of labour market behaviour gets pass marks. This is not to say that all aspects of behaviour can be explained by the model or that economics is the only relevant explanatory discipline. On the contrary, there are a number of important issues, particularly in the field of wage determination, which have not been satisfactorily resolved and will remain unresolved until we admit the influence of non-economic forces. Yet while this difficulty must be faced, and while it points to a need to escape from the narrow confines within which we have worked, it remains true that, in a number of areas we have examined, the tools of the economist appear to have a fairly sharp cutting edge. We have then to report on both the strengths and the weaknesses of labour market theory. This is done under three headings in Section 2—inter-plant wage differentials and labour turnover; recruitment and mobility; and internal mobility and internal wage structures. Section 3 deals with policy questions.

2. IMPLICATIONS FOR THEORY

(i) *Inter-plant wage differentials and labour turnover*

Our investigation of the relationship between wages and labour turnover did not reveal any systematic relationship between plant recruitment rates or net changes in employment on the one hand, and plant earnings levels or changes in earnings levels on the other. However, there was a systematic, negative relationship between earnings *levels* and quit rates—and as a result between earnings levels and total separation rates. Hence high-wage plants tended to have low wastage rates in exactly the manner predicted by economic theory. It must be admitted that the relationship was not always a strong one,[1] and was especially weak in the case of Birmingham males, but that it existed is not open to serious doubt. We would therefore take issue with Silcock's view that if labour turnover were responsive to wage differences: 'we should expect the firm providing higher earnings to retain the labour it attracts. The movement of labour should be, to some extent, a one-way traffic from lower to higher earning occupations. There is no evidence that this is so. . . . It is evident that the economic theory, so often advanced and so

[1] This may be partly due to our enforced reliance upon *average* earnings by plant and by occupation, but it is also likely to indicate that other variables apart from wages are important.

rarely supported by quantitative observation, is insufficient to to explain the observed facts.'[1]

Silcock's scepticism seems to derive from the failure of previous investigations to establish a strong relationship between earnings and turnover (or mobility). Certainly the results of empirical research have often been inconclusive, but we might be justified in taking refuge in the view that it is the dearth of reliable material rather than the quality of the craftsmanship which is to blame. Previous studies conducted at a highly aggregative level and using net changes in employment as the dependent variable do not constitute an ideal test of the theory. Here we have been able to examine labour turnover at the appropriate level of the plant and to separate the different aspects of such turnover. The findings help us to understand why the results of many previous enquiries have been inconclusive, and go some way to confirming the view that earnings are an important determinant of job choice.

Given this, we might wonder why the rate of recruitment and net changes in employment were not positively related to wage levels. After all, if high wages discouraged employees from quitting the plant, surely they should also have encouraged new recruits to join it? It could be that current employees *know* when wage levels are high whereas potential recruits do not, but we have found it unsatisfactory to regard imperfect knowledge as the *main* cause of the apparent asymmetry in employee behaviour. Thus, while detailed knowledge on wage levels and differences was difficult to obtain, certain plants remained high-wage units over extended periods of time and their identity was widely known throughout the market. We are therefore thrown back on the proposition, which appears consistent with the remainder of our evidence, that high-wage plants were not necessarily seeking to recruit a large number of employees or to increase the size of their labour force relative to other units. High wages, although they will assist the plant to meet either of these aims should it so desire, do not arise primarily for these reasons; they are not necessarily wage signals reflecting a need for additional labour. Two implications flow from this. First, a weak or non-existent relationship between wage levels and net changes in employment (or recruitment) need not be interpreted as evidence of the fact that employees are indifferent to wage differences or that they do not know that they exist. Second, changes in the distribution of employment between plants, or indeed between industries or areas, can in

[1] Silcock, *op. cit.*, p. 431. It is only fair to observe that Silcock does qualify this view immediately but he remains of the opinion that the economic argument 'appears to have been over-emphasised in the past'.

certain circumstances be accomplished with little or no change in wage differentials. Thus high-wage plants, industries or regions could attract more labour simply by making more job opportunities available. This is not, of course, a pure 'job opportunity' explanation of labour mobility, for it also follows from our argument that low-wage payers will have to offer job opportunities *and* raise wages to obtain a larger share of the labour force.

If one accepts the above interpretation then some degree of competitive pressure must exist in the labour market. It may be limited because high-wage units, which *could* attract more labour if required, are not normally attempting to expand employment rapidly. Yet as long as quit rates are negatively related to wage levels, a plant cannot ignore the actions of other units employing similar types of labour. We have also adduced as evidence of competitive forces the fact that the inter-plant wage structure displays considerable stability over time, and that plants attempt to maintain, if only approximately, their relative position in the wage hierarchy. However, having said this, we must recognize that substantial and persistent wage differentials found to exist in practice are difficult to reconcile with traditional theory.

This finding need not be inconsistent with the competitive model if higher wages are offset by non-pecuniary factors or if there are differences in the efficiency of labour. Unfortunately for the theory these explanations, either singly or jointly, do not seem to be adequate to explain away the wage differentials observed. There is simply no indication that the low-wage units in our study offered especially favourable conditions in other respects so that net advantages were equalized. Indeed the balance of evidence points to the conclusion that non-wage factors were likely to be most favourable in high-wage units and, even more important, the view that net advantages must be equalized is inconsistent with the finding that high-wage units tended to have low quit rates.

Differences in labour productivity, even where the type of labour considered seems homogeneous on the surface, seem to have been more important. High-wage units did appear to be able to apply and maintain higher hiring standards, so that inequalities in efficiency wages were smaller than inequalities in money wages. Yet while we lack conclusive evidence on this point, and while more sophisticated investigations of differences in labour quality are urgently required, it seems improbable that variations in hiring standards were substantial enough to bring about equalization of efficiency wages at any point of time or, indeed, that there was any tendency towards equalization for periods extending over eight years.

Because employees do respond to wage differentials where these are accompanied by job opportunities, a plant cannot set wages without regard to the actions of other establishments in the relevant market. At the same time, the plant is far from being a price-taker, merely rubber-stamping a wage set by impersonal market forces. Particularly if it is a high-wage unit, it has more freedom of action than the competitive model suggests. High-wage units will pay a money wage greater than that 'justified' by the higher quality of its labour force. What we are suggesting is that the labour force will benefit in the form of higher wages if the plant enjoys high profitability, economies of scale, efficient management or methods of production, monopoly elements in the product market which allow higher earnings to be passed on, and so forth. Such an outcome does depend on some modification of the assumption of strict economic rationality on the part of the manager. We must recognize other motives apart from profit maximization, and the existence of institutional pressures which can force the employer to adopt a strategy different from that which he might wish to follow. For example, the employer may be prepared to concede higher wages provided some satisfactory level of profitability is attained. As Slichter has suggested, 'wages, within a considerable range, reflect managerial discretion, [so] that where managements can easily pay high wages they tend to do so, and where managements are barely breaking even, they tend to keep wages down'.[1] We might add in parenthesis that, even where managements are unwilling to concede higher wages readily, they may find it difficult to resist wage claims which are likely to be pushed strongly where profits are high or where other conditions are favourable.

It would seem therefore that the conditions prevailing at plant level may have an effect on wages in a manner which is rigorously excluded from the traditional model. Economic rationality and competitive forces are not strong enough to result in a situation where each employer pays no more and no less than the market wage. Because competitive forces are present the concept of the market wage has its uses, but it also has severe limitations as a description of the realities which face the employer. In a world of imperfect knowledge, with powerful institutions and persistent differences in efficiency and profitability at plant level, market forces appear to set only the outside limits within which the wage bargain will be struck. These limits appear to be fairly wide, but their extent requires detailed investigation in a fashion which gives more explicit recognition of the factors we have referred to, but have hardly begun to analyse.

[1] Slichter, *op. cit.*, p. 88.

When we turn from a cross-sectional analysis of the relationship between plant wages and plant turnover, to an investigation of fluctuations in market quit rates over time, we can see that the voluntary quit rate was sensitive to changes in employment conditions as measured by registered unemployment and unfilled vacancies.[1] In other words, the market quit rate was inversely related to the unemployment rate and directly related to the vacancy rate. Moreover the effect of a change in unemployment conditions was quickly transmitted to quits. Quits responded to the unemployment conditions currently ruling in the market, and there was no indication of a lagged response to unemployment and vacancies, or, indeed, of an expectational effect based on the supposition that current trends would be continued in the future. The finding that the market quit rate is responsive to employment conditions is not novel, but it is a useful corrective to the view seemingly held by some sociologists that the volume of labour turnover is largely determined by the internal characteristics of the plant.[2] Economists have been guilty of ignoring the characteristics of the employing institution and have often tended to represent turnover as a phenomenon determined by purely external factors over which managements have little control. This is, of course, an over-simplification, and we shall argue later that managements could reduce turnover by greater attention to selection and induction procedures. Nonetheless, a plant's turnover will vary with employment conditions. As unemployment declines and the demand for labour rises quit rates will increase, and vice-versa. The employee is thus able to perceive changes in employment conditions and reacts in the manner predicted by the economic model. This is by no means the end of the matter, but it is a sound beginning to any study of labour turnover.

(ii) *Recruitment and mobility*

Labour turnover generally involves a change in jobs and therefore is usually part of a wider process involving a change in employer and often a change in industry, area or occupation. We have looked at this process from the standpoint of the employer and the employee because the preferences and activities of the two parties interact to

[1] In view of the extensive use of vacancy statistics in empirical studies an important finding of our research is that vacancies tended to understate considerably the 'true' demand for labour. However, as the main implications of this finding relate to policy, our discussion is postponed to a later section.

[2] The 'Tavistock School' particularly has paid remarkably little attention to the relationship between quits and employment conditions. See Rice, Hill and Trist, *op. cit.*, and J. M. M. Hill, *op. cit.*

establish the characteristics of labour recruitment and of labour mobility. Both of these aspects are of fundamental importance if we are to understand the labour supply position confronting the plant and the job choices made by individual employees. Our intention is therefore to look in turn at the industrial, occupational and geographical characteristics of recruitment and mobility, to establish the broad patterns which emerge, and to see whether these are shaped by factors which economists regard as important.

We have seen that the Glasgow establishments drew heavily on other engineering units to meet their labour requirements, and this dependence increased the higher the skill level of the recruit. In the other labour markets there was the same tendency for the proportion of recruits drawn from the engineering industry to rise with skill level, but, relative to Glasgow, a substantially higher proportion of recruits at each skill level was drawn from non-engineering establishments. One consequence of this was that outside Glasgow the percentage of total recruits obtained from engineering plants was much the same as that industry's share of total employment in the local market.

This appears understandable enough. Glasgow plants faced a slack labour market situation. Employment in the long-established and important engineering industries was declining and was not being easily absorbed by an expansion of employment in other industries, so that unemployment was substantially above the national average throughout 1959–66. As a result, Glasgow engineering plants found it relatively easy to obtain employees with previous experience of the engineering industry, and they naturally preferred such recruits because of the low training costs involved. On the other hand, Birmingham units were less able to satisfy their preferences for experienced and trained labour. They were forced to cast their recruitment net over a much wider range of industries because of the tight labour market situation. This tendency was even more noticeable in the Other Scottish Areas, and above all in New Town and Small Town, where the case-study plants expanded their employment extremely rapidly over 1959–66 in markets with very few reserves of engineering labour. The plants met their needs by recruiting from a considerable spread of industries within the local market. This naturally involved heavier training and induction costs than in Glasgow, but it was a perfectly rational response given the labour market situation facing the plants.

When this process is looked at from the employee's angle a coherent picture again emerges. Industrial attachment did not appear to have much effect on mobility into the engineering industry except in Glasgow. In the other areas the industrial distribution of recruitment reflected fairly accurately the underlying distribution of the employed population by industries, indicating that the walls between industries, and certainly between manufacturing industries, may in some cases be a statistical artefact rather than a serious obstacle to inter-industrial movement. However, the impression that industrial mobility is 'random', and that labour moves between industries at will, only holds if one considers *all males undifferentiated by skill* (and to a somewhat lesser extent all females). As soon as we delve beneath the surface and examine the behaviour of occupational groups, the process of industrial mobility takes on a character and shape closely linked to the skill of the group concerned, for in each area the proportion of skilled men recruited by the case-study plants from other engineering establishments was higher than the corresponding proportion for unskilled men. To some extent this must reflect the fact that plants were little concerned with the previous industrial experience of unskilled recruits, while the previous training and experience of men taken on for semi-skilled, and especially skilled jobs, was of greater relevance to hiring decisions. Yet it must also reflect the preferences of employees, which differed according to the level of skill required. Hence the patterns of industrial recruitment and mobility were largely set by occupational factors.

The bulk of persons recruited to skilled jobs with the case-study plants had previously been employed in that occupational group, but a much smaller proportion of those recruited to semi-skilled and unskilled production work were previously employed at the same skill level. As far as the plant is concerned this merely reflected a preference for hiring experienced labour for skilled jobs, to reduce the heavy training costs which would be involved in providing inexperienced recruits with the necessary skills. Training costs are much lower for semi-skilled and unskilled workers, and so plants could recruit such labour from a much wider range of occupations. However, while this distinction between occupational groups held within each market, the extent to which plants could exercise their preference varied from one market to the next. We have already observed one consequence of this in the characteristics of industrial recruitment in the separate markets which, indeed, merely reflected the differing extent to which occupational needs could be met by the direct recruitment of labour with the requisite skills. Tight labour market conditions in Birmingham, and the lack of experienced engineering labour in the Other Scottish Areas, forced plants to draw recruits at all skill levels from a wider range of occupations than was the case in Glasgow. Amongst other things, this naturally meant that the range of industrial recruitment was narrower in the latter area.

As we would expect from the above, we find when we turn to consider occupational mobility that skilled groups were least likely to change their occupation on changing their employer, while such mobility was greatest for the unskilled. This is, of course, in accordance with our *a priori* expectations. Relative to the unskilled, skilled workers have a greater propensity to remain within the same occupational group both on changing employers and within the plant, because they have invested more time and resources on acquiring training and experience. While of value to them in particular jobs (generally within the engineering industry) these are of little value in other activities In contrast, the unskilled change jobs (and industries) much more readily because their experience does not confer any substantial advantage in a specific field of activity.

It is true that the high degree of occupational mobility accompanying employer-shifts for skilled workers in Birmingham does not fit this explanation, but even there skilled males showed relatively little industrial mobility, and downward occupational mobility was largely into semi-skilled engineering jobs. Here their training and experience must have conferred on them some advantages and, given the wage structure of the engineering industry, may have been perfectly rational on economic grounds.

As far as the geographical dimensions of recruitment are concerned, most plants obtained the bulk of their recruits over the study period from within their local labour market, although it is apparent that recruitment policies could be, and were, adapted to meet prevailing employment conditions. Hence plants in Small Town and above all in New Town drew a relatively high proportion of their employees from other areas of the country in order to supplement the supply of local labour which lacked previous experience of engineering. To this end, the New Town plants put very heavy emphasis on advertising and the public employment exchanges as means of searching out possible job applicants, and made less use of informal methods of recruitment which, while cheap, were likely to be effective channels of communication only within the confines of the local labour market.

Yet while it was possible to trawl a wider area and to reach a greater number of potential job applicants by adopting more aggressive (and expensive) search methods, it is nonetheless clear that labour mobility was heavily influenced by the spatial characteristics of the market and by the skill, and level of earnings, of the employee. The great bulk of employer-shifts, especially in the established engineering areas and larger labour markets of Birmingham, Glasgow and North Lanarkshire, involved movement between plants less than

20 miles distant from each other, and relatively few shifts involved geographical mobility over distances greater than 5 miles. The notion that it is sensible and useful to discuss behaviour within a 'local' labour market context is, then, well grounded in fact. Most manual employees do seek work within a fairly restricted geographical area, and, correspondingly, the elasticity of labour supply to a plant is to a large extent a function of the potential labour force already resident within its vicinity. The boundaries of the local labour market can be pushed back, as we have seen, by more aggressive recruitment policies (e.g. advertising), but this does involve heavier recruitment costs, and higher wages may be necessary to induce potential employees to change residence or undertake longer travel-to-work journeys.

The bulk of employees had travel-to-work journeys of less than 5 miles, underlining our previous suggestion that the location of the plant relative to population settlements will be of crucial importance in determining its ability to meet its labour needs. Hence the higher wages generally offered by larger units must to some extent have reflected a need to compensate their employees for the longer average travel-to-work journeys they had to undertake. Moreover, the extent to which different groups of employees could trade off higher wages against the opportunity costs of travel-to-work journeys also differed according to the skill, and therefore the wage, of the group concerned. Unskilled males had shorter travel-to-work journeys than either skilled or semi-skilled males, and female employees were not prepared to undertake such long travel-to-work journeys as males. This is as might be expected, for it implies a positive relationship between wage levels and the length of travel-to-work journeys, which can be explained along the following lines. The wages received by females and unskilled males are relatively low, and in consequence the *absolute* differences between plant wage levels are smaller than for skilled males (see Tables 4.1 and 4.2, pp. 71–2 above). Because of this we can suggest, following Stigler,[1] that females and unskilled males will be unwilling to incur such heavy job-search costs as more highly paid groups. This being so, they will place greater emphasis on informal channels of information (see Table 13.3, p. 357 above) which, while cheap, are generally effective only within a restricted area. By an analogous argument we can demonstrate that more lowly paid workers will tend to seek work in the immediate vicinity of their homes, because the possible benefits of a higher wage elsewhere are less likely to offset the opportunity costs of a longer travel-to-work journey or residential mobility.

Summing up our findings through a study of the simple and com-

[1] *Op. cit.*

plex mobility accompanying employer-shifts, we find that the most common form of mobility involved a change in industry. Occupational and geographical mobility much less frequently accompanied employer-shifts. Indeed, if we wish to understand the factors which condition labour mobility then these are the aspects which are crucial. An analysis of mobility which emphasizes industrial attachment is not very meaningful unless it is related to occupational and spatial factors. Movement (or lack of movement) between industries is likely to be determined by their propinquity and by whether they have similar occupational structures and training requirements. Industrial attachment *per se* is of little significance.

We can also conclude that, when considering how the recruitment policies of employers, and the mobility of employees, are moulded, the economist is quite justified in emphasizing the importance of employment conditions, investment in training, the monetary and non-monetary costs involved in geographical mobility and so forth. When these factors are brought into the discussion much of the behaviour of the case-study plants, and of their employees, becomes explicable. The employer reacts to tight labour market conditions by widening his area of recruitment in geographical, occupational and industrial terms. These elements may not be present in all cases, and the mix adopted will vary according to the special circumstances of the market and of the plant. Again, employers may react slowly to changes in their environment and may cling to traditional policies when the conditions which justified them no longer exist. The adjustment process in the market may then be slow and strewn with errors of judgment, but in the end the plants do react to labour market pressures in a manner which by and large is consistent with our *a priori* expectations, and the same is true of the actions of employees.

(iii) *Internal mobility and internal wage structures*

Occupational mobility can, of course, occur within a plant as well as between employers, and the extent, the nature and the rules governing internal mobility are of importance in explaining the behaviour both of employers and employees. To the employer the encouragement of internal mobility may be one means of meeting recruitment difficulties encountered in the external market; to the employee, the rules and practices governing 'ports of entry' and seniority will influence his opportunities for promotion within the plant and his attitude towards changing employers. We found that the volume of internal mobility was responsive to employment conditions, so that upgrading was more frequent where experienced and skilled engineering workers were more difficult to recruit. The differences between

Glasgow and Birmingham in this respect were not as great as might have been anticipated, but internal mobility was more substantial in the latter area and was noticeably more important in the Other Scottish Areas where the supply of trained engineering labour was severely limited. In other words plants could and did turn to their internal markets to help to make good labour requirements which could not be easily satisfied through direct recruitment.

In view of this, the interest shown by researchers in the operation of internal labour markets is well placed. Yet the tendency to fall back on 'the operation of internal labour markets' as a convenient method of disposing of certain awkward aspects of labour market behaviour is less satisfactory. For example, it is extremely unlikely that one can attribute the substantial inter-plant wage differentials within a local labour market to procedures which discriminate in favour of existing employees, emphasize seniority, restrict ports of entry and so forth. The view that such rules and practices could explain the existence of inter-plant differentials because they inhibited mobility between plants was never very convincing, given the substantial amount of such mobility in any labour market. It is even less attractive if we bear in mind that the large and persistent plant differentials found to exist in this study were accompanied by a system of internal labour markets with little emphasis on seniority[1] and with many ports of entry, so that occupational mobility accompanying employer-shifts was more frequent than occupational mobility within the plant.

This is not to say that further study of internal labour markets would not be extremely valuable. However, it does appear that the stress on developing analytical models from first principles has tended to over-emphasize the degree to which labour market behaviour in general might be structured by behaviour within the plant. What is required is an improved understanding of how internal markets actually operate and further testing and development of existing hypotheses. The data from this study suggest that the volume of internal mobility, at least in established engineering plants, is rather restricted. As the traditional, competitive model assumes, job opportunities arising at each level of skill are fairly open to competition between employees and non-employees. Certainly in the British engineering industry we are some considerable distance from a situation where the wages of those some way up the occupational ladder are insulated from market forces, because hiring only takes place at

[1] Except in redundancy situations, and even here the application of the seniority principle was more restricted than commonly supposed.

the bottom rung and because progress up the ladder is determined purely by seniority.

We also require more detailed examination of the factors which shape the internal mobility which does occur, for these are extremely complex. For example, while tight labour market conditions create greater pressure on employers to promote internal mobility to make good shortages of trained labour, they may also increase the costs of such a policy, for the higher turnover rate resulting will inhibit investment in training programmes and reduce the stock of employees with sufficient acquired experience to warrant upgrading. This was one reason for a rate of internal mobility in Birmingham which was lower than expected. Plants in the Other Scottish Areas also faced difficulties in obtaining experienced engineering labour through external recruitment, but this was accompanied by very low turnover rates, so that on both counts internal mobility was encouraged. Yet internal mobility did not respond simply to external market forces. Institutional factors such as trade union pressures and the lack of attention to internal mobility displayed by many British plants relative to their American counterparts, inadequate on-the-job training, the nature of jobs and occupations created by the production technology, and the internal plant wage structure were all important influences conditioning the volume and the type of internal mobility which did occur.

As far as the internal wage structure is concerned few conclusions can be reached on the basis of cross-sectional analysis. All we can say is that the internal wage structure was extremely complex, and that it varied considerably from unit to unit both in the ranking and in the differentials enjoyed by different groups. This does not take us very far, for wages theory does not predict that intra-plant differentials will be the same for all plants, or even that the earnings of different occupational groups should have an identical ranking from one plant to the next. Complex and different internal wage structures could then represent non-economic behaviour or, conceivably, a careful and calculated response to differing economic pressures.

While the pattern of intra-plant wage differentials which emerges at any point in time is extremely complex and rather difficult to explain, a more orderly picture emerges over time. Where a plant was a high-wage payer for one group of workers it tended to be a high-wage payer for all employees, and if any group 'fell out of line' this was generally made good subsequently. A wage increase for one group of employees tended to be passed on in greater or lesser degree to all employees, so that the internal wage structure tended to move bodily upwards through time with all groups of workers in a plant doing better or worse than the average for the market as a whole. What this means in practice is that an employee's increase in earnings over any substantial period of time will depend more on the plant in which he is employed than on the external employment conditions for the particular skill he possesses. This is in direct contrast to the prediction of the competitive model, which is that the economic circumstances of the employing unit will be irrelevant in explaining an individual's increase in earnings, for this will be determined by the demand and supply conditions for his particular type of skill. It is extremely difficult to find support for this prediction in our evidence. We must conclude that, while competitive pressures are present, they do not set a unique wage, and that the economic circumstances of the plant in terms of efficiency, profitability and so forth, will affect the development of the internal wage structure just as they have an important influence on inter-plant earnings differentials.[1]

A further implication of our findings is that the internal wage structure has a certain inflexibility and does not respond readily to external pressures. A plant facing recruitment difficulties for turners may not then be able to meet this problem by increasing turners' wages without facing wage increases 'across the board'. Once we admit this we must surely also recognize that a theory of wage determination which makes no reference to institutional mechanisms and notions of 'equity and justice' is seriously incomplete. A wage increase for one group sets off a chain reaction within the plant, and its result will depend to a considerable degree on the process of bargaining itself. Also, the wage-payment system and the very complexity of the wage structure may themselves have an important influence on wages. The institutional framework within which piecework bargaining takes place in the British engineering industry makes wage drift difficult to control and probably impossible to eliminate. The complexity of many plant wage structures is a product of *ad hoc* responses to past events which offers hostages to future wage claims to 'restore relativities', 'eliminate inequities', and so on.

All this is overlaid by the considerable uncertainty which attends the process of wage determination. Thus far we have tended to play down the influence of imperfect knowledge on decision-taking which has been stressed by so many previous observers, for we do not believe that the labour market is quite as chaotic and disorganized as is often supposed. Nonetheless, the information possessed by all parties is deficient in many important respects. The concept of the

[1] A recent American study also emphasizes the importance of establishment variables in explaining wage differentials for manual workers. See A. Rees and G. P. Shultz, *Workers and Wages in an Urban Labor Market* (1970).

market wage must sound like a sick joke to the employer and the employee faced with a shadowy outline of large inter-plant differentials and a bewildering array of internal wage structures. Certainly the market provides some signals, but they are often very faint. Plants often know little in detail of competitors' wage structures and make mistakes accordingly. Again, the outcome of the system of wage determination can produce results which employers did not intend. Here the most noticeable example was the high downward mobility shown by skilled men in Birmingham who could often increase their earnings by taking up semi-skilled jobs.

Imperfect knowledge must also affect employees who have even less access to reliable wage information at the plant level. Given the complexity of the engineering wage structure, it must be extremely difficult for the job applicant to carry through a satisfactory search amongst alternative openings without actually experimenting in different jobs. Some part of labour turnover must then result from disappointed expectations. Employees are not completely ignorant, for the negative relationship between wages and quits suggests that they can recognize a good job when they find one. Yet finding a good job may, because of imperfect knowledge, be a difficult operation involving substantial costs for both parties. The labour market is, then, a good way short of being a perfect allocative mechanism. The 'invisible hand' does not provide the optimum solution without need of outside assistance. With this in mind, we can now take up our policy conclusions.

3. IMPLICATIONS FOR POLICY

Turning from theory to more immediate policy questions, we will again divide our remarks into three sub-sections. First, we shall look at employer wage policy and, second, at employer manpower policy. This may appear a peculiar distinction, for wage policy is, or should be, central to employer manpower policy in general. Yet while these areas are closely interconnected wage policy is of such fundamental importance that it requires special attention, and hence separate treatment. It is here that crucial weaknesses lie, and it is only proper that these weaknesses should be highlighted. Turning from the plant to a broader canvas, we take up those topics relevant to national manpower policy. It is possibly premature to talk of a 'national' manpower policy because, although there has been some groping in this direction, we are as yet a long way short of a coherent set of policy objectives and measures. The same criticism could be levelled at our remarks in this area which are far from comprehensive.

Nonetheless, we can look at a number of subject areas in which additional information is vital if such a coherent policy is to be developed.

(i) *Employer wage policies*
The British engineering industry is the outstanding example of what the Donovan Report referred to as the 'two systems of industrial relations'.[1] The first, the formal system at national level, is based on industry-wide collective bargaining. The second, the informal system at local or plant level, embraces a wider range of issues than national agreements, rests on 'understandings' and custom and practice as opposed to written agreements, and, most significant from our point of view, is of considerable importance in determining the level and structure of earnings. According to the Donovan Report, conflict between the two systems, and a failure to recognize that in many instances effective power has shifted to the plant level, has produced a situation where plant bargaining is often inefficient, 'usually takes place piece-meal and results in competitive sectional wage adjustments and chaotic pay structures'.[2] The Prices and Incomes Board, referring specifically to the engineering industry, echoed this view:

> 'Within any one plant or company the fragmentation of bargaining leads to a settlement for one group which gives rise to a sense of inequity on the part of another group . . . there ensues a struggle for an adjustment to remedy the inequity felt. That the adjustment is never in fact secured, that inequities persist, has been amply demonstrated. . . .'[3]

Internal wage structures in engineering cannot be based effectively on any single, simple principle, given the diversity of the industry and the fact that any wage structure has to satisfy a number of different objectives. Yet while there is a danger of assuming that complexity is equivalent to chaos, it must be said that all too often the intricacies of plant wage structures do not represent a rational response to economic pressures, but reflect inadequate control over payment systems and the lack of a systematic and comprehensive management wage policy. Here the findings of Chapter 5 are by no means new, but it is worth while underlining certain points which have to be borne in mind in any restructuring of collective bargaining.

[1] *Royal Commission on Trade Unions and Employers' Associations, Report, op. cit.*, para. 46.
[2] *Ibid.*, para. 1010.
[3] National Board for Prices and Incomes, *Report No. 49*, p. 40. See also *Report No. 104*.

The first concerns the widespread use of pieceworking in British engineering. The tendency for pieceworkers' earnings to increase over time was of particular concern to the case-study plants because most plants, at least in Birmingham and Glasgow, employed pieceworkers and hence faced the difficulty of meeting compensatory wage demands from timeworkers. Upward pressure on wage levels was therefore felt 'across the board', but in a fairly large number of plants this had not prevented the method of payment from swamping the effect of skill so that semi-skilled pieceworkers earned more than skilled timeworkers. No doubt this result was sometimes intended and justified, but such a reversal of skill differentials is seldom the result of conscious decision.

The consequences of the haphazard wage structures which can emerge were most apparent in Birmingham. Plants found skilled labour most difficult to recruit, yet the high earnings obtained by many semi-skilled workers relative to the skilled must have militated against the upward occupational mobility which would have met this difficulty. Wage structure problems are likely to have contributed to the unexpectedly low volume of internal mobility from semi-skilled to skilled jobs in Birmingham, and explain the pronounced tendency of skilled men to take up semi-skilled jobs on changing employers in that area. It is no solution to recommend the wholesale abandonment of pieceworking, as such a method of wage payment would not have been so widely adopted unless it satisfied some important need. If properly applied, pieceworking provides an incentive to greater productivity through relating pay to effort rather than to hours of work, but the question is how far pieceworking as currently practised does effectively relate effort and reward without creating undue difficulties of supervision, inflexibility, and cost inflation.

Whether or not it is appropriate to introduce pieceworking largely depends on the prevailing production technology. As the P.I.B. has suggested, a conventional payment-by-results system is likely to be most effective where four conditions are met. These are that there should be a steady flow of work, that the pace of work should be capable of being influenced by the effort of the employee, that the work must be measurable, and that it should be repetitive and not subject to frequent changes in methods.[1] If we apply these criteria to our case-study plants then certain interesting results emerge. Glasgow plants used pieceworking more extensively than plants in Birmingham or in the Other Scottish Areas. Yet the first and last conditions referred to by the P.I.B. were less often met in Glasgow. The bulk of engineering plants in that city were engaged in the manufacture of

[1] National Board for Prices and Incomes, *Report No. 65*, pp. 32–3.

producer durable goods, and the mix and volume of work through the shop showed considerable variation. Such a situation did, of course, raise an acute problem of supervising effort, and the solution adopted in many cases was to adopt a pieceworking system as a 'silent supervisor'. To meet the problems associated with operating a piecework system in such a situation, the Glasgow plants placed considerable reliance on work study and measurement; but the changing nature of production and the institutional framework within which bargaining took place made wage costs difficult to control despite the slack labour market situation which prevailed.

In Birmingham the nature of production appeared to lend itself more readily to pieceworking and yet it was less widely adopted than in Glasgow. Nonetheless, most Birmingham plants employed some pieceworkers, and the strains imposed on the internal wage structure were every bit as formidable as in Glasgow. Birmingham plants placed much less emphasis on work measurement, and the traditional system of a price per piece was still in widespread use. In turn, this probably reflected the flexibility of response that such a method of payment confers on management, in particular the ability to meet recruitment difficulties by offering more generous piecework prices. Labour market considerations may also be the crucial factor in explaining why in Birmingham there was little tendency to move away from pieceworking, whereas in Glasgow plants had, over the study period, shown a greater willingness to experiment with alternative payment methods such as measured day work and job evaluation.

However, while there appears to be some scope for a reconsideration of the use made of pieceworking, most plants are likely to employ some pieceworkers over the foreseeable future. The problem of relating pieceworkers' and timeworkers' earnings will therefore continue to be important, and the upward pressure on wage costs will not be diminished unless there is a movement away from the fragmented process of bargaining at individual and group level towards comprehensive agreements covering the establishment. The complexity of internal wage structures and the tendency to deal with each individual problem as it arises on an *ad hoc* basis result in real or perceived anomalies and inequities between groups, and create the conditions for pressure to be exerted at other points in the wage structure. The locus of power within management also requires examination. As it is, decisions on wage issues are usually taken by those directly responsible for production, and there is a considerable temptation to give way to pressure and to postpone any adverse consequences until tomorrow, so that production is not interfered with today. This is most apparent in British plants. In units under American control, expediency is given less weight and management is less willing to countenance exceptions to those principles on which the wage structure is erected. In short, American units are closer to the Donovan ideal that a consistent body of principles should be developed for dealing with wage issues on a plant-wide basis.

It must be observed, however, that any movement towards greater reliance on plant bargaining[1] will leave certain difficulties unresolved and may raise others. Two such difficulties may be mentioned briefly. First, one of the principal arguments used in the advocacy of plant bargaining is that it would allow plants to respond to local conditions by freeing them from the restraints imposed by national agreements. Amongst other things, it is held that such bargaining would create a framework within which the internal wage structure could adapt more easily to changes in the demand/supply relationships for different types of labour in the local market. It is doubtful whether this is so. Coercive comparisons are extremely strong within the plant, so that the internal wage structure is fairly inflexible over time. There is no reason to suppose that this will change, and while the adoption of plant-wide bargaining rather than the fragmented system prevailing at present may have advantages in other directions, it seems likely to increase rather than reduce the inflexibility displayed by internal wage structures. In this regard it is relevant to note that changes in the internal wage structures of American plants in the Other Scottish Areas were no more responsive to external employment conditions than those in British units.

Second, plant bargaining, whose main advantage would be to relate wages and effort more closely and to reduce the extent to which one group could be played off against another, might increase inflationary pressures in other directions. The existence of substantial inter-plant differentials, while not unknown to employers and employees, may be somewhat obscured by the continuance of national negotiations and the lack of regular and comprehensive information at establishment level. This can hardly survive a major shift towards plant-level bargaining. Unions and employees find 'comparability' a useful argument in most negotiations and are likely to seize on inter-plant comparisons as the most appropriate to plant bargaining. Such comparisons are, of course, already made, but they may well become more vital and important. This could heighten the pressure on low-wage establishments to bid up their wages relative

[1] This movement is already under way as a result of the 1968 agreement between the E.E.F. and the C.S.E.U. See *Incomes Data*, Reports 57 and 59–60.

to other units. This outcome is by no means certain, as in the United States substantial inter-plant differentials have continued to persist alongside plant or company bargaining. Yet it is a possibility which must be borne in mind, as is the need to provide the more detailed wage information necessary to make plant bargaining effective.

(ii) *Employer manpower policy*

Our primary interest lies in those aspects of employer manpower policy which are concerned with the manner in which plants anticipate and react to labour needs. Viewed in sequence, these are manpower forecasting, the nature of recruitment, selection and induction procedures and internal mobility. We shall pay particular attention to the methods by which policy can be adapted to meet different labour market conditions, but while we shall be mainly concerned with the current policies and practices of individual establishments, we shall, where it appears desirable, draw certain conclusions relevant to our last main field of enquiry—national manpower policy.

As manpower forecasting was little used as a basis for day-to-day recruitment activity, it was difficult to detect much difference in the actions of plants with and without forecasts. Both seemed equally uncertain as to what tomorrow might bring, so that new labour was generally sought only when the need for it became apparent on the shop-floor, and any advance warning was usually based on a rough evaluation of the labour content of forward orders rather than on a formal manpower forecast. Possibly it is unrealistic to expect forecasting to reach a degree of accuracy sufficient to provide a reliable guide to ordinary recruitment needs, for it could be argued that its main purpose is to establish long-run and major changes in labour requirements. However, even here results were disappointing. Plants with forecasts were as liable to experience redundancies as non-forecasters, and in both groups apprentice-training programmes were determined by reaction to current events rather than by an evaluation of future needs. Only in the new and growing plants in the Other Scottish Areas did top-level management pay much attention to forecasting, and even here the interest was short-lived, so that forecasting had little effect in labour market policy once the plant was firmly established.

None of the above implies that manpower forecasting has no useful purpose. Properly conducted, manpower forecasting can be useful in anticipating and meeting future developments, and might be especially valuable in weaning plants away from apprentice-training programmes which at present are largely based on customary practice, rule-of-thumb conventions and present circumstances rather than on some long-term view of future requirements for skilled labour. For this to happen it is first necessary to provide managements with a more developed and reliable tool. The science of forecasting had been little developed amongst our case-study plants, and we have no reason to believe that their activities in this field were atypical of British industry as a whole. At present the techniques adopted are extremely rudimentary, so rudimentary that forecasting often remains a paper tiger which has little impact on management policy decisions.

Our findings are also relevant to national manpower policy, for it is dangerous to base national decisions on forecasts in which the forecasters themselves place so little trust. The best-known example of such an approach was the ill-fated *National Plan* which was ultimately founded on the long-run forward projections of firms. However, as such an exercise is not likely to be repeated at least in the foreseeable future, a more relevant example is the series of reports which have tried to estimate future needs for scientific manpower on the basis of forecasts prepared by firms.[1] The provision of accurate forecasts of future labour needs is, of course, fundamental to manpower policy, but, as our previous remarks would suggest, the past record of those forecasts based on firm estimates is singularly unimpressive. This is not simply a matter of technique although that is obviously important; 'there are in addition biases towards optimism, since declining firms are reluctant to report their condition, and also towards conservatism, which appears in a tendency for the dispersion of anticipated changes to be consistently less than realised ones'.[2] We might conclude that national manpower planning should be based on centralized and not on plant forecasting unless there are clear and compelling reasons for the latter approach.

In the Birmingham and Glasgow conurbations the most interesting features of recruitment and labour mobility were their occupational and geographical characteristics. In Birmingham, we have seen that the shortage of skilled labour must to some degree have been attributable to the considerable amount of downward occupational mobility from skilled to semi-skilled jobs which accompanied employer-shifts. To the extent that such mobility arose from skilled workers moving into semi-skilled jobs in high-wage units such as the motor-car plants, it was possibly inevitable. However, downward

[1] The most recent example is the *Report on the 1965 Triennial Manpower Survey of Engineers, Technologists, Scientists and Technical Supporting Staff*, Cmnd. 3103, 1966.

[2] J. R. Crossley, 'Theory and Methods of National Manpower Policy', *Scottish Journal of Political Economy*, Vol. 17, 1970, p. 143.

occupational mobility was not solely confined to such instances, and there is a much broader problem of inter-plant and intra-plant differentials, to which we have already referred. As far as geographical mobility is concerned, it is apparent that, while more aggressive recruitment policies could widen the catchment area, plants were forced to rely on local sources for the bulk of their recruits, at least in the short run. Even within the conurbations we can distinguish sub-markets, and it is clear from travel-to-work patterns that the maximum utilization of the potential labour force must depend to a considerable degree on the location of factory and residential development. The costs of long travel-to-work journeys impose greater restrictions on job-choice for low-income groups and may cause married women to withdraw from the labour force altogether. If maximum use is to be made of the potential labour force available, it is therefore essential to pay particular attention to these groups and to carry out more detailed investigations of the relationship between wages and travel-to-work thresholds.

Our study of labour turnover raises a number of policy questions which are relevant to the selection and induction procedures applied to new recruits. In each market a high proportion of all leavers were persons who had completed only a short period of service with a plant and then left voluntarily or were dismissed because of unsuitability. The high rate of voluntary quits during the induction crisis could reflect inadequate prior knowledge of the conditions prevailing in the plant, to which the solution is the provision of better job information. Yet while this explanation has a substantial element of truth, there are other deficiencies which can be laid more directly at the door of the plants. Many managements have given little thought to the development of more effective selection and induction procedures. The overwhelming impression was that the hiring of labour was carried through in an extremely casual fashion. It is not then surprising that many square pegs were shoved into round holes and fell out thereafter. While a high degree of systematization is no substitute for the accumulated skill and experience of a good personnel officer, the unsatisfactory outcome of present methods is amply portrayed in its results. With the exception of a handful of plants, the engineering units applied hiring standards which were implicit and subjective. Those applicants who were manifestly 'undesirable' were excluded, and there was some evidence of a bias against older and coloured applicants. For the rest, little attempt was made to evaluate competence, attitude or interest. The initial selection procedures were extremely perfunctory, and the real mechanism of selection was the process of attrition after recruitment. Those who found the job uncongenial left, and others were dismissed. Yet while the drop-out rate amongst new recruits was high, most managements accepted it as a fact of life. They grumbled, but did little to develop more effective selection and induction methods.

A similar criticism may be levied against management attitudes towards internal occupational mobility. It seems clear that, while such mobility was used to make good some external recruitment difficulties, plants did not reap the maximum potential benefit from such movement. Occupational mobility within the plant was, therefore, fairly severely limited, but it is important to be clear why this was so. It was not primarily due to institutional restrictions, for although the importance of rules imposed by unions or work groups did vary between areas, technological considerations, training facilities and management attitudes were more important in determining inter-area differences in the volume of internal mobility. We would highlight our conclusion that, although the volume of internal mobility was generally greater where labour with the appropriate skills could not easily be recruited, British management, relative to its American counterpart, paid relatively little attention to promoting internal mobility through appropriate job design and training programmes. It might be objected that the American plants were located in the Other Scottish Areas where the lack of experienced engineering labour compelled greater reliance on the internal market, but even within these areas there was a distinct difference between the attitudes of British and American management because the latter made a virtue of necessity. The advantages this can convey are amply demonstrated by the ability of these plants to expand their labour forces rapidly in a market short of trained and experienced engineering labour. Internal mobility can, then, go a considerable way to satisfying labour needs which cannot be met through direct external recruitment. It requires, however, a positive attitude on the part of management if it is to become an integral part of the plant's manpower policy.

(iii) *National manpower policy*

An active national manpower policy, 'seeks to promote the availability, mobility and quality of the human resources needed by the economy and to assure the smooth adjustment by people to the changing geographical and occupational patterns of employment'.[1] The extensive subject matter embraced by this definition cannot be covered by our brief concluding remarks, but as such a policy

[1] O.E.C.D., *Active Manpower Policy; International Seminars 1964—1* (1965), p. 8.

must be based on some view of how employers and employees behave we can set some of the background against which the policy must operate. Our results are also useful in a more specific sense, for as we saw at the outset of this volume, there has been an increasing tendency to adopt particular policy instruments as a means of correcting the alleged inefficiency of the labour market. We can examine certain of these instruments—most notably the Industrial Training Act, the Redundancy Payments Act and the employment exchanges—to see whether their aims are likely to be met in current circumstances.

We begin on a more general level by looking at the relationship between employment conditions and labour turnover. As we have seen, tighter labour market conditions produced an increase in labour turnover so that the recruitment, selection and training costs incurred by plants were higher in Birmingham than in the remaining labour market areas. The crucial question is, therefore, whether these higher costs were offset by other benefits through an improved allocation of labour between different tasks. We cannot provide any definitive answer from our evidence, but it is interesting to note that the negative relationship between earnings and quits was *less* noticeable for Birmingham than for Glasgow males. The main result of a tight labour market is sharply to increase quits amongst short-service and younger employees, and previous research has suggested that much of this mobility is haphazard and wasteful. This is not to suggest that the level of unemployment should be increased in Birmingham to, say, the Glasgow level, for the reduction of income and welfare resulting would far outweigh any possible 'gains' from lower turnover. However, as far as strictly labour market considerations are concerned, there does seem to be some justification for a regional policy which attempts to prevent too great a disparity between markets in the relationship of labour demand to labour supply. This becomes all the stronger if we recognize that the reluctance to undertake geographical mobility, clearly displayed in this study, must introduce a considerable time-lag before labour supply adjusts to changes in labour demand conditions.

Here we immediately come up against the difficulty of measuring labour demand and supply. The chief problem concerns vacancy statistics which are used in many branches of labour market analysis as a measure of the demand for labour. Such data have always been regarded with some suspicion, which this enquiry shows to be well founded. On the positive side, all we can say is that as a rough, ordinal measure vacancy statistics do indicate those types of labour in greatest demand in the market as a whole, and when used in conjunction with unemployment data will indicate the approximate nature of the employment conditions facing different occupational groups. When one attempts to go any further than this one rapidly reaches dangerous ground, for, as a cardinal measure of labour demand and as a statement of a plant's actual labour requirements, vacancy statistics contain some serious biases.

The difficulty of matching a plant's vacancies to its actual labour requirements throws considerable doubt on the main function of vacancy information, which is to assist in the day-to-day placement activities of the employment exchanges. We shall, however, postpone our discussion of this problem for the moment and concentrate on the use made of vacancies for other purposes. We have found that plants notify only a small proportion of actual vacancies to the exchanges. Moreover, this degree of undercounting, especially for males, is likely to be more substantial on the demand side than on the supply side where the relevant measure is the number registered as unemployed. In any labour market, therefore, employment conditions are likely to be more favourable than vacancy and unemployment data indicate. Worse still, there is some evidence from our enquiry which suggests that the degree to which vacancies understate the 'true' demand for labour might increase as the market tightens. This certainly seems true in an occupational context, for amongst manual workers the proportion of vacancies notified to the exchanges was lowest for skilled workers, for whom recruitment difficulties were most severe. Again, the relationship between unfilled vacancies and the market quit rate was weak in Birmingham but much stronger in Glasgow. This observation is open to different interpretations, but one possibility is that the extent of undernotification of vacancies was more substantial in the tighter Birmingham market.

If this is so—and the reader is warned that further investigation is required before this remains anything but a tentative suggestion—it suggests that, in spatial as well as in occupational terms, the tighter the labour market the more will the true demand for labour be underestimated by vacancy data. The implication of this becomes obvious when we consider that two of the main issues confronting manpower policy in Britain are the shortages of skilled labour and the so-called 'regional problem'. In most periods over the last two decades the demand for skilled labour has outstripped the supply, while the reverse has held for unskilled labour. Similarly, certain labour markets have been characterized by low unemployment and a high demand for labour while other labour markets have been much slacker. Hence labour bottlenecks and inflationary pressure have subsisted side by side with unused resources. Our findings give reason to believe that the extent of these difficulties has been greater than supposed and that in the occupational, and possibly in the regional, context, structural imbalances in the labour market have been more serious than has been realized.

Some of our findings with regard to labour mobility are also relevant to regional policy. The engineering plants in New Town and Small Town were able to expand their labour forces rapidly over 1959–66 without experiencing acute recruitment problems; and this despite the location of the units in areas with meagre reserves of engineering labour. This expansion was accomplished by widening the geographical area of recruitment and by providing in-plant training for inexperienced labour. It is not surprising that plants in New Town drew a relatively high proportion of their new employees from other areas, but geographical mobility into Small Town was also high relative to Birmingham, Glasgow and North Lanarkshire and in all probability reflected return migration by those who had previously left the area. The plants also obtained labour from a wide range of industries and found the labour adaptable, always provided that training programmes were designed to provide the basic skills required.

The important conclusion for regional policy is that the plants in New Town and Small Town experienced remarkably few difficulties in securing the labour they required and were of the opinion that the supply of labour was adequate for any future foreseeable needs. Investigation of the industrial distribution of the insured population at the beginning of our study period would have given little confidence that a rapid expansion of engineering employment could be sustained. However, the constraints on a plant's labour supply are set not by industrial but by occupational and geographical factors. We may be able to infer the occupational distribution of an area's labour force from its industrial distribution, but even if the former appears to be unfavourable the plant can make good many skill shortages by the adoption of suitable training programmes. It is, of course, true that the plants in New Town and Small Town required a certain nucleus of skilled labour on which they could draw, but although this obviously imposes some constraints, the plants were able to expand employment extremely rapidly in areas without a previous history of engineering employment. In short, the available labour supply proved itself adaptable to new demands, and this suggests that, as far as purely labour supply considerations are concerned, plants may be able to choose from a much wider range of localities than often supposed. There is nothing in the experience of plants in the Other Scottish Areas which supports the contention that regional policy is based on a false premise and that the development areas have an inadequate supply of labour to allow industrial expansion.[1]

We can now turn from these more general remarks to look at three specific instruments of national manpower policy—the Industrial Training Act, the Redundancy Payments Act and, most important of all, the public employment exchanges. The effectiveness of the two statutory measures is difficult to evaluate because of the short time which has elapsed since their enactment, but we can obtain certain pointers which may help towards a more informed view of their likely effectiveness. For example, there is some evidence that the diagnosis on which the Industrial Training Act was based was correct in certain important respects, and that this Act, whether for good or ill, did bring about one of its stated objectives in Birmingham.

In Birmingham, in contrast to Glasgow, there was prior to the Act a high wastage rate amongst those who began apprenticeships, and a relatively high proportion of the costs of such training was borne by the bigger establishments which, *relative* to their stock of skilled employees, accounted for a large proportion of apprentices. It is difficult to demonstrate cause and effect, but it must be noted that in Birmingham there had also been a more or less continuous shortage of skilled labour over many years.[2] Yet there was no indication prior to the Act of plants attempting to increase the ratio of apprentices to journeymen which was much below the ratio in Glasgow and North Lanarkshire. All this changed from 1964 with a substantial expansion in apprentice-training, which was shared amongst a greater number of plants than hitherto. Hence while it is possible to find fault with the Act,[3] and with its implementation, it could be unwise to dismiss it prematurely.

The Redundancy Payments Act was introduced with the primary intention of easing the path of rationalization and redeployment. Its effectiveness in this regard is difficult to judge by the techniques

[1] This criticism of regional policy has been most strongly expressed by Davies, *op. cit.* For an opposite point of view see D. I. MacKay, 'Regional Labour Reserves: A Comment', *Oxford Economic Papers*, Vol. 20, 1968.

[2] In certain sectors such recruitment difficulties were more widespread. Thus in mechanical engineering, 'there is a chronic shortage of skilled labour . . . so pronounced in some trades that the shortage exists even at the bottom of the trade cycle'. National Economic Development Council, *Economic Assessment to 1972; Industrial Report by the Mechanical Engineering E.D.C.* (1970), p. 13.

[3] It is possible to argue that the Act, at least in its present form, should have been confined to apprentice-training, for a large proportion of the costs of training semi-skilled labour is likely to be incurred in specific training which *should* be borne by the firm providing the training. For evidence on this latter point see, B. Thomas, J. Moxham and J. A. G. Jones, 'A Cost–Benefit Analysis of Industrial Training', *British Journal of Industrial Relations*, Vol. 7, 1969.

employed in this study, but we can show why the payments made under the Act have been much heavier than expected,[1] so that the individual employer's direct share of the payment has been increased until it now stands at 50 per cent. As originally framed the Act seems to have assumed that the great bulk of redundant employees would have completed only a short period of service with their employer. In fact, redundant employees, even before the Act, had, on average, a much longer period of service than other types of leavers. This was partly a consequence of the way in which seniority criteria were applied by job or department rather than by plant, but also reflected the high proportion of all redundant employees who lost their jobs in a few large redundancies, and the tendency for such redundancies to be preceded by a period of low recruitment and continued high wastage amongst short-service employees. In consequence expenditure under the Act was bound to be high and, indeed, is five times as high as the sums spent on Government Training Centres, and at least thirty times as great as the payments made to encourage geographical mobility.[2] It is not at all clear that this distribution of expenditure can be justified on the grounds of efficiency or even of equity.[3]

What we are suggesting is that a result something like the above was bound to occur even if the Redundancy Payments Act had no effect on the length of service and age distribution of redundant employees. The situation may, however, have been further aggravated by the provisions of the Act, for there is some evidence, although admittedly somewhat shaky, which indicates that one unwanted side-effect of the Act might be to increase the proportion of long-service, and hence older, workers amongst those declared redundant.[4] As the age distribution of redundant employees was unfavour-

[1] In 1968, redundancy payments under the Act averaged £230 to some 264,000 redundant employees.

[2] See O.E.C.D., *Manpower Policy in the United Kingdom* (1970), p. 179.

[3] For a discussion of the general principles underlying redundancy payments see S. Please, 'The Economics of Redundancy Compensation', *District Bank Review*, March 1963, and J. Wiseman and K. Hartley, 'Redundancy and Public Policy', *Moorgate and Wall Street*, Spring 1965.

[4] The Act was originally designed to give the firm 'no incentive to retain or release older rather than younger workers, save by reference to their length of service with the firm' (Wiseman and Hartley, *op. cit.*, p. 46). This was accomplished by allowing the employer to recover from the central Redundancy Fund a higher proportion of any redundancy payments made to older workers. This no longer applies, and as the individual employer's share of the payment now amounts to 50 per cent in all cases he may have a financial incentive to retain older and longer-service employees. In our view, however, this change is unlikely to have much effect on employer behaviour.

able to redeployment even before the Act, this possibility, and the whole procedure of the selection of redundant employees, deserves more serious investigation. The only further light we can throw on this area is that in labour markets where engineering labour is difficult to recruit there does appear to be a tendency to 'hoard' skilled labour and to discharge semi-skilled and unskilled labour. Such a policy may be perfectly rational as seen through the eyes of management who wish to avoid recruitment difficulties in the future. It is less certain if it is in the 'national interest' as it may lead to underemployment of scarce skilled labour.[1]

The results of this study provide further confirmation of the long-established view that the bulk of manual workers seek new jobs through informal channels. However, the policy conclusions which follow are less self-evident. To some observers, the mere fact that informal channels are used seems sufficient to demonstrate inefficiency without need of recourse to further information on the costs and effectiveness of alternative methods of job-search. Thus, a recent O.E.C.D. report with scant regard for empiricism concluded: 'The "individualistic" approach to job seeking entails two obvious disadvantages: the period of "frictional" unemployment tends to be longer than it would otherwise be; and there is little assurance that deployment is carried out with a maximum of efficiency.'[2]

It is easy to criticize such unsubstantiated views, but it would be mistaken to take up the opposite and equally untenable extreme, that because informal methods of job-search are frequently used they are necessarily the most effective and efficient means of communicating job information. Previous studies have provided ample evidence that the employee's knowledge of job opportunities is limited and fragmentary, and that an improved flow of information would therefore make an important contribution towards improving the operation of the labour market. Hence, although we have suggested that knowledge is less imperfect than often supposed, we would not dissent from the view that steps could profitably be taken to improve the quality of job information.

This brings us to the employment exchanges, which were

[1] It is possible that the operation of the Redundancy Payments Act might increase the tendency to hoard skilled labour in tight markets. In Birmingham, skilled workers tend to have a longer average period of service than other manual groups and also, as in other areas, higher average earnings. On both counts, therefore, the average lump-sum benefit under the Act will be greater for skilled employees declared redundant than for less skilled employees. As the proportion rebated to the plant is the same in all cases, the employer might be more reluctant to discharge skilled labour.

[2] *Manpower Policy in the United Kingdom*, p. 163.

established for precisely this purpose and whose effectiveness in fulfilling their appointed tasks has been severely questioned in this and other countries.[1] As Daniel has suggested, the employment exchanges have an unfavourable image with both employers and employees. As a result: 'On the one side it has been seen mainly as dealing with undesirable employees, on the other side it is seen as dealing with undesirable jobs. A vicious circle is set up as many employers who have vacancies above the level of unskilled labour do not inform the office, and people seeking good jobs do not go through the office.'[2]

This conclusion accords with the findings of the present study. Only a small proportion of job vacancies were notified by plants to the exchanges in all areas with the exception of the rather special case of New Town, and the reluctance to use the exchanges was most marked for skilled labour. Partly in consequence of this, the penetration rate achieved by the exchanges was low, but even where skilled vacancies were notified, exchange placements were relatively rare. Thus, employers avoid using the exchanges when seeking skilled employees, and the latter, in their turn, by-pass the exchanges and use other methods of job-search.

The employers' attitudes to the exchanges are, then, unfavourable, and the placement work of the public service is further impeded by the lack of correspondence between the vacancies notified by a plant and its actual labour requirements. As long as the present organization of the exchanges remains unaltered there is little prospect of any major improvement in their effectiveness. To cut the Gordian knot will require a fundamental organizational change which emphasizes the services provided for the job-seeker and the job-provider. As has so often been suggested, this implies the separation of the employment services from the payment of unemployment benefit.[3] Until this is done, the role of the exchanges will remain one of providing for the unemployed, the unskilled, the handicapped and the social misfits, and their unfavourable image will limit their usefulness even in this area. Here, as in so many areas, the efficient operation of the labour market depends to a great extent on the attitudes and actions of employers. The employer is often critical of the ability of the exchanges to meet his recruitment needs, but his own actions are to a

[1] See Kahn, *op. cit.*; Wedderburn, *Redundancy and the Railwaymen*; and S. D. Anderman, *Trade Unions and Technological Change* (Swedish Confederation of Trade Unions, 1967).

[2] W. W. Daniel, *Strategies for Displaced Employees* (Political and Economic Planning, 1970), p. 33.

[3] There seems at last to be some movement in this direction. See Department of Employment and Productivity, *Future of the Employment Service; Consultative Document* (1970).

large extent responsible for the deficiencies of the exchanges, particularly his failure to provide exchanges with comprehensive and accurate information on job vacancies.

When considering national manpower policy, therefore, we are again forced back to consider the action of plants, and it is with a final word on the labour market policies adopted by plants that we would like to conclude this study. Much of the behaviour of the plants was rational and sensible enough even judged by the exacting standards which economists apply to others. The more notable weaknesses to which we have referred stem to a considerable extent from the fact that labour market policies often suffer from a lack of co-ordination. Responsibility for the several fields we have investigated —wages, turnover, recruitment and mobility, manpower and personnel policies—is often divided among a number of different individuals and departments. It is true that in the British engineering industry the key executive decisions are often made by those responsible for day-to-day production matters, but while labour market policies must be firmly rooted in the present, the pressure of events often prevents consideration of the longer-term implications of particular decisions. This is particularly true in the wages field, but also applies to most other aspects of labour market policies. For example even apprentice-training programmes are determined by the need of the present rather than by the needs of the future.

The effective utilization of the human capital embodied in a plant's labour force requires as much thought as the planning and execution of a major programme of investment in physical capital. This it seldom receives. A wide and longer-term appraisal is sometimes provided by the personnel manager, but in most engineering plants his executive powers and influence are limited, and in many his functions do not extend beyond labour selection, welfare work and the keeping of personnel records. The labour market policies of engineering plants are therefore seldom considered in their totality. Decisions are taken in one field without thought to the interconnections between, for example, wages and recruitment, labour selection and wastage, forecasting and training. Until this situation is remedied and labour management in its broadest sense is given the attention it deserves, many of the labour market problems to which we have referred are likely to remain as intractable as ever, and the task of national manpower policy that much more difficult.

APPENDIX

THE DESIGN, CODING AND INTERPRETATION OF THE DATA SHEET

The main source data for this enquiry were obtained from the personnel records of the case-study plants. As explained above (see pp. 43–5) a data sheet was completed for any person who at any time between January 1, 1959 and December 31, 1966 had been employed as a manual worker by one of the 66 case-study plants. The only exceptions arose in plants with 1,000-plus employees, where a sample was drawn by the method described and which we call the 'first sampling procedure'. As this process would have under-represented units with 1,000-plus employees the data sheets so obtained were grossed up by the appropriate factor. This was done in all cases except for the data which form the subject matter of Chapters 9 and 10 above. Here the process of coding information was extremely time-consuming, so that a 'second sampling procedure' was sometimes adopted (see A and E below).

The distinction between the 'first' and the 'second' sampling procedure is important and must be clearly drawn. The first sampling procedure was applied to the collection of information from plant personnel records, to produce a number of completed data sheets which, with the exception of certain items, were coded in their entirety and then analysed. Those data sheets for plants with 1,000-plus employees were grossed up to allow for the fact that they were under-represented in the sample initially drawn, and any bias which would have been introduced by the first sampling procedure was thereby corrected. There were, however, certain exceptions to this which affect the subject matter of Chapters 9 and 10. In this case, certain items of information, relating to the address of the employee and place of work, were extremely difficult to code. For this reason, a second sampling procedure was adopted which varied according to the item involved (see A and E below). This, together with a substantial 'not known' element which emerged for any item dealing with employment previous to the case-study plant (see E, F and G below), made it impossible to gross up the results on which Chapters 9 and 10 are based. As this means, in effect, that plants with 1,000-plus employees must be under-represented in these chapters, some bias may be introduced to the affected items where they are systematically related to size of plant. The appropriate reservations are noted below (see A, E, F and G) but they are not thought to be serious enough to invalidate the conclusions reached in Chapters 9 and 10.

The data sheet for each employee contained the following items of information:

1. Name of plant
2. Identification number
3. Sex
4. Marital status
5. Year of birth
6. Home address
7. Training

8 and 9. Details of apprentice-training

10. First engagement with plant (date of joining and leaving; job on joining; reason for leaving)
11. Second (or last) engagement with plant (details as in 10 above)
12. Number of times left plant after 1/1/59
13. Job at 1/1/59 (job and date started)
14. First change in job after 1/1/59 (details as in 13 above)
15. Second change in job after 1/1/59 (details as in 13 above)
16. Job at 31/12/66 (details as in 13 above)

17, 18 and 19. Name, address and business of previous employer

20. Job with previous employer
21. Date of leaving previous employer.

The codes adopted for items such as sex, marital status and similar items were straightforward as were those used for dates and the identification of plants and individual employees. No special explanation is, therefore, required. In other instances, however, the coding procedures were more complex and the basis of the information used requires clarification.

A. *Address* (Item 6)

The home address of each employee was entered on the data sheet. Coding was, however, restricted to a sample of persons employed at

31/12/66. A 100-per-cent sample was taken for units with less than 100 employees, a 1-in-2 sample for those in the size range 100–299 employees and a 1-in-4 sample for the remainder. For this sample the distance between home address and the location of the case-study plant was ascertained and coded to show distance in the following ranges: (i) less than 0·5 miles, (ii) 0·5 to 0·9 miles, (iii) 1·0 to 1·9 miles, (iv) 2·0 to 4·9 miles, (v) 5·0 to 19·9 miles, (vi) 20·0 miles plus.

In interpreting these data, two points should be borne in mind. First, the introduction of a second sampling procedure and the existence of a small not-known category made it impossible to gross up the results obtained. Hence, the data for travel-to-work journeys contain a relatively high proportion of observations taken from units of less than 1,000 employees. Second, a number of establishments did not systematically 'update' information on home address, so that the address recorded on the personnel record is generally that which applied when the individual first took up employment with the plant. The effect of this is uncertain, but there is no *a priori* reason for supposing that it invalidates the travel-to-work comparisons made for the different labour market areas. In each market area, however, where we have summed the results for all plants this is likely to slightly understate the proportion of employees with long travel-to-work journeys. This is so because long travel-to-work journeys were most frequent in plants with 1,000-plus employees (see p. 251 above) and these units are under-represented for the reasons stated.

B. *Job* (Items 10, 11, 13, 14, 15 and 16)
Following examination of the Department of Employment and Productivity's 'Classification of Occupations', and discussion with D.E.P. officials, representatives of the Scottish and West Midlands Engineering Employers Association, plant management and officials of the Amalgamated Union of Engineering and Foundry Workers, a list of engineering jobs was compiled. The list contained 86 job codes which were used to classify all jobs held with the case-study plants. These 86 job codes were arranged to yield the following occupational groups:

(i) apprentices, (ii) skilled engineering and allied occupations, production and non-production, (iii) intermediate group, skill not specified, (iv) semi-skilled production workers, (v) semi-skilled non-production workers, (vi) unskilled production workers, (vii) unskilled non-production workers, (viii) staff, (ix) job not known.

Groups (ii), (iv) and (vi), on which most of our occupational analysis is based, were so arranged to yield categories similar to those applied by the D.E.P. and by the Engineering Employers Federation in their occupational wage returns (see pp. 69–70 above).

The design of the data sheet provided for a limited number of job changes with a case-study plant. For example, if an individual joined a case-study plant before 1959 and did not leave subsequently, the maximum number of jobs which could be recorded was five (four if he joined after 1/1/59). When more than five jobs were held, the procedure was to record the first four and the last known jobs. Some undercounting of internal mobility may result, but inspection of the data sheets shows this to be of only marginal significance.

Two more important difficulties arise in interpreting job data. In the first place, some plants may not have recorded all job changes by employees, and administrative practice did vary from one establishment to the next. Job changes did, however, appear to be generally recorded where they involved a change in the rate of pay, a move between departments or a significant change in the type of work done. Hence, it was usually possible to identify shifts between occupational groups as defined above, and this formed the basis of the analysis of internal mobility in Chapter 11. Job changes which did *not* involve movement between occupational groups were, however, likely to be undercounted. This is not serious, as there was no evidence of any systematic error affecting the comparisons made between different labour market areas, and our analysis is only very rarely dependent on job shifts *within* occupational groups.

The second difficulty arises from the differing terminology and job classifications adopted by the case-study plants. This problem was tackled by relating the list of job codes to the system of classification adopted by the plants through discussions with plant management. It is not claimed that this yielded complete uniformity of treatment, and in most establishments some problems of allocation arose. Nonetheless, this procedure did minimize as far as possible inconsistencies in job classification. Once again the results are likely to be most accurate when the analysis is in terms of occupational groups rather than of individual job codes.

C. *Recruitment to and separation from the case-study plants* (Items 10, 11 and 12)

The data sheet allowed for the possibility that an individual might enter or leave a case-study plant on two occasions. When there were more than two engagements, details of the first and last engagement only were recorded. This will have resulted in some undercounting of recruitment and separation rates.[1] To ascertain whether this was significant, space was provided (item 12) to record the number of separations in the period of the study. The percentage of male and female employees who left the same case-study plant more than twice in the period 1959–66 is shown for each labour market area below. It can be seen that the percentage is extremely low in all cases, so that the underestimation of separation rates is not important. The same comment applies where measurement relates to the recruitment rate.

Percentage with more than two separations, 1959–66

	Birmingham	Glasgow	North Lanarkshire	New Town	Small Town
Males	0·5	0·6	0·1	0·0	0·0
Females	0·6	1·2	0·8	0·1	0·8

D. *Reason for leaving* (Items 10 and 11)

With the exception of a few Birmingham units (see p. 133 above) this information was generally recorded accurately and comprehensively by plants, for it may be useful if an individual re-applies for employment at a subsequent date. Thus it may be relevant to be able to distinguish those applicants who were previously employed by the plant and who were dismissed, declared redundant or left voluntarily. In a number of establishments the personnel records attempted to draw further distinctions between various categories of voluntary leavers. These subdivisions were based on unsubstantiated information supplied by the individual, who may well have had cause to conceal the real factors which motivated his decision. For this reason no attempt was made on the data sheet to distinguish between various categories of voluntary leavers.

[1] It will also have resulted in some undercounting of the number of individuals employed by the case-study plants at any date. The above discussion demonstrates that this must have been of minor importance.

E. *Address of previous employer* (Items 17, 18 and 19)
For employees for whom data on previous employer were available, the name and address of that employer was recorded on the data sheet. From this the place of previous employment was established and, by comparison with the location of the case-study plant, the distance between the two locations was ascertained. Coding was restricted to a sample of employees. In this instance the second sampling procedure was based on those who had joined a case-study plant during 1959–66 and were still employed at 31/12/66. The codes showed distance in ranges equivalent to those in A above, except that the first two ranges were amalgamated to show all distances less than 1 mile.

It should be observed that the personnel records of case-study plants were least comprehensive when they related to the employment experience of the employee prior to his joining the establishment. There is, therefore, a fairly substantial not-known element as regards the above item of information, and this varies from one unit to the next in a manner which is not related to the size of the plant. For this reason, it is not possible to gross up the results of the above analysis to allow for the fact that the first sampling procedure produced relatively few data sheets for units with 1,000-plus employees. Hence the results presented for area of recruitment and geographical mobility will contain a relatively small proportion of observations for the largest plants. This has much the same result as in A. The catchment area of a plant tends to increase with the size (see p. 243 above), so that the under-representation of the larger units is likely to slightly understate the amount of geographical mobility when the results for a market area are summed for all units.

F. *Business of previous employer* (Items 17, 18 and 19)
The business of previous employer was coded to yield industry of previous employment on the basis of the Main Order Headings of the 1958 Standard Industrial Classification. To this classification additional categories were added as follows: (i) school leavers, (ii) armed services, (iii) garages, (iv) unemployed, (v) Government Training Centre.

As the study was concerned with the pattern of industrial recruitment and mobility over 1959–66, coding was undertaken only for those persons recruited by the case-study plants in that period for whom the relevant information was available. As in E above, there was a significant proportion of not-knowns, this proportion varying from one establishment to the next in a manner independent of the size of the case-study plant. Hence it was not possible to gross up the results of the analysis. In this case, however, size of plant did not appear to exert any influence, so that the qualification is of little significance.

G. *Job of employee with previous employer* (Item 20)
The 86 job codes derived as in B were applied where the previous employer was engaged in the engineering industry. However, where the individual had previously been employed outside the engineering industry, the 86 job codes were not used except where the type of work done was similar to that encompassed by one of the 86 job codes. This was most often the case where unskilled work was involved but also in other instances such as maintenance electricians, joiners, welders, etc.

Most jobs held with previous employers not engaged in engineering production had, however, to be dealt with by the provision of supplementary job codes. Provision also had to be made for other categories such as school-leavers, unemployed, married women entering the labour force, etc. A further nine codes were therefore devised to supplement the 86 job codes in B. These were used to classify previous activity outside the labour force or outside the engineering industry.

The job with previous employer of all employees who were engaged by case-study plants over 1959–66 was coded in the above manner where the relevant information was available. Because the data come from the same source as E and F above, there was a significant proportion of not-knowns, so that it was not possible to gross up the results to counteract the underweighting of large plants due to the first sampling procedure. As in F above, this does not appear to be of much consequence, as the occupational pattern of recruitment does not appear to be much affected by size of plant.

INDEX